Manufacturing Engineering and Materials Processing

Manufacturing Engineering and Materials Processing

Edited by **Casan Anderson**

WILLFORD PRESS

New York

Published by Willford Press,
118-35 Queens Blvd., Suite 400,
Forest Hills, NY 11375, USA
www.willfordpress.com

Manufacturing Engineering and Materials Processing
Edited by Casan Anderson

International Standard Book Number: 978-1-68285-030-5 (Hardback)

The publisher's policy is to use permanent paper from mills that operate a sustainable forestry policy. Furthermore, the publisher ensures that the text paper and cover boards used have met acceptable environmental accreditation standards.

Trademark Notice: Registered trademark of products or corporate names are used only for explanation and identification without intent to infringe.

Printed in the United States of America.

Contents

Preface

Scientists and engineers across the globe, from different engineering disciplines, are constantly trying to design and build integrated systems and processes for developing new materials. Computational data management techniques, advanced engineering design frameworks, creating infrastructure for innovations in materials manufacturing, application of advanced materials in different manufacturing sectors, etc. are some of the diverse topics covered in this book. The aim of this text is to present researches that have transformed this discipline and aided its advancement. Students and researchers in search of information to further their knowledge will be greatly assisted by it.

This book has been the outcome of endless efforts put in by authors and researchers on various issues and topics within the field. The book is a comprehensive collection of significant researches that are addressed in a variety of chapters. It will surely enhance the knowledge of the field among readers across the globe.

It gives us an immense pleasure to thank our researchers and authors for their efforts to submit their piece of writing before the deadlines. Finally in the end, I would like to thank my family and colleagues who have been a great source of inspiration and support.

Editor

Phase-field modelling of microstructure evolution during processing of cold-rolled dual phase steels

Jenny Rudnizki, Ulrich Prahl[*] and Wolfgang Bleck

* Correspondence: ulrich.prahl@iehk.
rwth-aachen.de
Department of Ferrous Metallurgy,
IEHK, RWTH Aachen University,
Intzestr 1, 52056, Aachen, Germany

Abstract

Cold-rolled dual-phase steels, which belong to the advanced high strength steels, have gained much interest within the automotive industry. The formation of dual-phase microstructure, which provides an optimal combination of strength and formability for automotive applications, occurs during intercritical annealing of cold-rolled strip. Variations in the chemical composition as well as in the heat treatment parameters influence very strongly the microstructure development and therefore the final mechanical properties of the strip. Thus, the precise control of the microstructure evolution during full processing route is required for the achievement of essential mechanical properties. The current work is focused on a through-process model on a microstructural scale for the production of dual-phase steel from cold-rolled strip, which is based on Phase-Field Method and combines the description of ferrite recrystallisation and all phase transformations occurring during intercritical annealing. This approach will enable the prediction of final microstructure for varying composition and processing conditions, and therefore, can be used for the process development and optimisation.

Keywords: Phase-field modelling, Phase transformation, Dual-phase steels

Background

The mechanical properties of dual-phase steels are provided by a microstructure which consists of two different phases with different hardness, i.e. hard martensitic islands in a soft ferrite matrix. Thereby, not only phase fractions but also distribution and morphology of both phases are responsible for the mechanical properties [1,2]. The formation of dual-phase microstructure from cold-rolled strip occurs during intercritical annealing which involves different metallurgical phenomena, like recrystallisation and phase transformations [3,4]. Each of these processes contributes to the establishment of the final microstructure. Thus, the control of the microstructure evolution during the whole processing of dual-phase steels is therefore the key for the processing of steels with predetermined mechanical properties. Therefore, several detailed experimental studies of DP microstructure evolution have been performed e.g. [5-7].

Beside of that, the numerical investigation which is an alternative to complex and expensive experiments can be very helpful. Nowadays, computational materials science is a powerful approach for understanding physical mechanisms and their interactions during industrial processing allowing the prediction of microstructure as well as of mechanical properties [8,9]. It can be applied for the improvement of process parameters and optimisation of chemical composition.

Up to now, different mechanisms along the annealing route of cold-rolled strips in dual-phase processing, as there are recrystallization in cold formed ferrite, austenite formation, grain growth in ferrite and austenite, formation of ferrite and martensite, have been investigated separately. However, the current trend in computational materials science moves towards ICME (Integrative Computational Materials Engineering) where individual approaches on the different scales are combined in order to describe a process chain completely [10,11]. Such combination of models describing single process steps allows the prediction of final product properties from the knowledge of chemical composition and processing conditions.

At the time, no through-process model for the processing of dual-phase steels, which covers all metallurgical phenomena on a microstructural scale in one integral approach, is available. Setting up such through-process model is expected to lead to significant improvement in the prediction of microstructure development and therefore, to enable robust production of dual-phase steels with predefined mechanical properties. Hereby, the Phase-Field approach represents a powerful method for the calculation of microstructure evolution in multi-phase multicomponent systems which can be applied to the establishment of through-process model. The potential and ability of this approach for the prediction of several phase transformations have been already show in many works [12-15].

The main focus of this work is the formulation of a through-process model on a microstructural scale for the production of dual-phase steels from cold-rolled strip by means of Phase-Field approach. 2D- and 3D-simulation results of the microstructure evolution during intercritical annealing will be shown and compared to experimental results.

Methods

Experimental

Material characterisation

The material used in this study is industrially produced cold-rolled steel sheet of 1 mm thickness and 60% cold rolling reduction. The chemical composition of the steel is listed in Table 1. As in a typical DP-steel, the main alloying elements are C, Si, Mn, and Cr.

According to metallographic analysis, the initial microstructure of investigated material consists of two phases: the bright ferrite matrix and dark pearlite grains, Figure 1. Both types of grains have an elongated shape, due to cold-rolling. The average pearlite content was determined to be 16% by means of image analysis. In order to determine the distribution of the alloying elements in the initial structure, a microprobe (ESMA) analysis was performed. The concentration of C, Si, Mn, and Cr was measured along the arrow in the zone between two indents that were used to mark the selected area.

Table 1 Chemical composition (in wt.-%) of the investigated steel

C	Si	Mn	P	S	Cr	Al	N	Nb	Ti	V
0.098	0.24	1.65	0.017	0.007	0.55	0.04	0.004	0.001	0.002	0.008

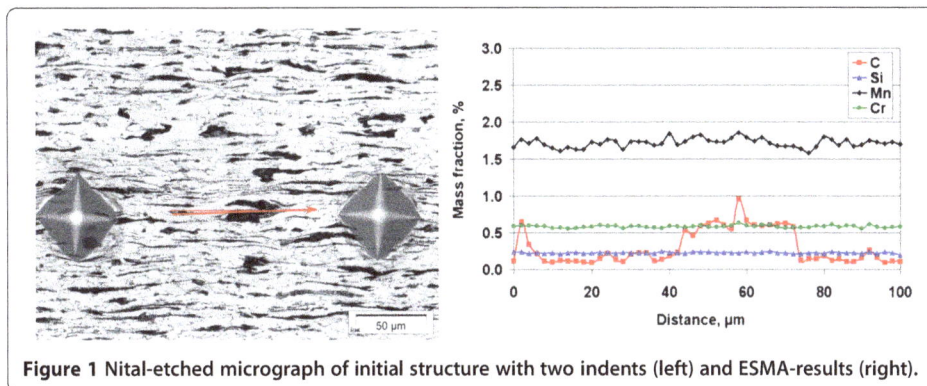

Figure 1 Nital-etched micrograph of initial structure with two indents (left) and ESMA-results (right).

According to the microprobe analysis, carbon concentration of pearlite is about 0.60 wt.-%. In parallel, Thermo-Calc® calculations were performed taking into account C, Si, Mn and Cr. This confirms as well the eutectic concentration of investigated steel at approximately 0.60 wt.-% of carbon. The C concentration in ferrite is much lower. However, the exact determination of the carbon content in ferrite by means of microprobe analysis is limited due to the very small concentration. No difference was found between the concentrations of Si, Mn and Cr in ferrite and pearlite in the initial structure: the content of the substitutional elements in both phases is equal to the total content in the steel. Thus the initial structure contains 16% of pearlite with 0.60 wt.-% C, 0.24 wt.-% Si, 1.65 wt.-% Mn, and 0.55 wt.-% Cr in a ferritic matrix with the same content of substitutional elements. Knowing the carbon content and phase fraction of pearlite, the carbon concentration in ferrite was calculated to be 0.002 wt.-% using total carbon content and mass balance.

The stored energy distribution, which is also an important characteristic of material, was obtained based on EBSD analysis assuming that the misorientations within deformed grains reflect the distribution of a small angle grain boundary and therefore, the stored energy. The approach to identify the stored energy from misorientation maps has been developed by Zaefferer and can be found at [16,17]. The average value of the stored energy for the investigated steel was evaluated to be 4.5 J/cm^3. Additionally, the EBSD analysis is applied in this work for the determination of ferrite grain size distributions.

Experimental process simulation

In order to investigate the microstructure evolution during intercritical annealing, tests were performed on Baehr Dilatometer DIL-805A/D. For these experiments, the samples with 1 mm thickness, 4 mm width and 8 mm length were cut parallel to the rolling direction of the ingot. The dilatometric tests have been carried out with a dilatometer type DIL-805A/D fabricated by Baehr Thermoanalyse GmbH. Experiments were realised by a low-pressure environment of the order of 10^{-5} MPa in order to protect the samples from oxidation. During experiments temperature was recorded by a thermocouple. Figure 2 represents intercritical heat treatment performed in dilatometer which corresponds to a simplified industrial processing of dual-phase steel. According to this heat treatment, the samples were heated in two-steps to the intercritical temperature of 800°C, held 60 s at 800°C and afterward cooled down slowly with 1°C/s to 600°C.

Along the intercritical heat treatment samples were quenched at various stages as can be found in Figure 2 using a helium atmosphere. Thus, the microstructure at the

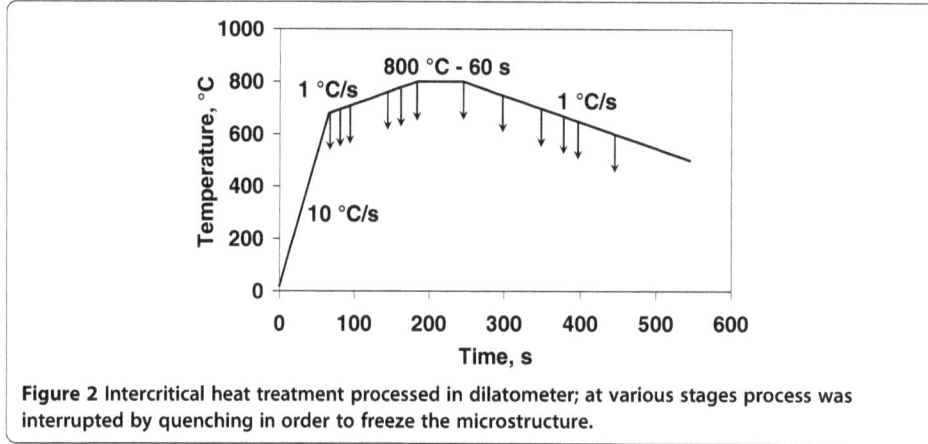

Figure 2 Intercritical heat treatment processed in dilatometer; at various stages process was interrupted by quenching in order to freeze the microstructure.

chosen temperatures was frozen and afterwards studied by means of light optical microscopy. The corresponding average values of the ferrite fraction were obtained by an image analyser of four nital-etched micrographs per sample at magnification 1000x.

Modelling

Multicomponent multiphase-field approach

The Phase-Field approach is developed for the modelling of the phase transformations in multicomponent systems [18-20]. For simulation work in this paper the commercial software package MICRESS® is used which allows the calculation of phase fractions during solid-solid transformations in multicomponent steels as well as the description of the corresponding microstructure evolution [21].

Multiphase-Field is a computational approach which describes the evolution of multiple Phase-Field parameters $\varphi_i(\to x, t)$ in time and space. These fields map the spatial distribution either of different grains with different stored energy or of phases with different thermodynamic properties. At the interfaces, the Phase-Field variables change continuously between 0 and 1 over an interface thickness η which can be chosen to be large compared to the atomic interface thickness, but small compared to the microstructure length scale. The time evolution of the phases is calculated by a set of Phase-Field equations formulated by minimization of the free energy functional [19]:

$$\dot{\varphi}_i = \sum_j M_{ij}(\overrightarrow{n}) \left(\sigma_{ij}^*(\overrightarrow{n}) K_{ij} + \frac{\pi}{\eta} \sqrt{\varphi_i \varphi_j} \, \Delta G_{ij}(\overrightarrow{c}, T) \right) \tag{1}$$

$$K_{ij} = \varphi_j \nabla^2 \varphi_i - \varphi_i \nabla^2 \varphi_j + \frac{\pi^2}{\eta^2} \left(\varphi_i - \varphi_j \right) \tag{2}$$

In Eq. (1), $_{ij}$ is the mobility of the interface as a function of the interface orientation, given by the normal vector n$\overrightarrow{}$. σ_{ij}^* is the effective anisotropic surface energy (surface stiffness), and K_{ij} is related to the local curvature of the interface. The interface is driven by the curvature contribution $\sigma_{ij}^* K_{ij}$ and by the thermodynamic driving force ΔG_{ij}. ΔG_{ij}, which is a function of the local composition $\to c$, couples the Phase-Field equations to the diffusion equations:

$$\dot{\vec{c}} = \nabla \sum_{i=1}^{N} \varphi_i \, \vec{D}_i \, \nabla \, \vec{c_i},$$

(3)

where $\rightarrow D_i$ is the multicomponent diffusion coefficient matrix for phase ι. $\rightarrow D_i$ is calculated online for the given concentration and temperature using the Fortran TQ® interface to Thermo-Calc® and the mobility databases MOB2 [22]. In case of recrystallization modelling the thermodynamic driving force is replaced by stored energy.

Input parameters

2D-Modelling The performed 2D-Phase-Field simulation is carried out based on the real microstructure of cold-rolled steel, Figure 3. Here, the left image represents the nital-etched micrograph of the cold-rolled sample, where the ferrite is bright and pearlite and grain boundaries are dark. The centre image shows the phase arrangement reproduced by simulation software. The size of this 2D-simulation domain is about $110 \times 80 \ \mu m^2$. Similar to the nital-etched micrograph, in the replicated pattern, ferrite is white while pearlite and grain boundaries are black. The right image shows the carbon distribution in black-and-white scale, wherein black corresponds to zero and white to the maximum carbon concentration 0.7 wt.-%.

There, the important material characteristics, i.e. distribution of alloying elements and stored energy, are taken from the microprobe and EBSD analyses. In the simulation C, Si, Mn and Cr were included, which are the main alloying elements in this DP600 grade. Thus, according to the microprobe results, it is taken that the substitutional elements are homogeneously distributed in the structure whereas the carbon content in pearlite is about 0.60 wt.-% and in ferrite 0.002 wt.-%.

The stored energy of ferritic grains which is the driving force for the recrystallisation is set between 3 and 7 J/cm^3, so that the average value corresponds to the EBSD results. As pearlite is a much harder phase compared to ferrite, the stored energy of pearlite is set to zero and thus, recrystallisation occurs only in ferrite and is finalised before austenite forms.

Recrystallisation is followed by the nucleation and growth of new grains. Due to the high heating rate in industrial annealing lines, recovery has been neglected in this study. Additionally to the stored energy, which is the driving force for this process, the

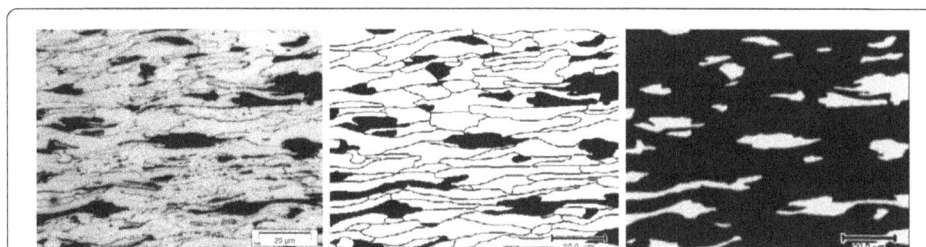

Figure 3 Nital-etched micrograph of the cold-rolled sample (left), phase arrangement reproduced by simulation software (center) and representation of carbon distribution in black-and-white scale (right).

main input parameters for the simulation of recrystallisation are nucleation density and grain boundary mobility. These parameters determine the rate and characteristics of the whole process.

To approximate the real transformation behaviour, 2D-simulation of recrystallisation is performed following a local maximum stored energy criterion assuming continuous nucleation. The reason for this is the fact that in the 2D-simulations each grain emerging on the simulated surface should be assumed to be a newly-generated. Thus, for the 2D-modelling it is important during the whole process to introduce new grains which reflect and compensate growth of grains from the underlying level in the considered 2D-section. The total number of nuclei appearing during the simulation is set to 250 based on the experimental grain size distribution after recrystallisation. This approach is similar to that used by [23] combining nucleation scenario assumptions with mobility adjustment to replicate the experimental data.

The grain boundary mobility is typically given by the Arrhenius equation describing its temperature dependence. The activation energy for α/α grain boundary mobility is fixed to 140 kJ/mol as this is the most widely used value in the literature [24]. The pre-exponential factor is set to 3.2×10^{-6} m^4/(Js) based on the comparison of the simulated growth of the single grains by the variation of the pre-exponential factor with the experimental grain sizes. As the simulation focuses on micro scale, all grain boundaries are assumed to be high angle grain boundaries and will be kinetically treated identically.

The modelling of phase transformation (austenite formation from ferrite pearlite and ferrite formation from austenite) will be performed separately and the results will be combined in one through process simulation. For the simulation of pearlite dissolution a simplified approach is used. Pearlite in this approach is considered as an effective pseudo-phase with mixed properties of ferrite and cementite. The thermodynamic driving force for dissolution is calculated from overheating obtained by a linearization of the phase diagram. The interactions of pearlite with other phases are defined by assumed slopes which restrict the two phase regions [13]. Based on the metallographic observations, nucleation of austenite during heating is assumed to take place only within pearlitic grains. The number of austenite nuclei is set to 100 so that in the large pearlitic grains several austenite nuclei could form, similar to the metallographic finding.

The thermodynamic driving force for the modelling of ferrite-to-austenite and austenite-to-ferrite transformations is obtained by the direct coupling to the thermodynamic database via the Gibbs energy minimisation software Thermo-Calc®. The modelling of these phase transformations are performed taking into account redistribution of substitutional alloying according to LENP (Local Equilibrium Non Partitioning) conditions using special model for solute redistribution [12].

The diffusion parameters are taken from the kinetic database MOB2. The mobility for the ferrite/austenite and pearlte/austenite grain boundaries are defined by an activation energy, of 140 kJ/mol and a pre-exponential factor of 1.5×10^{-6} m^4/(Js). The other input parameters which have to be mentioned here are interfacial energy the interfacial energy σ_{ij}^{*}, of 0.4 Jm^{-2} for ferrite/austenite interface, 0.9 Jm^{-2} for austenite/pearlite interface, 0.7 Jm^{-2} for austenite/austenite interface [13], the grid spacing Δx, which is taken to be 0.2 μm, and the interface thickness η, which in this work equals to 1 μm.

3D-Modelling Due to the fact that a 3D-simulation demands high computational capacity as well as being very time consuming, a small simulation domain of $30 \times 30 \times 30$ μm^3 is selected for 3D-simulations of microstructure evolution during processing of cold-rolled dual-phase steel. Figure 4 shows this created structure consisting of about 16% of pearlite in a ferritic matrix as a 3D-volume domain. Again, in this representation ferrite is white, while pearlite as well as the grain boundaries are black.

However, the direct comparison of 3D-structures with experimental data is hindered due to the fact that grain size distribution obtained from the volume element and from the surface of the same structures could be very different [25-27]. For the comparison of the simulated 3D- and 2D-structures the 3D-volume element has been be subjected to a series of cuts in order to get sufficient cumulative area of 2D-cuts. Figure 5 shows nine 2D-cuts of the 3D-volume presented in Figure 4 and comparison of the obtained ferrite grain size distribution with experimental data from EBSD-analysis.

The grain size distribution of the created volume domain obtained from the XY-sections reflects an elongation of the real microstructure due to cold-rolling. Moreover, the ferrite grain size distribution of 3D-structure obtained from 2D-cuts is in fair correlation with the experimental data. The average grain size of created 3D-structure equals to 11.5 µm and diverges only by 2–3 µm from the experimentally-obtained grain size value 8.6 µm.

As in the case of 2D-structure, the content of alloying elements is taken according to the microprobe results and the stored energy of ferritic grains in the created 3D-structure is set corresponding to the EBSD results. Most simulation parameters such as behaviour of substitutional alloying elements, mobility, interfacial energy and diffusion parameters are taken from 2D-approach. However, the nucleation parameters are adjusted due to the fact that continuous nucleation proposed for 2D-simulations assumes not only nucleation of new grains but also the growing up of grains from neighbour level. This is therefore not suitable for 3D-simulations. Thus, it was established that the site saturation nucleation is much more applicable for the 3D-modelling of recrystallisation due to the fact that according to the experimental data, recrystallisation under the investigated heating parameters takes only 20–30 s. In order to reflect the experimentally-obtained average grain diameter after recrystallisation, which is determined by the nucleation density, it is supposed that 50 nuclei form at 660°C. The grain junctions and interfaces are addressed as probable nucleation sites. Considering austenite formation, as in the case of 2D-simulation it is assumed that nucleation takes

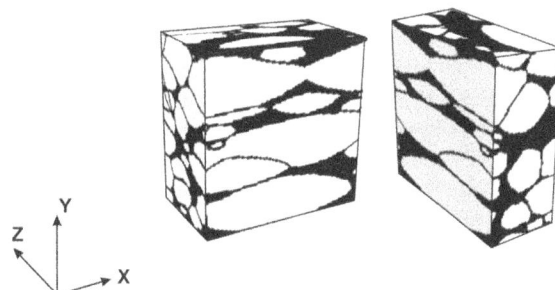

Figure 4 The generated initial structure for 3D-simulation.

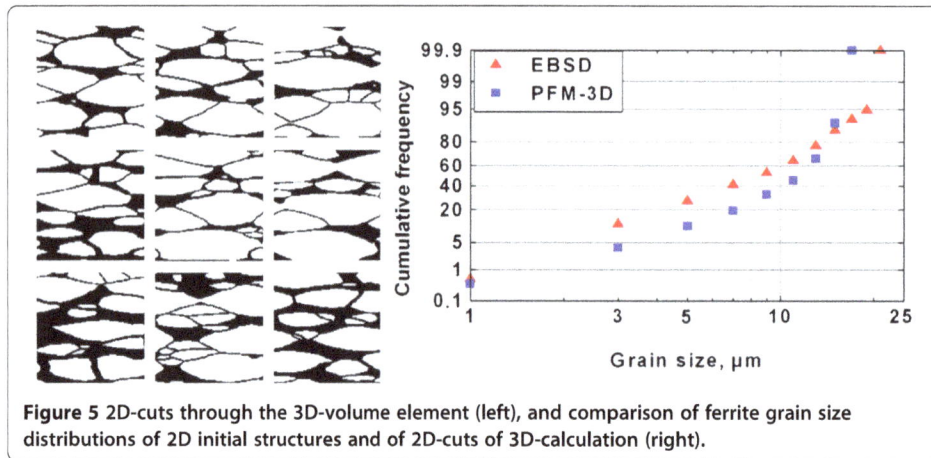

Figure 5 2D-cuts through the 3D-volume element (left), and comparison of ferrite grain size distributions of 2D initial structures and of 2D-cuts of 3D-calculation (right).

place only in pearlite. The number of nuclei is fixed to 25 so that in each pearlitic grain several austenite nuclei could form.

Results and discussion

In the following the results of the 2D- and 3D-modelling for the microstructure evolution during intercritical annealing with the process parameters according to Figure 2 will be presented.

Figure 6 represents 2D-simulated evolutions of transformed fractions. The data repeat quite well the experimental results for the same process conditions obtained by image analysis of corresponding micrographs. The analysis of the data shows that the evolution of transformed phase fractions during intercritical annealing as predicted by the 2D phase field simulation coincides with all experimental points.

The ferrite fraction predicted by the 3D-simulation during intercritical annealing with selected process parameters also reflects the experimental results quite well, Figure 7. The other phases are not shown in this diagram in order to keep the clearness in the representation. Beside the good agreement with the experimental data, evolution of the ferrite fraction from 3D-simulation reproduces well the 2D-resuts. Finally, the ferrite

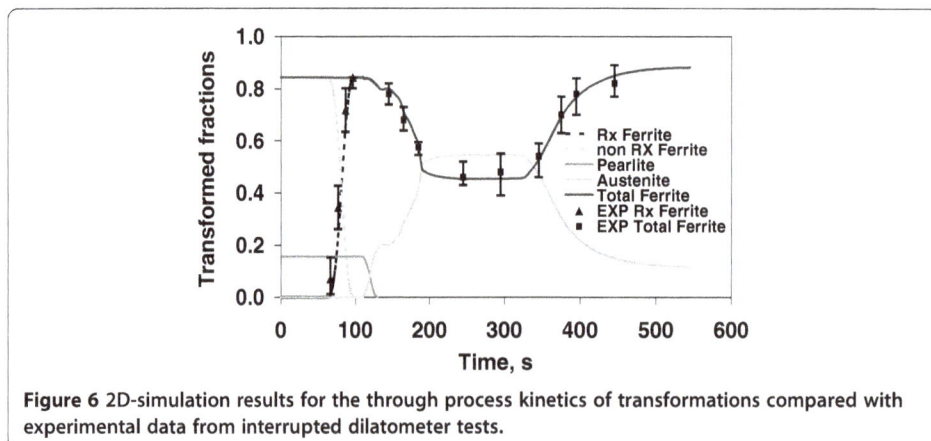

Figure 6 2D-simulation results for the through process kinetics of transformations compared with experimental data from interrupted dilatometer tests.

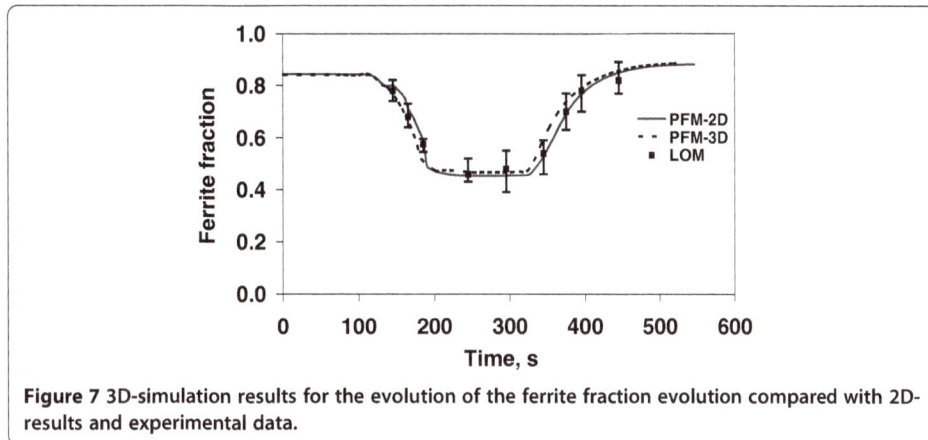

Figure 7 3D-simulation results for the evolution of the ferrite fraction evolution compared with 2D-results and experimental data.

amount simulated in 3D and 2D is inside the scatter bars obtained by metallographic analysis.

The comparison of 2D- and 3D-simulated and experimental grain size distributions of ferrite after intercritical annealing is presented in Figure 8 at 650°C. Both simulated grain size distributions matches the EBSD results quite well. Moreover, the average ferrite grain sizes at 650°C obtained from the 2D- and 3D-simulations versus experiment are 7.9 µm, 9.8 µm and 8.0 µm, respectively.

Mecozzi et al. have shown that the predicted ferrite grain size distribution depends in detail on the assumed nucleation behaviour [23]. Further, it is suggested that the apparent mobility in 2D is lower than in 3D to match a reference kinetics for a given nucleation scenario. In the present simulation 2D and 3D mobilities are the same, while in 2D continuous nucleation has been assumed and in 3D all nuclei form at the same temperature (site saturation). The apparent agreement of 2D and 3D mobilities might be accepted to be a compensation effect of 2D vs. 3D growth geometries and increased nucleation spread in 2D as compared to 3D simulations.

For the more efficient evaluation of modelling results, the quality of the phase transformation simulation is verified by comparing micrographs and simulated microstructures at selected temperatures following cycle from Figure 2. Figure 9 represents the micrographs and outputs from 2D-simulation at 710°C (after recrystallization), 800°C (heating to intercritical temperature) and 650°C (subsequent cooling after intercritical

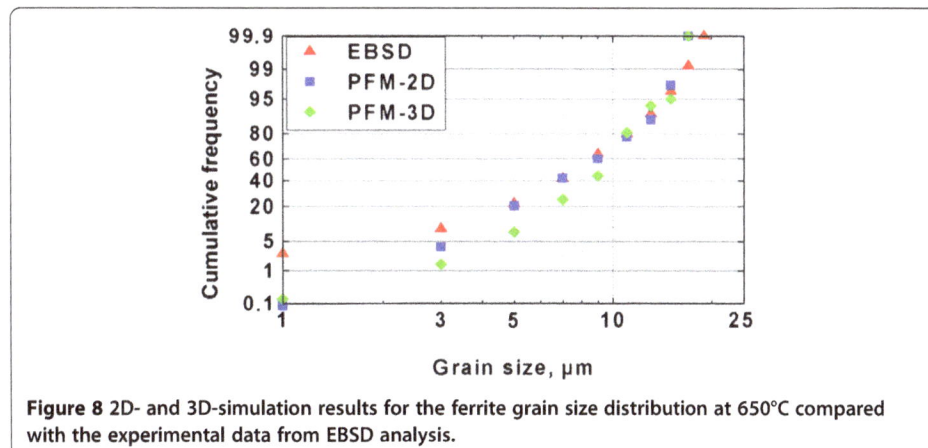

Figure 8 2D- and 3D-simulation results for the ferrite grain size distribution at 650°C compared with the experimental data from EBSD analysis.

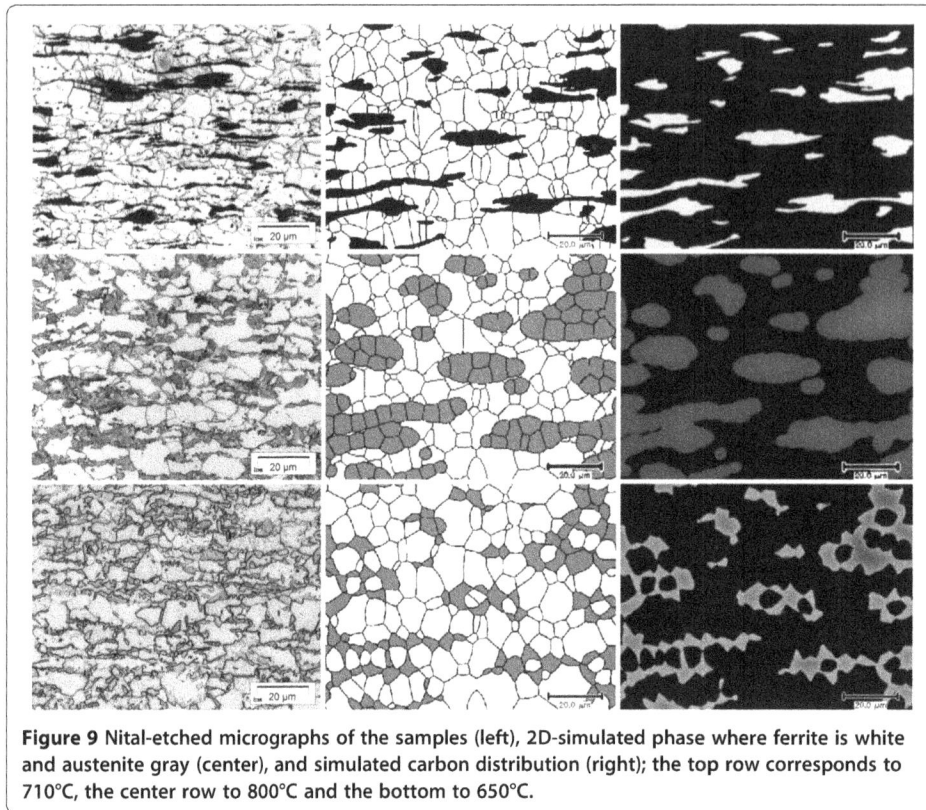

Figure 9 Nital-etched micrographs of the samples (left), 2D-simulated phase where ferrite is white and austenite gray (center), and simulated carbon distribution (right); the top row corresponds to 710°C, the center row to 800°C and the bottom to 650°C.

annealing). The left images represent micrographs of the nital-etched samples interrupted at the corresponding temperatures and quenched to Room temperature. In these micrographs, ferrite is bright, and martensite (which is assumed to have been austenite before quenching) as well as grain boundaries are dark. The central and right images represent simulated phase and carbon distribution; here the same colour scale is used as in Figure 1.

The visual comparison of the phase distributions and the grain sizes at 710°C shows that 2D-results are quite similar with real micrograph. Both, metallographic and simulation results show that the recrystallisation has been completed. At corresponding temperature the deformed elongated ferritic grains are substituted by the round recrystallized grains. Nvertheless, there is still room for improvement. One aspect that is not considered in the simulations is banding due to segregation effects that is quite obvious from the micrograph shown in Figure 1.

At 800°C, the calculated austenitic fraction is about 41%, which agrees with the experimentally observed 42%. The grain distributions in the simulated structure and in the micrograph are similar. An elongated arrangement of phases can be observed in both as a result of the stretched arrangement of pearlite in the initial structure. Due to the fact that the ferrite to austenite phase transformation is controlled by carbon diffusion, a redistribution of carbon during the progress of phase transformation occurs. At this temperature, the austenite contains about 0.24 wt.-% carbon.

The simulated structure at 650°C with about 77% of ferrite and 23% of austenite also corresponds well to the real dual-phase microstructure. The experimental results with

78% of ferrite and 22% of martensite confirm the simulation. Although the phase arrangement and distribution of ferritic grains in both structures are very similar, the network of austenitic/martensitic grains in the simulated structure seems to be somewhat coarser. This could be a consequence of the fact that the final dual-phase microstructure still reflects the initial distribution of pearlite. Though the performed simulation based on the real micrograph, very small pearlitic grains (smaller than 3 grid element) have not been reproduced in the starting structure. The absence of these grains may lead to the somewhat coarser martensite network in the final simulation structure.

The carbon distribution map shows the enrichment of carbon fraction in the remaining austenite with decreasing austenite fraction. The average carbon concentration in the austenite at 650°C reached a value of about 0.41%.

Figure 10 shows the phase and carbon distributions at 710°C, 800°C and 650°C predicted by the 3D-simulation with the same colour scale as in the 2D-results in Figure 9. As mentioned above, the size of ferritic grains from 3D-simulation in general agree 2D-results and experimental data, Figure 10. The 3D-results concerning phase arrangement, seem to be similar to the 2D-simulation and experimental data as well. Moreover, the average carbon concentration in the austenite at 650°C according to 3D-results is about 0.46 wt.-% that also matches 2D-simulation result.

Outlook

From the information of the carbon content in austenite, the M_s-temperature at each point of structure can be calculated. Thus, from the structure simulated by means of Phase-Field Method, ferrite-martensite structure can be obtained by the application of Koistinen-Marburger [28] or an alternative approach. This will enable coupling to the already available models for the modelling of mechanical properties, such as for example RVE-FE Method for the prediction of flow behaviour [29-31].

Conclusions

2D- and 3D-simulations of microstructure evolution during processing of dual-phase steels from cold-rolled strips have be realized by means of Phase-Field Modelling approach. It allows to describe all metallurgical phenomena occurring on a microstructural

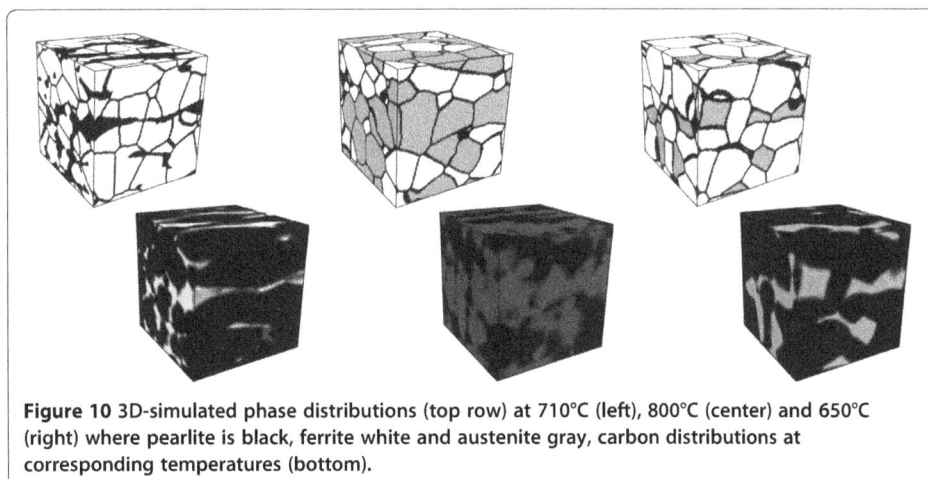

Figure 10 3D-simulated phase distributions (top row) at 710°C (left), 800°C (center) and 650°C (right) where pearlite is black, ferrite white and austenite gray, carbon distributions at corresponding temperatures (bottom).

scale during intercritical annealing, i.e. recrystallisation, austenite formation, ferrite formation as a function of chemical composition, starting microstructure and process parameters. The accurate definition of input model parameters yields both 2D- and 3D-approaches to simulate the microstructure evolution successfully. The comparison of 2D- and 3D-simulated results with experimental data demonstrates an overall agreement of the predicted evolution of phase fraction and grain size distribution.

Moreover, evolution of carbon distribution, especially the carbon distribution in the final structure, is an important output of Phase-Field simulations. Carbon dissolved in martensite determines its hardness, and therefore, mechanical properties of the investigated steel. This can be used for the following prediction of the microstructure evolution. It will allow determining martensite start temperature in case of the subsequent quenching as well as further modelling of flow behaviour of material.

The advantage of 2D-simulation compared to the 3D-approch is the achievement of fast results and directly comparison with experimental data enabling rapid revelation of the influencing parameters. Therefore it can be utilised for the optimisation of process parameters to achieve the essential microstructure. In contrast, the 3D-simulations are much more time-consuming and need high computational capacity but can be applied easily for the following coupling with the models for the prediction of mechanical properties.

Competing interests
The author(s) declare that they have no competing interests.

Author's contributions
JR did the experiments, the calculations and prepared the first version of article text. UP discussed and analysed the experimental and simulation results, contributed during text and figure preparation and corrected the article. WB analysed and discussed the final results and conclusions and the article text. All authors read and approved the final manuscript.

Acknowledgement
This research was carried out under project number MC5.06257 in the framework of the Research Programme of the Materials innovation institute M2i (www.M2i.nl). The authors acknowledge the financial support of M2i as well as the fruitful discussion with Henk Vegter, Piet Kock (both now at Tata Steel Europe) and Jilt Sietsma.(Delft University of Technology).

References
1. Bag A, Ray KK, Dwarakadasa ES (1999) Influence of martensite content and morfology on tensile and impact properties of high-martensite dual-phase steels. Metallurgical and Materials Transactions 30A **5:**1193–11202
2. Tomita Y (1990) Effect of morphology of second-phase martensite on tensile properties of Fe-0.1C Dual-phase steels. Journal of Materials Science **25:**5179–5184
3. Militzer M, Poole WJ (2004) A critical comparison of microstructure evolution in Hot-rolled and cold-rolled dual-phase steels. AHSSS Proceedings :219–229
4. Huang J (2004) Microstructure evolution during processing of dual phase and TRIP steels. PhD Thesis, University of British Columbia
5. Peranio et al (2010) Microstructure and texture evolution in dual-phase steels: competition between recovery, recrystallization, and phase transformation. Materials Science and Engineering A **527:**4161–4168
6. Calcagnotto et al (2008) Ultrafine grained ferrite/martensite dual phase steel fabricated by large strain warm deformation and subsequent intercritical annealing. ISIJ International **48:**1096
7. Calcagnotto et al (2011) and fracture mechanisms in fine- and ultrafine-grained ferrite/martensite dual-phase steels and the effect of aging. Acta Materialia **59:**658–670
8. Bäker M (2002) Numerische methoden in der materialwissenschaft. Braunschweiger Schriften des Maschinenbaus 8, Braunschweig ISBN 3-936148-08-2
9. Raabe D, Roters F, Barlat F, Chen LQ (2003) Continuum scale simulation of engineering materials. Weinheim Wiley-VCH Verlag, Weinheim
10. National Research Council (2008) Integrated computational materials engineering: a transformational discipline for improved competitiveness and national security. National Academic Press, Washington D. C
11. Gottstein G (2007) Integral materials modeling. Weinheim Wiley-VCH-Verlag, Weinheim

12. Rudnizki J, Böttger B, Prahl U, Bleck W (2011) Phase-field modelling of austenite formation from a ferrite plus pearlite microstructure during annealing of cold-rolled dual-phase steel. Metallurgical and Materials Transactions A 8:2516–2525
13. Thiessen RG (2006) Physically-based modelling of material responce to welding. PhD Thesis, TU Delft
14. Mecozzi MG (2007) Phase-field modelling of the austenite to ferrite transformation in steels. Ph.D Thesis, TU Delft
15. Militzer M (2011) Phase field modeling of microstructure evolution in steels. Current Opinion in Solid State and Materials Science 15(3):106–115
16. Zaefferer S, Konijnenberg P, Demir E, Woodcock T (2010) Progress in 3-dimensional EBSD-based orientation microscopy: New software tools for 3-dimensional materials characterization. Materials Science and Engineering MSE 2010, Darmstadt
17. Eshelby JD, Read WT, Shockley W (1953) Anisotropic elasticity with applications to dislocation theory. Acta Metallurgica 1:251–259
18. Steinbach I, Pezzolla F, Nestler B, Seeßelberg M, Prieler R, Schmitz GJ, Rezende JLL (1996) A phase field concept for multiphase systems. Physica D 94:135–147
19. Eiken J, Böttger B, Steinbach I (2006) MultiPhase-field approach for multicomponent alloys with extrapolation scheme for numerical application. Phys Rev E 2006:066122
20. Steinbach I (2009) Phase-field models in materials science; a tutorial review. Modelling and Simulation in Materials Science and Engineering 17:073001–31
21. MICRESS – The Microstructure Evolution Simulation Software http://micress.de
22. Thermo-Calc Software http://www.thermocalc.com
23. Mecozzi MG, Militzer M, Sietsma J, van der Zwaag S (2008) The role of nucleation behavior in phase-field simulations of the austenite to ferrite transformation. Metall Mater Trans 5:1237–1247, A 39
24. Krielaart GP, van der Zwaag S (1998) Simulations of pro-eutectoid Ferrite Formation using a Mixed Control Growth Model. Material Science and Engineering A246 1998:104–116
25. Giumelli AK, Militzer M, Hawbolt EB (1999) Analysis of the austenite grain size distribution in plain carbon steels. ISIJ International 39:271–280
26. Calcagnotto M et al (2010) Orientation gradients and geometrically necessary dislocations in ultrafine grained dual-phase steels studied by 2D and 3D EBSD. Mater Sc Engin A 527:2738
27. Zaefferer S et al (2008) Three-dimensional orientation microscopy in a focused ion beam-scanning electron microscope: A new dimension of microstructure characterization. Metal Mater Trans A 39A:374–389
28. Koistinen DP, Marburger RE (1959) A general equation prescribing the extent of the austenite-martensite transformation in pure iron-carbon alloys and plain carbon steels. Acta Metallurgia 7:59–60
29. Rodriguez R, Gutierrez I (2004) Mechanical behaviour of steels with mixed microstructure. Proceeding of TMP'04, B-Liege 363:356–363
30. Thomser C, Uthaisangsuk V, Bleck W (2009) Influence of martensite distribution on mechanical properties of dual phase steels: experiments and simulation. Steel research international 80(8):582–587
31. Uthaisangsuk V, Prahl U, Bleck W (2009) Failure modeling of multiphase steels using representative volume elements based on real microstructures. Procedia Engineering 1(1):171–176

DREAM.3D: A Digital Representation Environment for the Analysis of Microstructure in 3D

Michael A Groeber[1][*] and Michael A Jackson[2][*]

*Correspondence:
michael.groeber@wpafb.af.mil;
mike.jackson@bluequartz.net
[1] Air Force Research Laboratory,
2230 Tenth St, 45433, WPAFB, Ohio,
USA
[2] BlueQuartz Software, 400 S.
Pioneer Blvd, 45066, Springboro,
OH, USA

Abstract

This paper presents a software environment for processing, segmenting, quantifying, representing and manipulating digital microstructure data. The paper discusses the approach to building a generalized representation strategy for digital microstructures and the barriers encountered when trying to integrate a set of existing software tools to create an expandable codebase.

Keywords: Electron back-scatter diffraction; Synthetic microstructure; HDF5; 3D Microstructure reconstruction; Programming library; Open-source

Background

In recent years, two major initiatives have been introduced that promise to affect how the materials science community integrates with the larger system design process. These initiatives, known as Integrated Computational Materials Engineering (ICME) and the Materials Genome Initiative (MGI), are built on the ability to represent materials digitally, both in a structural and performance context. Under the ICME construct [1], materials engineering can be treated as a series of models (empirical or physical) that link a processing history to a suite of properties (mechanical, optical, electromagnetic, etc.). In the most general terms, processing models predict the internal structure of materials under some processing conditions, either directly or through a correlation with continuum state variables like thermal history and strain path. Similarly, property models predict a material's performance under some operating conditions, given a description of its internal structure. Thus, it becomes obvious that the natural link between these models is the internal structure of the material that is output from one and input to the other. The internal structure of nearly all materials is complex, multi-scale and not easily defined by a small number of parameters. As such, there exists an opportunity in materials engineering to advance the quantitative description of internal structure and move further away from ad-hoc, word-based descriptors (i.e. equiaxed, acicular, basket-weave, etc.). Historical efforts have been made to quantify selected aspects of microstructure (ASTM grain size, etc.), but generally the metrics chosen stopped at average quantities, which in part has been driven by the limited description of microstructure in models. The MGI has challenged the materials community to develop a framework for describing materials in a consistent and quantitative way [2], more similar to the approach applied to sequencing the human genome.

Two critical themes in these initiatives are the move to the digital basis and the call for tools with clear and understandable inputs/settings (be they software or hardware). There are 'easy-to-use' software tools that exist in both the processing (ProCast, Deform, etc) and property (Darwin, Abaqus, Deform, etc) modeling regimes. However, there is a lack of easy-to-use software tools that exist to process, quantify and represent microstructure in a general sense, especially in three dimensions (3D). This becomes a problem if one is attempting to validate the predictions of processing models or provide property models with accurate input. The work discussed in this paper is aimed at developing a software architecture that is both open and scalable to address the growing needs for quantitative, digital analysis of microstructural data. The ultimate goal of this effort is to fill the gap in the ICME chain with respect to 'easy-to-use' microstructure quantification and representation tools across all material classes and length scales. Another important goal of this work is to standardize the format of material microstructure data, so that the increasing demand for access to scientific research data can be met [3].

It should be mentioned that the initial focus of DREAM.3D was far less general and pervasive than the ideas discussed in this paper. It was only during this initial development effort that the authors encountered the difficulties that will be discussed here and subsequently broadened the scope and vision of DREAM.3D. This broader vision is in line with efforts in the biological community [4,5] and the authors see a potential for further integration with that community. Many of the examples in this paper reflect the personal experiences of the authors and within this context, we highlight the path needed for the advancement of digital microstructure analysis. Note that microstructure is used throughout this paper as a general term for the internal structure of materials and does not refer to a specific length-scale.

Barriers to integration/development

At the outset of this work, many computational tools existed for treating various aspects of microstructure quantification, post-processing/clean-up, data visualization, etc. However, these tools remained disjointed and generally non-transferable between researchers. It became clear, both to the authors of this work and authors of many of the disjointed tools, that a larger integrated environment was needed to be developed to fully realize the utility of any of the individual tools. Integrating computational codes into a larger ecosystem presents many barriers, for example: storage format of the data, usability of the codes (Graphical User Interface (GUI) vs. Command-Prompt), documentation and Intellectual Property (IP) rights, to name a few. In the typical case, each of these issues needs to be addressed in order for the code to be widely usable by other researchers. A critical component for efficiently solving these issues in a consistent way is having a long-range vision for how the software will be used and any possible growth opportunities. In the following subsections, some of the most critical barriers are highlighted and how the authors addressed them in the development of DREAM.3D will be discussed.

Data structure and storage

One of the critical barriers to integration of software codes is ensuring that downstream algorithms can properly interpret the data produced by upstream algorithms.

Misunderstanding how data is structured can lead to a significant barrier that prevents algorithms from being integrated. During the development process, often the researcher is mainly focused on the correct implementation of the algorithm and gives less time to designing a data and file format that is both efficient from a computational standpoint and shareable with other researchers. This leads to input and output files that are not written in any standardized format. Subsequently, substantial effort may go into manipulating the output files of one algorithm so that the next algorithm can use the data as an input, which is highly inefficient whether done manually or by a computer. A basic example of this problem is an algorithm that stores data as a comma separated list of values and another algorithm that reads values from a space separated list of values. In order for these algorithms to work seamlessly together, one or both of the codes would need to be modified or commonly a third program would be created to 'translate' between the data structures. Developing software this way hinders the reusability of the codes and will present a barrier to the adoption of the codes in the greater community. DREAM.3D aims to use a widely available open-source format to store both archival and processed data. However, this only addresses the external, or resting format of data. Another important issue is how the data is represented internally, which can significantly impact how easily algorithms are able to share data and information. DREAM.3D utilizes a scalable organization to describe data at all dimensionalities.

Ease of use

As mentioned previously, research grade codes are often developed with little emphasis given to the usability of the codes by other researchers. Generally, this is because the code is never intended for use beyond the author. Many codes typically run from a command prompt or terminal environment and offer little information about the required number and types of input parameters. Worse still is when the author stops actively developing the code and the knowledge of how to use the algorithm and the sensitivity to its input parameters is lost. When this happens, the code becomes effectively unusable. DREAM.3D has tried to mitigate these situations through the use of formal coding protocols. These protocols dictate how the documentation for a filter is written, how the user interface is created and how the filter will interact with the rest of the system. Collectively these formal design patterns are used to ensure that the filter can be employed by researchers in the field simply by reading a documentation file and/or following a simple example. Some of the items that go into integrating a filter into the DREAM.3D system include, but are not limited to, the following items

- Filter is documented including required input parameters and data, output data created and an explanation of the algorithm (including citations if needed).
- Input parameters are written to and read from a native DREAM.3D file.
- Required inputs are enumerated using native DREAM.3D data structures.
- Outputs are clearly defined and relayed to the DREAM.3D internal data structures.

Intellectual property

During the early stages of DREAM.3D a conscious decision was made to structure the codes in such a way as to allow the use of external libraries that may contain proprietary algorithms. This allows academia, industry and government institutions to contribute algorithms and still protect their intellectual property. These various institutions can elect to release the source code to the open community or keep the source private and only release a precompiled library that is compatible with the current release of DREAM.3D. As other institutions begin to contribute to DREAM.3D they can make their own decision as to what is the best model to distribute their specific computational tools.

Long-term maintainability

The maintainability of DREAM.3D has several facets of discussion. From a programmer's perspective the authors strive to use best practices when developing the various algorithms, reusing algorithms and creating reusable software objects that can be applied or adapted to new algorithms. A suite of unit tests are continually developed to ensure the behavior of the public functions is not altered when bug fixes and algorithmic enhancements are added. Currently, all the external libraries that DREAM.3D is built on top of are all open-source, thus giving the development team complete access to the entire code base that is used to build DREAM.3D. Another perspective to consider is the source of funding for DREAM.3D development. The current development of DREAM.3D has been essentially exclusively funded by U.S. Government sources. This funding enabled building the integrated core infrastructure that enabled much of the critical aspects already discussed. However, the core of DREAM.3D should become relatively static, with only minor 'usability' additions in the near future. It will be at this point, which has already begun to occur, that academia and industry will begin to drive the growth and development of DREAM.3D. This growth will likely be focused almost entirely on filter development and expansion. It is the belief of the authors that the materials community, possibly with government support (either directly or through academic funding), will view the core as an enabling tool that will be in the 'best interest' to update as needed with a small overhead on filter development efforts. Finally, it should be noted that since DREAM.3D is currently open-source, the current instance of the core will always be available. Any filter additions that can operate with the current design can always be used. Also, any user can download the current core and extend it to address their research needs.

Methods
Data representation and file format

Material microstructures come in many different sizes and shapes and the features of interest have different dimensionalities. Data describing attributes of microstructure can be obtained from many sources (Scanning Electron Microscopy, Transmission Electron Microscopy, Optical Microscopy, Electron Backscatter Diffraction, Energy Dispersive Spectroscopy, Wavelength Dispersive Spectroscopy, 3D Atom Probe, Atomic Force Microscopy, etc.). Unfortunately, during the development of these experimental methodologies, no common data structure was developed and as such, combining data from multiple sources is difficult. Further, the tendency to link the data with a material class (metal, ceramic, composite, polymer, etc.) has stunted the development of a unified method for describing microstructure data. During development of DREAM.3D, the vision of a unified representation of all digital microstructure data for all material classes and length-scales presented a challenge. As discussed in [6], when writing code or designing data structures that operate on or represent a variety of features and dimensions, it is critical to establish a proper abstraction layer to ensure transferability. In the case of DREAM.3D and microstructure, the authors believe the proper abstraction layer is to work with all features of structure as geometrical objects. By abstracting the materials interpretation of the features and focusing only on how the feature is described digitally, DREAM.3D has been able to institute a general, unified structure for digital data that assumes no prior knowledge of length-scale or material class. The following subsections will discuss this generic data structure and illustrate its direct use in a wide range of materials applications.

Geometric mesh element construct for holding digital microstructure data

Spatially-resolved digital data, of which most material microstructure data is a subset, are simply information or attributes that are associated with discrete geometrical elements. These elements can be pixels in an image, points in a probe scan, line segments in a digital model, etc. At this level, all digital microstructure data can be treated/organized similarly within a computer. Meshes of appropriate dimension can be created and data can sit on the mesh element(s) that they describe. For example, the mass-to-charge ratio of an atom in an atom probe dataset is information associated with a point, while the misorientation across a boundary in a electron backscatter diffraction (EBSD) dataset is information associated with a surface. As such, any given dataset has an associated mesh dimensionality equal to the highest dimension of feature its data describes. It should be noted that the mesh dimensionality may be different from the dimensionality of the dataset. For example, the atom probe dataset consists of 3D locations having (x,y,z) coordinates, but represents microstructure features that are treated as a 0-D point.

DREAM.3D organizes/stores mesh data (and subsequent feature and ensemble data discussed in the next section) in a structure called a "data container". DREAM.3D uses four types of data containers for the different possible data dimensionalities (Vertex = 0D, Edge = 1D, Surface = 2D, Volume = 3D). Figure 1 illustrates the different data containers and the data they can hold. As Figure 1 shows, lower dimensional geometrical objects bound higher dimensional objects and a given data container can store data on mesh elements of a lower dimension. For example, in a 3D EBSD dataset, the collected orientation data is generally treated as belonging to a cell, but the misorientation between neighboring cells could also be stored on the face shared by the cells and the edges and vertices of the cells could store the coordination number of different features they belong to (i.e. triple line or quadruple point). An example dataset of each type of data container can be found in the supporting material. The examples include a Vienna Ab initio Simulation Package (VASP) input structure (Vertex data container - Additional file 1), a ParaDis output structure (Edge data container - Additional file 2), a grain boundary mesh of a synthetic polycrystalline microstructure (Surface data container - Additional file 3) and a synthetic polycrystalline microstructure (Volume data container - Additional file 4).

The mesh that represents the data locations is unique to the dataset itself. While the mesh can be altered via smoothing, regridding or other processing steps, it is generally defined by the data collection or generation protocol/settings. Furthermore, the mesh itself is not influenced by the material class and can exist at any length-scale. The mesh is solely the physical location of all data elements and their associated attributes.

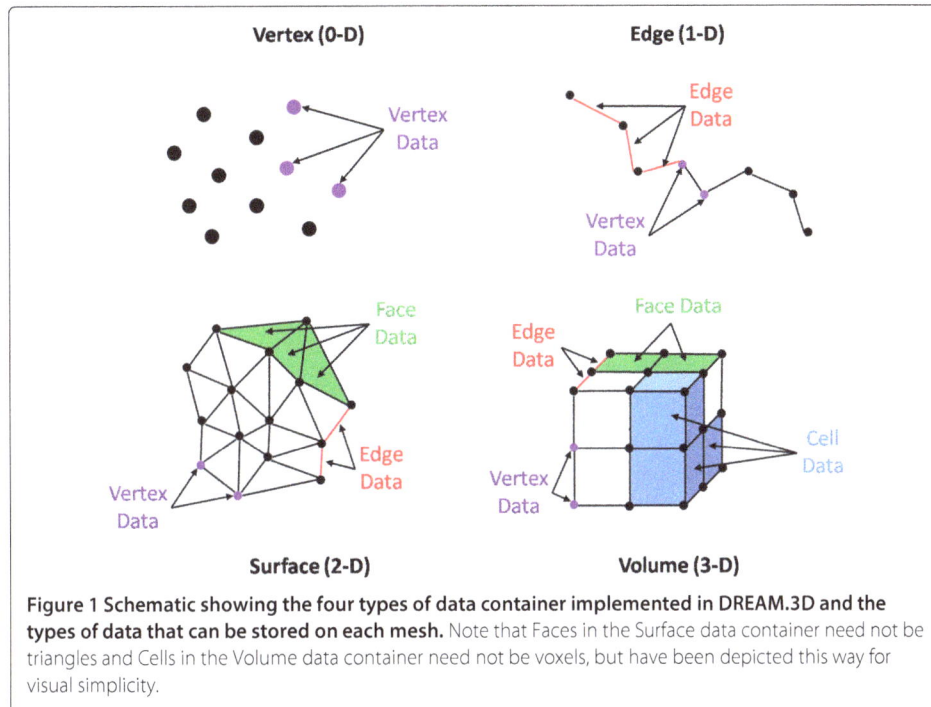

Figure 1 Schematic showing the four types of data container implemented in DREAM.3D and the types of data that can be stored on each mesh. Note that Faces in the Surface data container need not be triangles and Cells in the Volume data container need not be voxels, but have been depicted this way for visual simplicity.

Hierarchical grouping for feature and ensemble representation

A given material's microstructure can be thought of as being constructed using building blocks called "features" such as grains, fibers, pores, magnetic domains, corrosion pits, dislocations, individual atoms and many other possibilities. Though these features are very different in the "real world" material's sense, digitally they are all simply groups of discrete mesh elements. It is the user's prerogative to group the mesh elements in whichever way makes most sense for their uses, which imparts a certain uniqueness to the data set. It is the human interpretation of what the features represent that links the data to a specific material class and/or length-scale. DREAM.3D utilizes a software engineering technique where all of the domain specific groupings can be represented by a generalized data structure. This is commonly referred to as an "Abstraction Layer" in the software engineering field and allows the DREAM.3D system to grow and adapt to new domains.

From the perspective of the computer, the act of assigning elements to a given feature is still material class and length-scale independent. Mesh elements are simply noted to belong to a given feature for a given segmentation/grouping protocol. For each grouping/segmentation protocol, all elements are set to belong to one and only one feature. It is possible that a user would want to group mesh elements by multiple protocols. For example, mesh elements could be grouped by common orientation and then by common chemistry if a data set had both orientation and chemical information. If multiple grouping protocols are used, then each mesh element would have a vector of feature IDs listing which feature it belongs to in each grouping.

After features are defined, attributes such as size, shape, etc. can be calculated and stored associated with each feature. The structure of how these attributes are stored will be discussed in the next section. Also, it may be desirable to the user to group features together to establish "ensembles". Ensembles are groups of features that the user has linked for some reason. Similar to each mesh element having one (or more) feature IDs to list what feature it belongs to, each feature has one (or more) "ensemble IDs". For example, a group of features could be linked because they are all the same phase, because they are the largest 10% of features, etc. Similar to features and individual elements, attributes describing ensembles such as size distribution, average feature curvature, orientation distribution function (ODF), etc. can be calculated and stored associated with each ensemble.

Scalable layout for information storage

At all levels, from the individual mesh elements to features and ensembles, the method of how information is stored must be dynamic. In order to be a flexible software environment that can work with data from multiple sources and treat microstructures from all material classes, it is not reasonable for DREAM.3D to predefine what attributes can be associated with a mesh element, feature or ensemble. As such, a matrix-style container is needed for holding information of this type. For example, in an EBSD scan, each pixel has an Euler angle set, a phase ID, a coordinate in space and a list of values associated with the indexing approach of the commercial software that collected the scan. These attributes, as a set, are called a 'property vector' in DREAM.3D and define the pixel with which they are associated. These property vectors are shown as columns in Figure 2. The rows in Figure 2 are the lists of single attributes for all pixels and are called "attribute array". Given this container structure, it becomes clear that as filters are applied to the data, more attribute arrays are generated and each property vector grows.

At each level (mesh element, feature, ensemble), attribute matrices can exist. Only one matrix exists at the element level because there is no user grouping at that level and as such there is only one definition or instance of the mesh. However, at the feature and ensemble levels, many attribute matrices can coexist. In an attribute matrix, every property vector is the same size and every attribute array is the same size. This is because filters calculate attributes and filters must loop over all members in the attribute matrix for which the attribute is being calculated.

HDF5 File structure

The Hierarchical Data Format Version 5 (HDF5) is an open-source library developed and maintained by "The HDFGroup" [7] that implements a file format designed to be flexible, scalable, highly performant and portable. HDF5 allows each application to organize

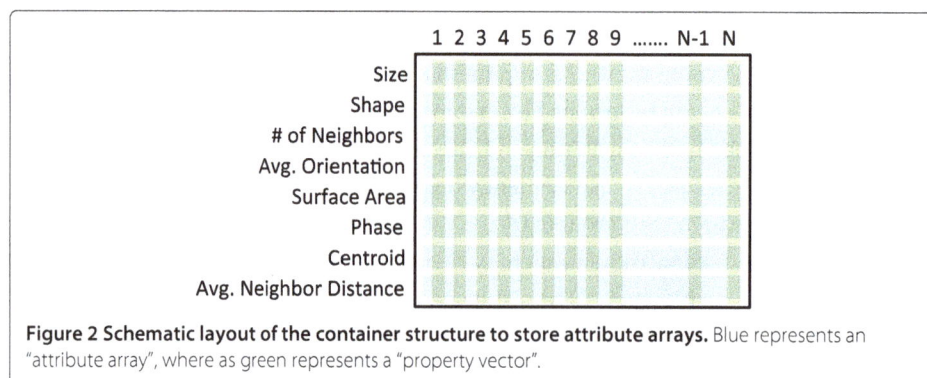

Figure 2 Schematic layout of the container structure to store attribute arrays. Blue represents an "attribute array", where as green represents a "property vector".

its data in a hierarchy that makes sense for the application. Virtually any type of data, from scalar values to complex data structures, can be stored in an HDF5 file. Scalability has been a design consideration from the outset and HDF5 can handle data objects of almost any size or dimensionality. The library has also been designed to be efficient at querying, reading and writing data objects, and including utilizing parallel I/O when needed. One of the most important aspects of HDF5 is its portability across all the major computing operating systems. HDF5 has support for C, C++, Fortran and Java as its native implementations; many higher-level programming languages also have direct support for HDF5, including IDL (Interactive Data Language), MATLAB and python.

HDF5 files can be thought of as 'a file system within a file'. Data can be stored as datasets (analogous to files) and arranged inside groups (analogous to folders) all within the HDF5 file. This structure is well-suited for storing the organized data from DREAM.3D. The organization of a typical DREAM.3D file is shown in Figure 3. At the 'root directory' or highest level in the file, two groups exist for holding 1) the processing pipeline and 2) all data containers of the dataset. Inside the pipeline group, there are subgroups for each filter and within each subgroup there are datasets for each of the input parameters of the filter. The subgroups are titled as their numerical order in the processing pipeline, but have attributes stored on the group listing the name of the filter and its version number. The datasets inside the subgroups are titled as the name of the input parameter they hold and the contents are the value(s) of the input parameter. Inside the data container group are subgroups for each data container that exists in the dataset. The subgroups are titled as the name the user gave to the data container. Within each subgroup there are multiple groups (the number depending on the dimensionality of the data container).

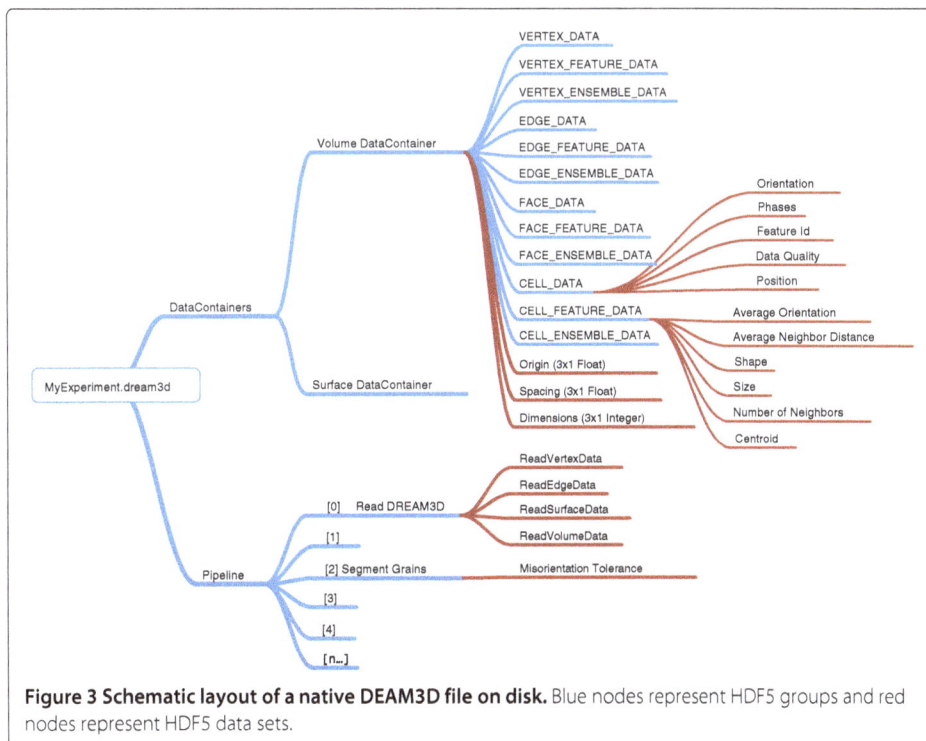

Figure 3 Schematic layout of a native DEAM3D file on disk. Blue nodes represent HDF5 groups and red nodes represent HDF5 data sets.

Each group at this level is associated with an attribute matrix described in the previous section. For example, if the data container was a vertex data container, then there would be a group for the vertex mesh element attribute matrix and there could be multiple groups of feature and ensemble attribute matrices depending on the number of grouping schemes employed by the user. In the example in Figure 3, the dataset contains a single volume data container. Within an attribute matrix group, each dataset represents an attribute array (or row from Figure 2). The name of the dataset is the name of the attribute array and the contents are the entire attribute array in order from object 1 to N.

The structured layout of HDF5 and the DREAM.3D file also offer potential for databasing of datasets. The ability of HDF5 to query the existence of datasets and groups without reading the entire file is well-suited for determining if data meets a specified criterion, whether it be a specific processing path, attribute array, etc.

Pipeline concept

DREAM.3D's pipeline workflow is designed around the concept of signal processing. In this analogy, the signal is the 'raw' data and the individual algorithms/programs in DREAM.3D are filters that process the signal. It is for this reason that DREAM.3D refers to each individual program as a filter. It should be noted that unlike typical signal or image processing, many of the filters in DREAM.3D do not change data/attributes existing on each element, feature or ensemble, but rather create new data/attributes to be stored. The intent of modular pipeline workflows is to separate the two critical aspects of data processing: algorithms and order of operations. When designing a pipeline, the user is solely focused on the latter while using an existing set of algorithms. Each algorithm can be treated as a module that can be modified or replaced if it is not generating the desired results. The following subsections will discuss the user interface of the workflow and how it is linked to the data.

Visual programming workflow

In a typical high level programming environment, such as MATLAB or IDL, the user must manually type in the proper commands to build the desired pipeline/workflow and ensure that all data is available during the execution. Many times missing data can cause the systems to crash at worst or give a cryptic error message in the best case. With DREAM.3D a visual approach to designing the workflow was engineered. Each filter can still be thought of as a pre-packaged subroutine like the functions in MATLAB or IDL, but in DREAM.3D a visual linking of the filters/subroutines is more analogous to programming environments like LabView. Each filter has the knowledge of every piece of data that is required before it will execute. As each filter is placed into the pipeline area the workflow is executed in a "preflight" step where each filter dynamically checks to make sure it will have the required input data to operate (also similar to LabView). If any inputs are not correct or there is missing data an error message is displayed for the user to correct. Once all the errors are corrected the pipeline will be allowed to be executed. In this respect DREAM.3D presents a very high level and simple programming model that is easy and straight forward to learn. An example of the DREAM.3D GUI with a pipeline containing errors is shown in Figure 4 to illustrate the layout of the software.

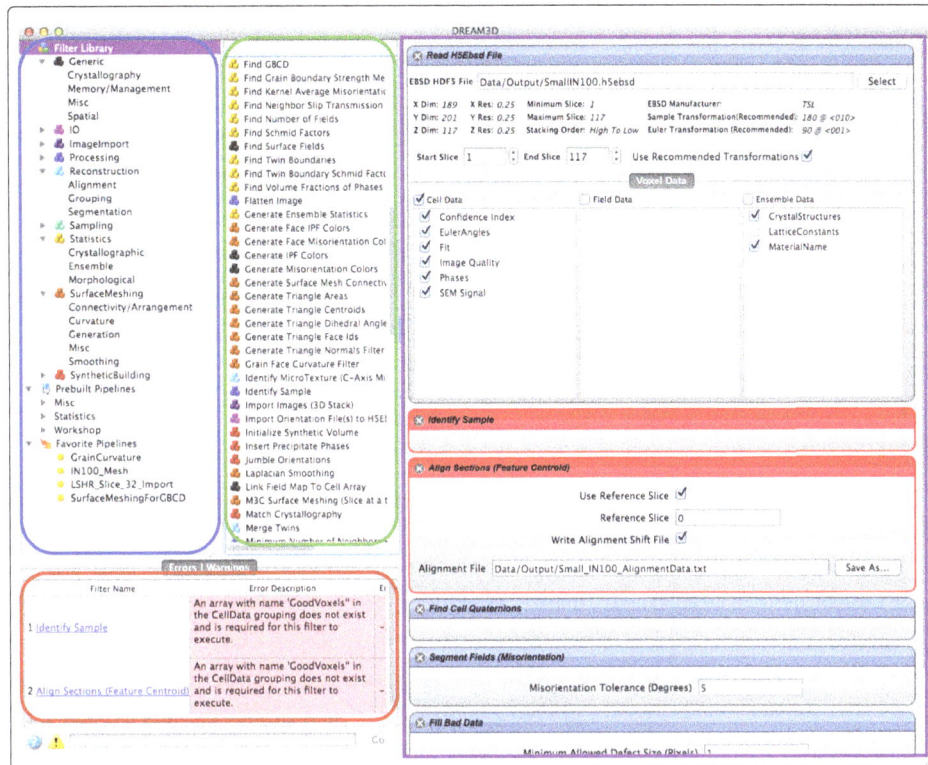

Figure 4 DREAM.3D GUI with example pipeline. The region highlighted in blue contains the Filter Library, Prebuilt Pipelines and Favorite Pipelines, where filters are grouped and common pipelines are stored. The region highlighted in green contains the Search Window, where filters can be searched or viewed by group. The region highlighted in purple contains the Pipeline Area, where filters are inserted and ordered to build a pipeline. The region highlighted in red contains the Errors and Warnings Window, where messages regarding the feasibility of the pipeline are displayed to the user.

Working file format and complete data provenance

Continuing the analogy of signal processing, one can think of the digital data, raw or otherwise processed, as existing in a discrete processing-step domain. Between all filter steps in the processing pipeline, the data exists as a 'snapshot'. DREAM.3D terms these points as a 'digital instance' of the microstructure. Any digital instance should be reproducible by beginning with the raw data and following the same processing pipeline to the point the instance was captured. It is this realization that led to the differentiation between a working file and an archival file in DREAM.3D. An archival file, which is discussed in more detail in (Jackson, Groeber, Rowenhorst, Uchic and DeGraef: "h5ebsd: An archival data format for electron back-scatter diffraction data sets.", submitted), contains the unaltered data from the collection instrument along with meta data that may help infer the inherent artifacts of digitizing the true 'analog' microstructure. DREAM.3D terms the unaltered data as step = 0 in the processing-step domain. A working file implies the contained microstructure is beyond step = 0 in the processing-step domain and some filtering has been applied to the data. During the execution of a processing pipeline in DREAM.3D, the user can export/save an instance of the microstructure at any point by writing all of the in-memory data arrays to a data file using the HDF5 format. These files are organized by each data container that is being used within the active pipeline. In addition to saving the complete set of in-memory data arrays to a file, the complete processing history of that data is also saved in the form of the user's pipeline up to the point of saving the instance. If the user selects to start a pipeline by reading a DREAM.3D file, the previous pipeline stored in the starting file is transferred to any newly written DREAM.3D file, thus keeping the entire provenance of the data safely stored in the same file. The pipeline information is written such that the specific version of DREAM.3D that was used during the processing is attached to each filter's inputs values. This type of meta-data attachment can help researchers independently recreate results of past experiments and is fast becoming an important aspect of scientific publishing.

Growth and scalability of DREAM.3D

For DREAM.3D to realize the vision of a material class and length-scale independent analysis environment, careful thought had to be given to the creation of a method for implementing new filters with a low barrier to entry. No one researcher, group or laboratory has the knowledge diversity or time to implement all useful processing or analysis filters for even one material class, let alone all. For this reason, an environment where collaboration and competition of ideas/algorithms/implementations can occur on a common basis is critical to the development of standards for microstructure analysis.

Plugin Architecture

In order to allow DREAM.3D to grow organically through the addition and integration of new algorithms, a plugin architecture has been designed and implemented. This allows researchers with some programming experience to expand and enhance the capabilities of DREAM.3D. The plugin architecture allows entities such as government labs and commercial businesses to create DREAM.3D compatible binary plugins and retain full rights to their specific sources. Plugin developers can distribute their tools in a number of different ways. First, the developer(s) can contribute their filters directly to DREAM.3D and have them compiled with the core. The core of DREAM.3D is simply the previously discussed internal data management classes, macros to facilitate filter-to-filter communication, the GUI and a set of libraries for common operation like math and I/O. This path is only possible if the developer(s) release their plugin as open-source, as the core of DREAM.3D is open-source. A second option is for the developer(s) to compile their plugin themselves and then distribute their plugin as a library. Under this second option, the developer(s) retain multiple avenues for dissemination. The plugin can be offered as freeware, can be licensed or can be provided with the source (albeit disjointed from the DREAM.3D core). Offering these options is intended to help drive adoption of DREAM.3D across a diverse set of materials science domains and industries. This type of programming model leverages contributions of different organizations to allow the entire DREAM.3D system to grow larger, thus spreading the development cost among all of those different organizations.

Documented interface protocol and common libraries

DREAM.3D has been developed with a concerted focus on lowering the barrier for future developers to contribute codes. Common libraries for math, I/O and internal data management are supplied with the core of DREAM.3D. DREAM.3D also supplies libraries for dynamically creating the visual presence of a filter, provided a relatively simple interface required by the filter. This limits the amount of 'low-level' computer science knowledge the developer must have. Furthermore, DREAM.3D can be compiled to build an accompanying program that will generate all necessary files for a user that is making a new plugin. The shell files created contain all the required functions of new filters along with examples of how to add input parameter calls to the user and how to request and add data to the scalable attribute matrices. This allows the developer to focus on their algorithm, while operating in a 'put-your-algorithm-here, list-your-input-requirements-here, list-your-output-details-here' type of environment.

Interface with external software

Future growth of DREAM.3D is also likely to be tied to the ability to integrate DREAM.3D with other software packages. For example, the authors made a conscious decision early in the development of DREAM.3D to not invest time and effort into generating a visualization package within DREAM.3D. Instead, a link was built to interface with ParaView [8], an open-source visualization environment developed by Kitware with Department of Energy (DoE) funding. ParaView is a powerful visualization package with many, many man-years of development already invested. Using the HDF5 file structure already discussed, coupled with an XML description (Xdmf format), DREAM.3D files can be opened and viewed within ParaView. As such, new developments to ParaView are indirectly developments to DREAM.3D.

Results and Discussion
Case studies

This section will demonstrate the workflow and data structure of DREAM.3D in a set of case studies. Due to the historical focus of the software tools that evolved to become DREAM.3D, many of the filters currently in DREAM.3D are related to processing and analysis of polycrystalline metal datasets with 3D EBSD data. The case studies presented here show a subset of current DREAM.3D functionalities, but should not be viewed as an exhaustive list of current or future capabilities. Furthermore, the final results of the various pipelines may not be different than previous codes or similar analysis software packages. The major differentiating factor with DREAM.3D is the time and manual effort to get results. The use of a visual representation of the workflow reduced the learning curve greatly, which makes DREAM.3D an approachable software suite for all levels of user.

Reconstruction and Meshing (3D EBSD)

A dataset consisting of 117 serial sections through a polycrystalline Ni-based superalloy with EBSD data on each section was collected in [9]. DREAM.3D was used to reconstruct, segment, clean-up and mesh the features of the dataset. The pipeline used to accomplish these tasks, which is listed in Table 1, will be discussed briefly. In the interest of brevity, the details of each individual filter will not be discussed here, but can be found in the documentation of DREAM.3D. Many of the steps in the pipeline are also discussed in [9] and [10]. It should be noted that the results presented in [9] and [10] were generated prior to the existence of DREAM.3D. The total processing/analysis time in the previous work

Table 1 3D EBSD reconstruction pipeline

Filter #	Filter name	Reason for use
0	Read H5EBSD File	Loads raw EBSD data
1	Multi Threshold (Cell Data)	Allows user to define which voxels are 'good'
2	Find Cell Quaternions	Converts voxel Euler angles to Quaternions
3	Align Sections (Misorientation)	Rough alignment of sections by minimizing misorientation between sections
4	Identify Sample	Adjusts the 'good' voxels assuming one contiguous block of 'good' data
5	Align Sections (Feature Centroid)	Secondary alignment of sections assuming sample is parallelepiped
6	Neighbor Orientation Comparison	Checks orientation of 'bad' voxels against neighboring 'good' voxels
7	Neighbor Orientation Correlation	Second check of 'bad' voxels against neighboring 'good' voxels
8	Segment Features (Misorientation)	Identifies features of similar orientation
9	Find Feature Phases	Determines the phase of each feature
10	Find Feature Average Orientations	Calculates average orientation of each feature
11	Find Feature Neighbors	Determines list of neighbors for each feature
12	Merge Twins	Merges features misoriented by 'special' sigma3 relationship
13	Minimum Size Filter	Removes small features and fills gaps with neighboring features
14	Find Feature Neighbors	Determines neighbors after removing features
15	Minimum Number of Neighbors Filter	Removes features with few neighbors
16	Fill Bad Data	Fills in 'bad' data with neighboring 'good' data if 'bad' data regions are small
17	Erode/Dilate Bad Data	Shrinks any remaining 'bad' data regions
18	Erode/Dilate Bad Data	Grows back any remaining 'bad' data regions
19	Write DREAM.3D File	Writes attribute matrices and pipeline to file

Table listing the filters in the reconstruction pipeline.

took approximately 24 hours and involved moderate manual interaction between multiple software codes. The current processing/analysis time was reduced to approximately 5 minutes and required effectively no user interaction (beyond setting up the pipeline). In both cases, the times quoted reflect use of a standard desktop PC. The resulting digital instance is shown in Figure 5 and is attached as supporting material in the form of a DREAM.3D file (Additional file 5).

Statistical analysis

The previous section discussed the reconstruction and segmentation of a polycrystalline Ni-based superalloy dataset with 3D EBSD data. Upon reconstructing and segmenting the data to obtain features, those features and ensembles of those features can be measured and statistically described. Table 2 lists the pipeline used to calculate a number of morphological and crystallographic attributes of the features and ensembles within the dataset. The list of features and all attributes calculated to describe them can be found in a comma separated value (.csv) file in the supporting material (Additional file 6). Some of these results were also presented in [10] (using tools prior to DREAM.3D). After determining the attributes of the individual features, distributions of those attributes can be calculated for ensembles of the features. For this dataset, the material was treated as single-phase and all grains were said to belong to a single ensemble. The distribution of sizes, shapes, numbers of neighbors, orientations and misorientations

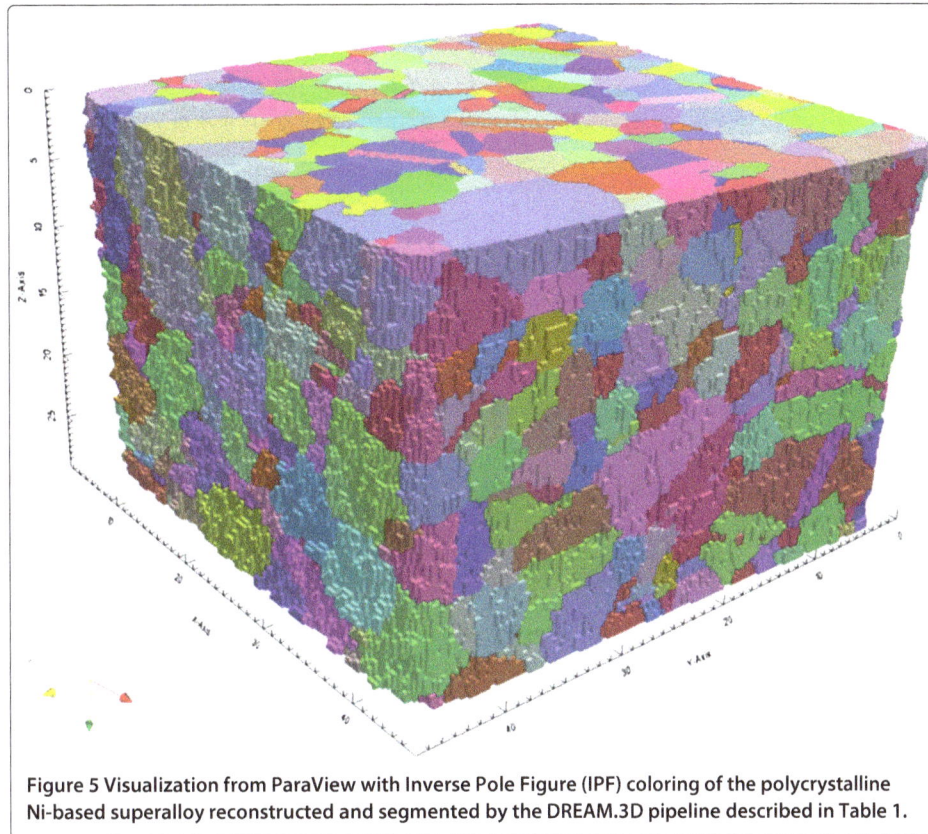

Figure 5 Visualization from ParaView with Inverse Pole Figure (IPF) coloring of the polycrystalline Ni-based superalloy reconstructed and segmented by the DREAM.3D pipeline described in Table 1.

Table 2 Statistics pipeline

Filter #	Filter name	Reason for use
0	Read DREAM.3D File	Loads reconstructed and segmented dataset
1	Find Feature Centroids	Determines the centroid locations of each feature
2	Find Feature Sizes	Determines the volume of each feature
3	Find Feature Shapes	Determines aspect ratios and omega3 of each feature
4	Find Feature Neighbors	Determines the number and list of contiguous neighbors for each feature
5	Find Feature Neighborhoods	Determines the number and list of features within one diameter of each feature
6	Find Euclidean Distance Map	Determines the distance each voxel is from the nearest grain boundary, triple line and quadruple point
7	Find Feature Average Orientations	Second check of 'bad' voxels against neighboring 'good' voxels
8	Find Feature Average Orientations	Calculates average orientation of each feature
9	Find Feature Neighbor Misorientations	Determines the misorientation for each contiguous neighbor of each feature
10	Find Schmid Factors	Determines the Schmid factors of each feature
11	Find Feature Reference Misorientations	Determines the misorientation between each voxel and a reference orientation for the feature it belongs to
12	Find Kernel Average Misorientations	Determines the average misorientation between each voxel and its neighbor voxels
13	Write Feature Data As CSV	Outputs attributes of features to CSV file
14	Write DREAM.3D File	Writes out all attribute matrices and pipeline to file

Table listing the filters in the statistical analysis pipeline.

Table 3 Synthetic structure generation pipeline

Filter #	Filter name	Reason for use
0	Initialize Synthetic Volume	Loads goal statistics and creates empty volume
1	Pack Primary Phases	Generates set of grains and places them inside of volume
2	Find Feature Neighbors	Determines the number and list of contiguous neighbors for each feature
3	Find Number of Features	Determines the number of features in the volume
4	Match crystallography	Assigns orientations to match the ODF and MDF
5	Write DREAM.3D File	Writes out all attribute matrices and pipeline to file

Table listing the filters in the synthetic structure pipeline.

were all calculated. The following section will discuss one use case for applying this information.

Sythetic structure generation

DREAM.3D has filters to generate synthetic digital microstructures with a goal set of statistics as input. The synthetic generation process is discussed in detail in [11]. Using the statistics calculated by the pipeline in the previous section, a 'statistically-equivalent' microstructure was generated using DREAM.3D. The pipeline used to generate this microstructure is listed in Table 3 and the DREAM.3D file corresponding to the synthetic volume is attached as supporting material (Additional file 4). A visualization of the resultant synthetic microstructure is shown in Figure 6. It should be noted that the synthetic microstructure generated was created using statistics calculated without the twin features

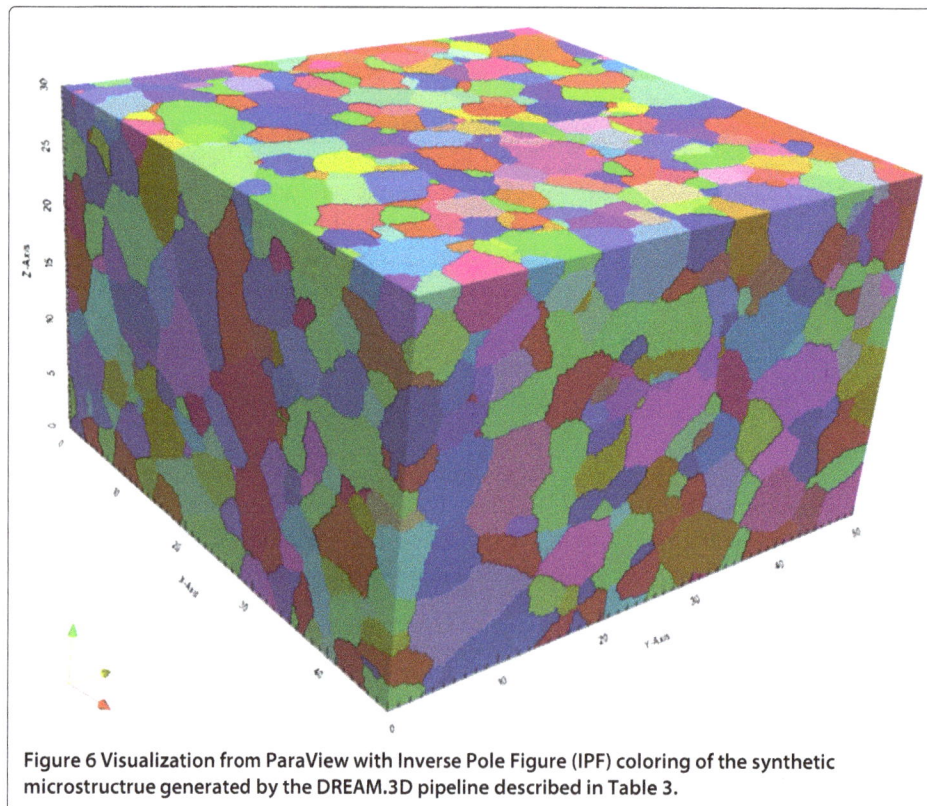

Figure 6 Visualization from ParaView with Inverse Pole Figure (IPF) coloring of the synthetic microstructrue generated by the DREAM.3D pipeline described in Table 3.

in the microstructure and as a result may look slightly different that the experimental microstructure in Figure 5.

Conclusion

DREAM.3D is an open-source software package focused on creating a high-level programming environment to process, segment, quantify, represent and manipulate digital microstructure data. DREAM.3D's central goal is to enable the move of microstructure quantification to a digital basis with easy-to-use software tools. The core of DREAM.3D implements a standardized approach to working with and storing digital microstructure data. Additionally, protocols are included to allow independently-developed filters and plugins to interface with one another. The DREAM.3D environment is constructed in a way that small research groups, government laboratories, start-up companies and major industrial corporations can collaborate and leverage each other's work. It is the belief of the authors that DREAM.3D will reduce the time and cost to conduct microstructural characterization, due to the ability to leverage community-wide developments and bring disjointed research areas into a common environment for development.

Additional files

Additional file 1: Example vertex data container. This file is an example of a vertex data container containing a Vienna Ab initio Simulation Package (VASP) input structure.

Additional file 2: Example edge data container. This file is an example of an edge data container containing a ParaDis output structure.

Additional file 3: Example surface data container. This file is an example of a surface data container containing a grain boundary mesh of a synthetic polycrystalline microstructure.

Additional file 4: Example volume data container. This file is an example of a volume data container containing a synthetic polycrystalline microstructure.

Additional file 5: Polycrystalline ni-based superalloy 3D EBSD reconstruction. This file contains a reconstructed and segmented experimentally measured polycrystalline Ni-based superalloy microstructure.

Additional file 6: Grain statistics. This file contains the attributes calculated in the Statistics pipeline for all features identified in the 3D EBSD reconstruction pipeline.

Competing interests
The authors declare that they have no competing interests.

Authors' contributions
The authors contributed to this paper equally and would like to be considered as joint First Authors. MJ generally contributed more of the computer science vision and MG generally contributed more of the materials engineering vision. All authors read and approved the final manuscript.

Acknowledgements
MAJ would like to acknowledge financial support from the U.S. Air Force Research Laboratories contracts # FA8650-07-D-5800 and # FA8650-10-D-5210 and U.S. Naval Research Laboratories contract # N00173-07-C-2068. The authors would like to acknowledge the following people:

- Chris Woodward for his strong belief in the need to integrate software tools that initiated this effort
- Marc DeGraef, Sukbin Lee, Greg Rohrer, Anthony Rollett, David Rowenhorst, and Joseph Tucker for sharing their codes and knowledge to help create the initial functionality of DREAM.3D and for setting a standard for what collaboration should be in these types of ventures
- Michael Uchic and Dennis Dimiduk for their daily discussions and consistent belief and support
- Joseph Kleingers for his efforts to improve the usability of DREAM.3D
- Jeff Simmons for his vision beyond the materials community that helped crystallize the views now central to DREAM.3D
- Paul Ret for his unwavering support for DREAM.3D's vision
- The early adopters of DREAM.3D for their patience and suggestions that have helped to grow and strengthen the software.

References

1. Committee on Integrated Computational Materials Engineering (2008) Integrated Computational Materials Engineering: a Transformational Discipline for Improved Competitiveness and National Security. National Research Council, The National Academies Press, Washington D.C.
2. National Science and Technology Council (2011) Materials Genome Initiative for Global Competitiveness. Executive Office of the President, Washington D.C
3. Holder J (2013) Increasing Public Access to the Results of Scientific Research. https://petitions.whitehouse.gov/response/increasing-public-access-results-scientific-research.
4. OME. http://www.openmicroscopy.org/site/.
5. OSS. http://loci.wisc.edu/software/oss/.
6. Pietzsch T, Preibisch S, Tomancak P, Saalfeld S (2012) ImgLib2 - generic image processing in Java. Acta Materialia 28(22): 3009–3011
7. HDF5. http://www.hdfgroup.org/HDF5/.
8. Kitware. http://www.paraview.org/.
9. Groeber M, Haley BK, Uchic MD, Dimiduk DM, Ghosh S (2006) 3d reconstruction and characterization of polycrystalline microstructures using a fib-sem system data set. Mater Charac 57: 259–273
10. Groeber M, Ghosh S, Uchic MD, Dimiduk DM (2008) A framework for automated analysis and simulation of 3d polycrystalline microstructures. part 1: Statistical characterization data sets. Acta Materialia 56: 1257–1273
11. Groeber M, Ghosh S, Uchic MD, Dimiduk DM (2008) A framework for automated analysis and simulation of 3d polycrystalline microstructures. part 2: Synthetic structure generation. Acta Materialia 56: 1257–1273

3

Development and application of MIPAR™: a novel software package for two- and three-dimensional microstructural characterization

John M Sosa*, Daniel E Huber, Brian Welk and Hamish L Fraser

* Correspondence:
sosa.12@osu.edu
Center for the Accelerated
Maturation of Materials, The Ohio
State University, 1305 Kinnear Rd.,
Columbus, OH 43212, USA

Abstract

Three-dimensional microscopy has become an increasingly popular materials characterization technique. This has resulted in a standardized processing scheme for most datasets. Such a scheme has motivated the development of a robust software package capable of performing each stage of post-acquisition processing and analysis. This software has been termed Materials Image Processing and Automated Reconstruction (MIPAR™). Developed in MATLAB™, but deployable as a standalone cross-platform executable, MIPAR™ leverages the power of MATLAB's matrix processing algorithms and offers a comprehensive graphical software solution to the multitude of 3D characterization problems. MIPAR™ consists of five modules, three of which (Image Processor, Batch Processor, and 3D Toolbox) are required for full 3D characterization. Each module is dedicated to different stages of 3D data processing: alignment, pre-processing, segmentation, visualization, and quantification. With regard to pre-processing, i.e., the raw-intensity-enhancement steps that aid subsequent segmentation, MIPAR's Image Processor module includes a host of contrast enhancement and noise reduction filters, one of which offers a unique solution to ion-milling-artifact reduction. In the area of segmentation, a methodology has been developed for the optimization of segmentation algorithm parameters, and graphically integrated into the Image Processor. Additionally, a 3D data structure and complementary user interface has been developed which permits the binary segmentation of complex, multi-phase microstructures. This structure has also permitted the integration of 3D EBSD data processing and visualization tools, along with support of additional algorithms for the fusion of multi-modal datasets. Finally, in the important field of quantification, MIPAR™ offers several direct 3D quantification tools across the global, feature-by-feature, and localized classes.

Keywords: Three-dimensional; Software; Characterization; Image processing

Background

The emergence of 3D characterization tools has permitted significant advancement in the field of materials characterization. Various data acquisition techniques exist across length scales [1-6], each with their own strengths and weaknesses. However, data collection is only one in a sequence of steps required for 3D characterization. In fact, the majority of the effort is spent post-collection, with the quality of the reconstructed data critically dependent on these processing steps. The typical processing sequence for 3D characterization is as follows:

Acquisition

The first step in any 3D characterization effort is the collection of three-dimensional data. As stated above, multiple techniques are available. Which one is chosen depends on the questions one wishes to answer and at what length scale they are being asked. For example, if one wishes to accurately quantify the morphology of 1 um precipitates, DualBeam™ FIB/SEM serial sectioning is well suited. In this technique, material is iteratively "sliced" off the edge of a small cantilever using a focused ion beam of Ga^+ ions, while subsequent images are acquired with a scanning electron beam [2]. On the other hand, if 100 μm precipitates are of interest, a larger scale technique such as Robo-Met.3D™, where slices are removed via mechanical polishing followed by optical imaging, may be ideal. At the other size-scale extreme, if 10 nm precipitates are the target features, electron tomography would likely be performed in a transmission electron microscope (TEM). A non-destructive technique, X-ray tomography, has garnered much interest in recent years. In this method, multiple X-ray scans are acquired at various sample tilts [3]. This technique can produce a wealth of information, with signal generated from multiple sources.

While each of these techniques is well suited for different length-scales, there is some overlap. For instance, consider the first example of 1 um precipitates. Although DualBeam™ FIB/SEM serial sectioning may be the intuitive choice, if sampling statistics or full precipitate reconstruction are not required, electron tomography offers superior spatial resolution, and in some cases, yields stronger image contrast depending on the image formation signal. Furthermore, despite X-ray tomography's lower spatial resolution, its non-destructive nature may be a paramount factor, thus rendering it the technique of choice. Regardless of the chosen technique, the steps that follow data acquisition are of equal, if not greater significance to the efficacy of 3D characterization.

Alignment

Image registration, or image alignment, is the process by which similar images are shifted relative to one another in order to maximize agreement of their spatial intensity distribution. As with data collection, multiple techniques are available [7-9]; however, most techniques rely on two steps: image transformation and similarity quantification. That is, the image to register is subjected to an iterative process of image transformation of some class (e.g. rigid, similarity, affine, etc.) followed by similarity measurement (e.g. correlation coefficient, mutual information, etc.) with respect to the reference image. Optimum registration is defined as the image transformation parameters for the given class which yield the maximum similarity value. The pairing of a transformation class with a similarity metric defines the particular technique. The most common technique is known as cross-correlation. Cross-correlation typically performs successive rigid image transformations (translation and/or rotation) while attempting to maximize the normalized correlation coefficient (i.e. dot-product) of the two images. While cross-correlation is quite effective at registering highly random microstructures without the need for fiducial marks, it can be rather inaccurate given a set of similarly spatially oriented features. In these cases, artificial fiducial marks are required. Additional registration techniques involving metrics such as mutual information have been employed [10] and are well suited for the fusion of multi-modal datasets (i.e. those involving multiple collected signals). Selection of the optimum alignment technique strongly influences the success of subsequent processing steps.

Pre-processing

These steps are defined as any which serve to manipulate the raw pixel intensities, typically on the grayscale spectrum, in an effort to improve the accuracy of their eventual segmentation. Such steps include levels adjustments (i.e. brightness/contrast enhancement), noise-reduction filters, and FFT filtering for sectioning artifact removal. Details of the latter will be further discussed in the later pre-processing section.

Segmentation

Segmentation is formally defined as the separation of data into disjoint regions [11]. It is perhaps the most critical step to extracting useful quantitative data from a two- or three-dimensional dataset. In more complex datasets, such as those acquired from titanium alloys, segmentation can involve multiple stages. The most familiar stage is phase segmentation, where pixels are labeled according to the phases which they are deemed to belong. In the case of titanium alloys, a single phase can exist in various morphologies. Therefore, it is often necessary to perform a second stage of segmentation where pixels are assigned to each morphology. Finally, a third stage may involve the discretization of individual microstructural features such as particles or plates. As with data collection and alignment techniques, there exists a multitude of algorithms [12,13], each suited for overcoming the various challenges of image segmentation. The most common sub-set of segmentation tools are binary, that is, they employ only two classes and assign either a 0 or 1 to each pixel in the dataset. While some view binary segmentation algorithms to be limited in their applicability to multi-class datasets, they offer reduced algorithm complexity and can be readily complemented by a variety of cleanup techniques. The later section on segmentation discusses a data storage framework and user interface which leverages the simplicity of binary segmentation while overcoming many of its multi-class dataset limitations.

Visualization

Visualization was once regarded as the ultimate goal of 3D characterization. As 3D characterization has been applied to a wider problem scope, quantification has become the primary focus of most experiments. However, integrating visualization tools into the processing and quantification framework of any 3D analytical software is paramount to maximizing its characterization potential.

Quantification

The final stage, quantification, is often the purpose of a 3D characterization effort. Its efficacy depends entirely on the metric and corresponding algorithm, as well as the accuracy of the data's segmentation. The various quantification tools can be divided into three classes: global, feature-by-feature, and localized. Global quantification involves an extraction of a single quantity such as volume fraction, surface area density, or mean linear intercept. In contrast, feature-by-feature quantification extracts a metric such as volume, surface area, or diameter from each feature of interest. This type of quantification is quite common in 3D datasets as many of these individual feature metrics cannot be accurately determined from two-dimensional images. Finally, localized quantification is performed at each point, or vertex, on a reconstructed surface. Such metrics can include local curvature, surface roughness, and thickness. Examples of several localized quantification results are presented in the localized quantification section.

Methods

To be truly robust, any 3D characterization software package must possess a broad array of tools capable of subjecting a dataset to each step described in the previous section. Many powerful software programs have been developed for 3D characterization, both commercial and open source [14-17]. However, few can equip users with extensive toolsets in the areas of alignment, pre-processing, segmentation, visualization, and quantification, with equal attention given to each. A fully integrated toolset, designed by material scientists, would provide an attractive 3D characterization platform. Along that vision, Materials Image Processing and Automated Reconstruction (MIPAR™) has been developed. The software was written and developed in the MATLAB™ environment, with the MATLAB™ compiler enabling MIPAR™ to be executed as a standalone application on Macintosh, Windows, and Linux platforms. MIPAR™ is based upon a modular construction. These modules may be launched from a global launch bar, as well as from within one and another. The modules were designed as standalone programs, each suited for different tasks, and each capable of communication with other modules. MIPAR™ consists of five total modules; three of which are critical for 3D characterization of most materials. The following sub-section will discuss the capabilities and purpose of each of these salient modules, as well as describe MIPAR's conventional 3D characterization workflow.

Image processor

Nearly all image-processing efforts will originate in the first module known as the Image Processor. This module provides an environment for users to develop a sequence of processing steps known as a recipe. The specific steps and parameters constituting a recipe will vary based on the user's intent, but the ultimate goal of most recipes is the same: to segment grayscale intensity into a binary image. Much like building an action or macro in Adobe® Photoshop™, a user's process and parameter selections are recorded real-time, permitting subsequent parameter editing as well as process removal or insertion. Perhaps the Image Processor's most useful feature is the ability for process and parameter tweaks to propagate down through all subsequent recipe steps. Indeed, parameter optimization is one of most critical aspects of image segmentation, and given the complexity of most recipes, such optimization can be quite time-consuming. The auto-update feature greatly facilitates parameter optimization, increasing the likelihood that users produce near-optimum recipes. Figure 1 reveals the layout of the Image Processor along with lists of many of the included image-processing algorithms, each of which are paired with their own graphical user interface for efficient parameter selection.

Batch processor

Following completion of a recipe, it may be saved and loaded into the second module, the Batch Processor. This module functions to automatically apply the recipe to a series

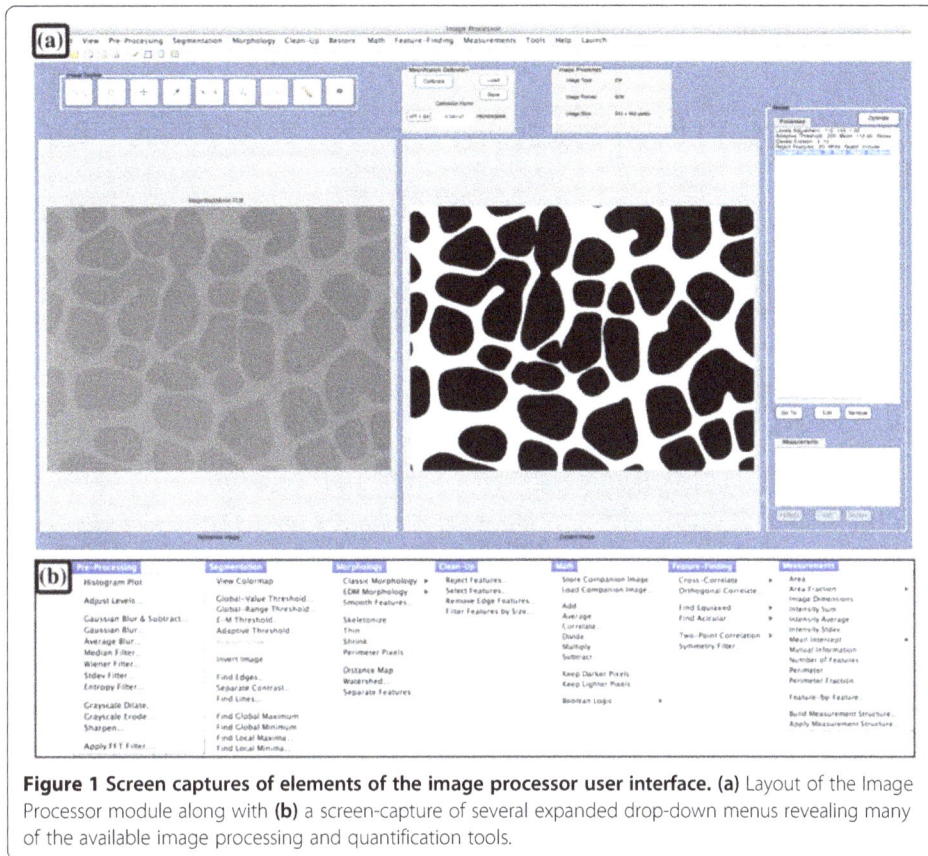

Figure 1 Screen captures of elements of the image processor user interface. (a) Layout of the Image Processor module along with **(b)** a screen-capture of several expanded drop-down menus revealing many of the available image processing and quantification tools.

of images, in this case, those collected during a serial sectioning experiment. In addition, tilt-correction, image alignment, and volume reconstructions may all be performed. The Batch Processor was designed to provide a single environment where each of these steps could be performed in one sequence, transforming a set of unaligned raw images into both a stack of aligned slices as well as a segmented, reconstructed volume. The layout of the Batch Processor is shown in Figure 2.

3D Toolbox

This module provides a host of tools for interacting with the image stack and reconstructed volume output from a batch process. The left side of the user interface is dedicated to viewing, manipulating, and exporting the image stack output from the Batch Processor. The image stack is merely a sequence of frames used to examine the slice-to-slice alignment of the raw data. The stack can be viewed with or without the negative space resulting from such alignment. Additionally, the quantitative slice-to-slice translations can be viewed as scatter plots. If severe translations were necessary to align certain slices, resulting in a significantly reduced aligned volume, these slices can be removed and the alignments automatically recalculated. Any desired cropping of the image stack can be performed interactively. Once completed, the modified image stack can be exported as an image sequence for re-processing and segmentation.

The right side of the 3D Toolbox is used for processing, segmenting, cleaning, interacting with, visualizing, and quantifying the voxelized 3D reconstruction. In some

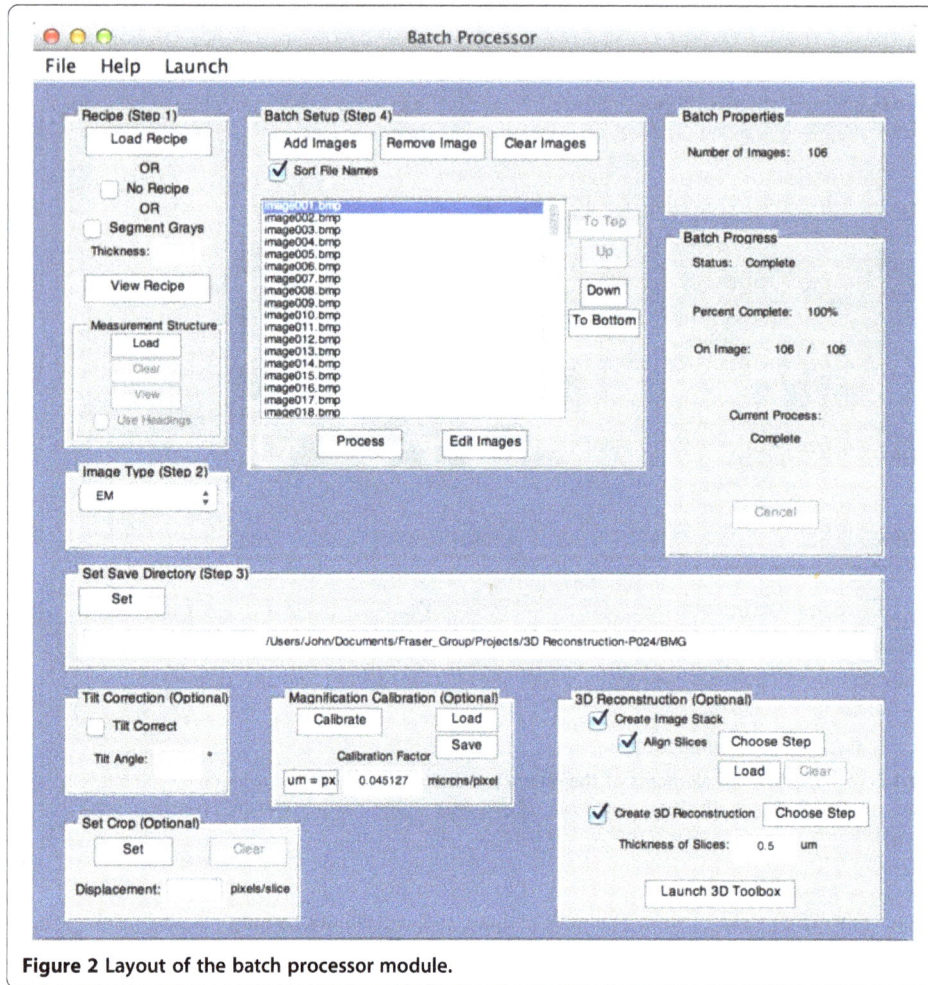

Figure 2 Layout of the batch processor module.

cases, a direct 3D segmentation can be superior to conventional slice-by-slice segmentation. In such cases, only the alignment stage would be performed in the Batch Processor, and all pre-processing and segmentation would be performed within the 3D Toolbox directly on the aligned three-dimensional data. If slice-by-slice segmentation *was* required, the 3D Toolbox can still offer many binary noise reduction, erosion/dilation, and smoothing tools – all performed directly on the 3D segmentation.

On the visualization front, while MIPAR™ does not possess some of the powerful visualization capabilities of commercial applications (e.g. Avizo®), it offers convenient tools for interactively visualizing parts or all of a 3D segmentation as either surface reconstructions or volume renderings. This provides users with direct feedback on their segmentation quality and illustrates relationships between two-dimensional feature cross-sections and the 3D structure to which they belong. Should higher-end visualization be required, the reconstruction may be output in a variety of formats which are compatible with a number of 3D visualization packages, both commercial and open-source. Finally, the 3D Toolbox offers a host of quantification tools in the global, feature-by-feature, and localized categories discussed in the introduction. Example applications of many of these tools are further discussed in the section on quantification tools.

A layout of the 3D Toolbox is shown in Figure 3 along with screenshots of the drop-down menus which offer the pre-processing, segmentation, and quantification tools. Additionally, several interactive tools are labeled which offer functions such as 2D/3D vector and angle measurement, plane indices determination, and single-feature quantification and visualization.

Results and discussion
Pre-processing
As described in the introduction, pre-processing tools are those that manipulate the intensity values of raw data pixels or voxels. Such tools aim to improve image contrast and reduce noise, thus improving the efficacy of subsequent segmentation steps. The following section discusses a unique pre-processing tool included in MIPAR™.

Frequency domain filtering
One issue that can interfere with image segmentation of DualBeam™ FIB/SEM serial section datasets is a milling artifact known as "curtaining". It receives its name from the surface modulations normal to the incident ion-beam which result from either incomplete or excessive removal of material [18]. These modulations can exist as either surface protrusions or surface relief, both of which are deleterious to image segmentation. In many cases, curtaining can be minimized or eliminated by proper sample preparation, milling parameter selection, and/or advanced techniques. However, for certain

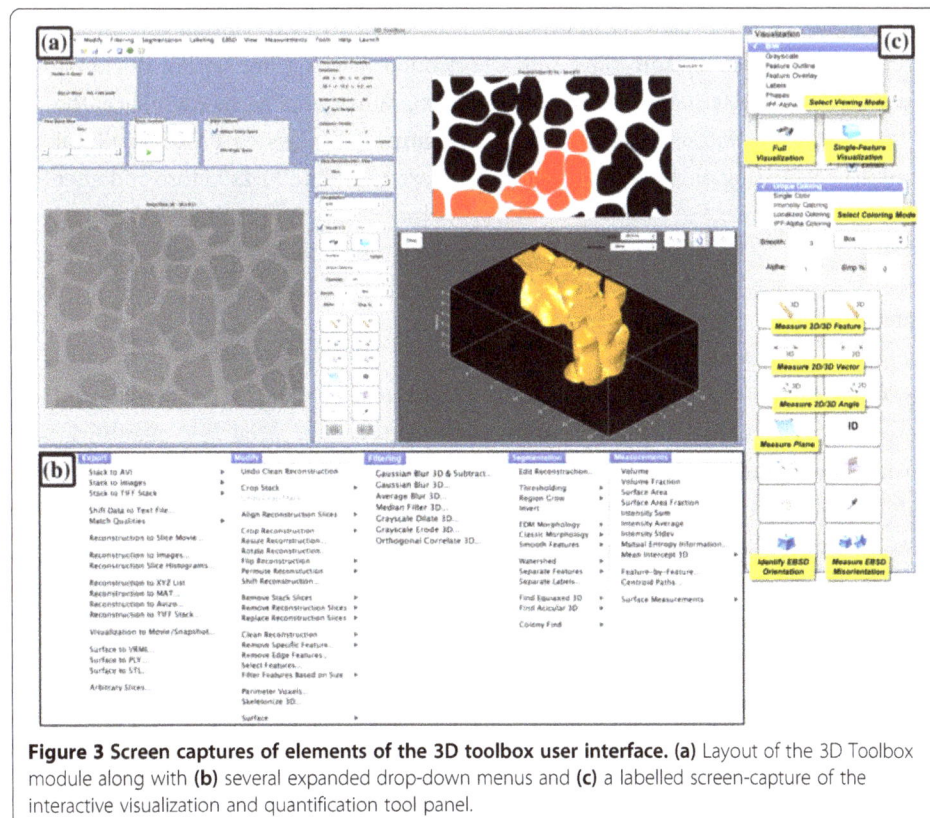

Figure 3 Screen captures of elements of the 3D toolbox user interface. (a) Layout of the 3D Toolbox module along with **(b)** several expanded drop-down menus and **(c)** a labelled screen-capture of the interactive visualization and quantification tool panel.

sample geometries and microstructures, curtaining is unavoidable, in which case, post-acquisition image filters are the only recourse.

Once such filter is the frequency domain filter (i.e. FFT filter). The fast-Fourier transform (FFT) is a classical algorithm which expresses spatial data as a collection of frequencies. When applied to a two-dimensional image, abrupt intensity transitions are expressed as high frequencies and gradual transitions as low frequencies. The spatial orientation of these transitions determines the vector along which the frequencies are plotted relative to the FFT origin. Once an FFT is computed for a given image, filters may be constructed which either discard or retain certain frequencies. Milling curtains are typically oriented parallel to the milling direction (vertical in most images) and thus exist as a collection of frequencies oriented horizontally within the image As such, curtaining tends to manifest as frequencies which lie along the x-axis of the FFT. By discarding such frequencies and inverting the FFT to recover the filtered image, the intensity of most curtains can be significantly reduced and in some cases removed. Additionally, FFT filtering is effective at reducing the influence of similarly oriented scratches in 2D and 3D datasets as they manifest similar to milling curtains as frequency bands in an FFT.

MIPAR's Image Processor offers a graphical environment for the construction and application of FFT filters including low-pass, high-pass, annular, and custom filters. In the case of curtain reduction, a custom filter can be created as a thick line which discards all frequencies which lie along the x-axis. It should be noted that frequencies in close proximity to the origin should be retained as these low frequencies contribute to a majority of the image contrast. As examples, slices from two DualBeam™ FIB/SEM serial sections: a hydrogen blister in a tungsten alloy and a bimodal microstructure in [α + β]-processed Ti-6Al-4 V (wt%) (Ti-64) were subjected to FFT filtering for curtain reduction. This particular custom filter was a horizontal masking line with a thickness of 6 pixels (i.e. 110 cycles/micron in frequency units). When applied to each slice, the segmentation of both datasets vastly improved. Figure 4 displays MIPAR's graphical interface for FFT filtering along with the slices and corresponding FFTs before and after filtering.

Segmentation

Optimization via mutual information

Perhaps the most critical, yet often unexplored areas of 2D and 3D segmentation are those of objective parameter selection and segmentation accuracy quantification. In fact, nearly all aspects of most segmentation algorithms have become fully automated with the exception of parameter selection. The influence of these parameters on subsequent quantification can be profound. In some cases, metrics such as volume fraction acquired from a 2D image can vary by several volume percent from an under- or over-segmentation of a single pixel. In a 3D dataset, segmentation fluctuations can result in even more dramatic volume fraction variations. Dataset to dataset, these fluctuations can result from variations in image quality and contrast, as well as user bias.

Prior to exploring a method for objective segmentation parameter selection, one must first define the concept of segmentation accuracy. Consider a volume of material

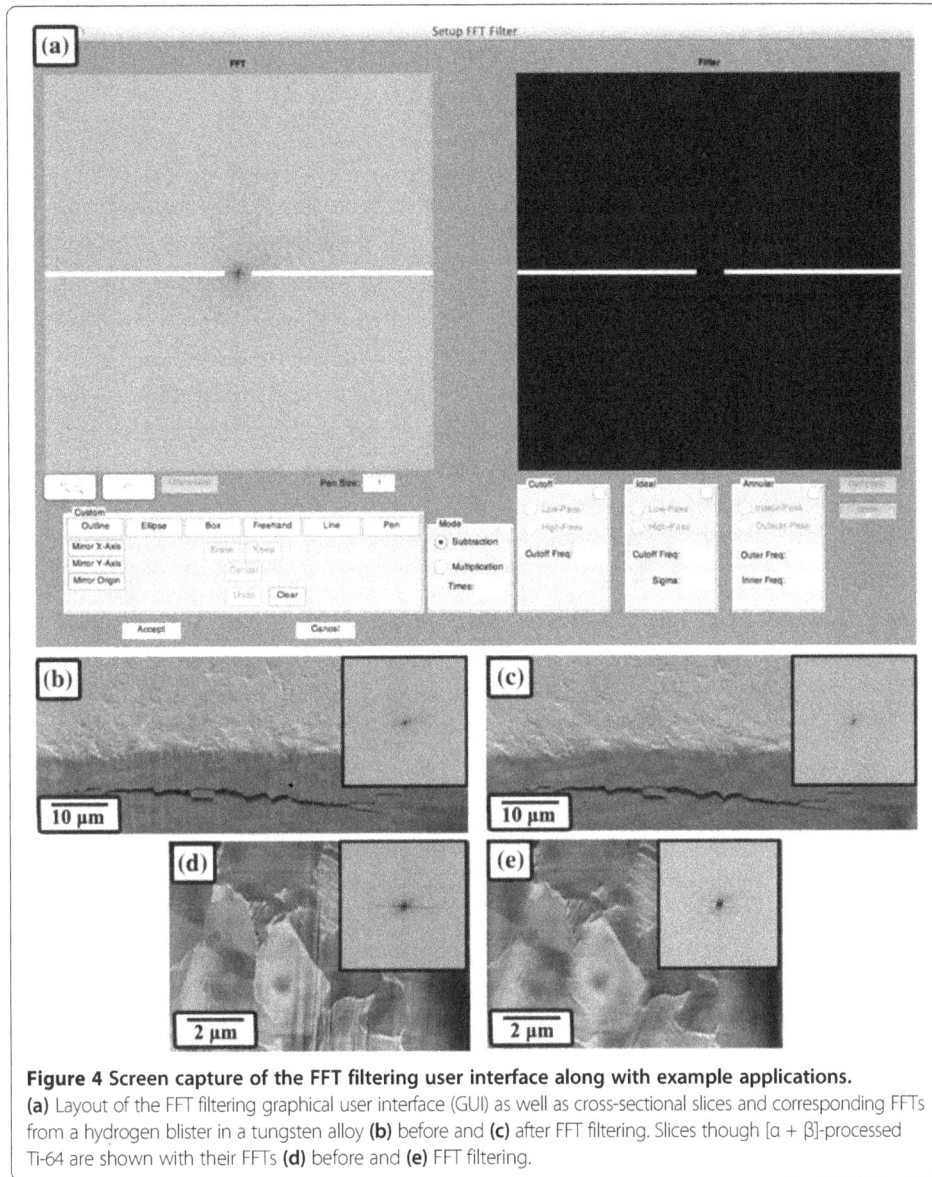

Figure 4 Screen capture of the FFT filtering user interface along with example applications.
(a) Layout of the FFT filtering graphical user interface (GUI) as well as cross-sectional slices and corresponding FFTs from a hydrogen blister in a tungsten alloy (b) before and (c) after FFT filtering. Slices though [α + β]-processed Ti-64 are shown with their FFTs (d) before and (e) FFT filtering.

containing precipitates of a certain shape and size. An image acquired of such precipitates is a product of sample preparation and image formation physics. Any method that attempts to quantify the "accuracy" of the image's segmentation, would be doing so relative only to the *image*, and is therefore entirely dependent on how faithfully that image represents the actual microstructure. Therefore, when discussing the notion of segmentation accuracy, one *must* consider the segmentation-to-image comparison separate from that of image-to-microstructure. The image-to-microstructure comparison has historically been left to the judgment of the experimenter. Recent advances in the field of forward modeling; however, have the potential to greatly contribute to this area [19,20].

The segmentation-to-image comparison is one in which information theory tools can contribute. One such tool is mutual information. By definition, mutual information describes the similarity between two variables, in this case, two images [21]. Eq. 1 displays a mathematical expression for mutual information:

$$I(X,Y) = \frac{H(X) - H(X|Y)}{H(X)} \tag{1}$$

In this expression, $H(X)$ represents the entropy, or uncertainty, of the original image X. $H(X|Y)$ represents the conditional entropy of the original image X when the segmented image Y is known. Their subtraction therefore describes the reduction in uncertainty of original image X upon knowing its segmentation Y, or in other words, how well the segmentation describes the original image. Normalizing by the entropy of the original image yields a range of 0 to 1 as possible values of mutual information. A mutual information of 0 implies that the segmentation in no way describes the original image, while a mutual information of 1 implies that the image is perfectly described the segmentation.

Using mutual information as the quantifying metric, a mechanism has been developed and incorporated into MIPAR™ which allows users to objectively determine the optimum parameters for a given image-processing algorithm. Algorithms such as global thresholding, adaptive thresholding, erosion and dilation, feature rejection, and watersheding are all candidates for optimization. When either of these processes is selected within a recipe, an "Optimize" button becomes active at the top of the recipe panel. Selecting this button spawns a series of windows allowing users to select the parameter ranges from which to identify the optimum parameter set. Using either a sub-area or the entire image, the grayscale data is segmented over all possible parameter sets within the specified range. The mutual information between each segmentation and the original image is computed, and the parameter set which yields the maximum mutual information is chosen as optimum. This tool is currently applicable to two-dimensional processing algorithms with a maximum of two parameters; however, extension of this tool to three-dimensional volumes and multi-dimensional parameter space is an ongoing effort.

As an example, the adaptive thresholding of an SEM micrograph of secondary Ni_3Al γ' precipitates is optimized. Briefly, adaptive thresholding assesses each pixel relative to a statistic (typically the mean) of its local neighborhood. This algorithm involves two parameters: the window size which defines a pixel's neighborhood, and the threshold by which a pixel's intensity value must exceed its local mean in order to be selected. The threshold may also be expressed as the percentage of the local mean that a pixel's intensity must meet or exceed. In this example, the window size was fixed at 15 pixels, and the threshold value was allowed to vary from 50% to 110% in steps of 1%. Figure 5 displays the result of the optimization along with several candidate segmentations determined during the process.

Although this technique does not address the fidelity between image and microstructure, it does provide a method for consistently and objectively determining processing parameters across images and users. Additionally, if a recipe step has been optimized in MIPAR's Image Processor, the optimization process is repeated for every image to which that recipe is applied in the Batch Processor. In this way, slice-to-slice contrast variations within a 3D dataset may be overcome by allowing parameters of several recipe steps to dynamically adjust to a given image's intensity profile. Other research

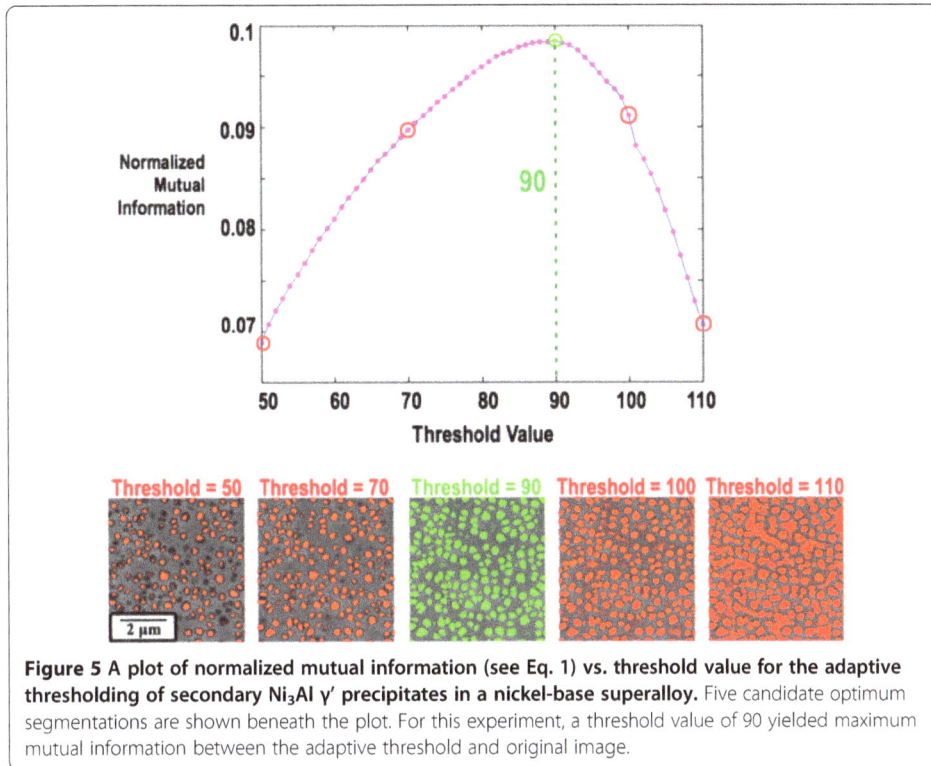

Figure 5 A plot of normalized mutual information (see Eq. 1) vs. threshold value for the adaptive thresholding of secondary Ni₃Al γ' precipitates in a nickel-base superalloy. Five candidate optimum segmentations are shown beneath the plot. For this experiment, a threshold value of 90 yielded maximum mutual information between the adaptive threshold and original image.

has been performed on the topic of optimizing segmentation via mutual information [22,23]. However, to the best of the author's knowledge, MIPAR™ exhibits the first incorporation of this method into a graphical software interface. This will permit its application to a variety of problems and microstructures so that the extent of its strengths and limitations can be thoroughly explored.

Feature-sets

Up to this point, this paper has discussed segmentation exclusively in a binary sense, where each pixel or voxel in a grayscale dataset is labeled as either a 0 or 1. Historically, this has worked well for the segmentation of two-phase microstructures, or for the isolation of a certain phase or feature type. As a result, the image processing community has developed an extensive library of binary segmentation and cleanup routines [24], all of which are computationally efficient. However, in many image-processing frameworks, binary segmentation can limit the type of microstructures that can be processed since any with more than two phases of interest are incompatible with binary classification. For two-dimensional images, this is not a great inconvenience since a separate binary segmentation can be performed for each phase interest with minimal overhead. However, the size and storage demands of 3D datasets make this approach less practical. Furthermore, by discretizing each phase or feature type into an independent segmentation and file, any metrics that quantify one phase or feature type with respect to another are quite difficult.

Rather than develop MIPAR™ as a framework to handle multi-class 3D segmentations, a data structure was developed wherein separate binary segmentations of the same dataset could be stored as layers in a multi-dimensional space. Furthermore, this structure can effectively store multi-modal datasets where raw data is collected from several techniques (e.g. secondary/backscatter electron imaging, EBSD, EDS, etc.). A schematic of this multi-dimensional structure is shown in Figure 6. Under this format, every voxel receives a set of multi-dimensional coordinates which identify the voxel's placement in 3D Cartesian space as well its position in other dimensions such as data type and data level. A voxel's location in the "ith" dimension (see Figure 6) defines the binary segmentation to which it belongs. The different segmentations are termed feature-sets.

The initial segmentation of the original data produces the first feature-set, whether performed slice-by-slice in the Batch Processor, or three-dimensionally in the 3D Toolbox. A dropdown menu allows users to add additional feature-sets, as well as switch to, remove, rename, or merge existing feature-sets. By treating multi-class segmentations as a set of concatenated binary segmentations, interacted with using a simple interface, the simplicity and computational efficiency of binary data is preserved, while the complexity of multi-phase datasets and inter-phase quantification can be handled.

An example application of the feature-set framework is presented in Figure 7. This particular dataset from a nickel-base superalloy consisted of FCC γ grains, carbides, and twins [25]. Thus, three binary segmentations were necessary to fully characterize a single grain within the microstructure. Under the feature-set framework, three fairly complex segmentation recipes were carried out independently, yet their results combined into a single data structure. A fourth feature-set was then generated from the merging of the first three, where each voxel was labeled according to the feature-set from which it originated. This fourth feature-set was employed to visualize the multi-phase microstructure in three dimensions.

3D EBSD and data fusion

The emergence of high-speed electron backscatter diffraction (EBSD) cameras has increased the popularity of three-dimensional EBSD. These datasets permit a deeper investigation of 3D microstructures and have advanced the understanding of microstructural evolution. While robust commercial software has long existed for the analysis of *two-dimensional* EBSD scans, analogous *3D* software has been somewhat underdeveloped. Recently developed software programs (e.g. DREAM.3D™ [16]) offer unique collections of 3D EBSD data processing tools housed in graphical interfaces. These programs also provide tools for both 3D EBSD quantification and synthetic microstructure generation. However, few software titles offer tools for the fusion of 3D multi-modal

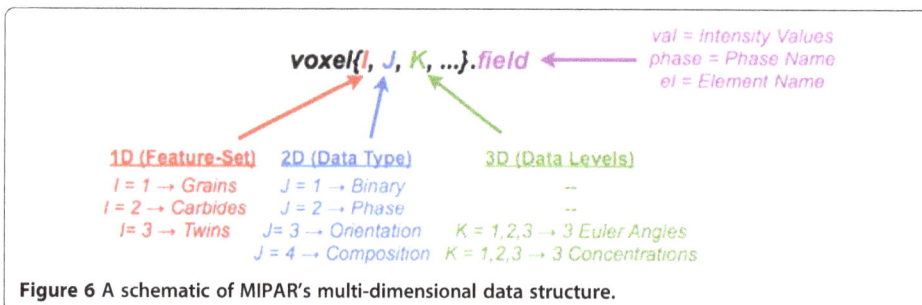

Figure 6 A schematic of MIPAR's multi-dimensional data structure.

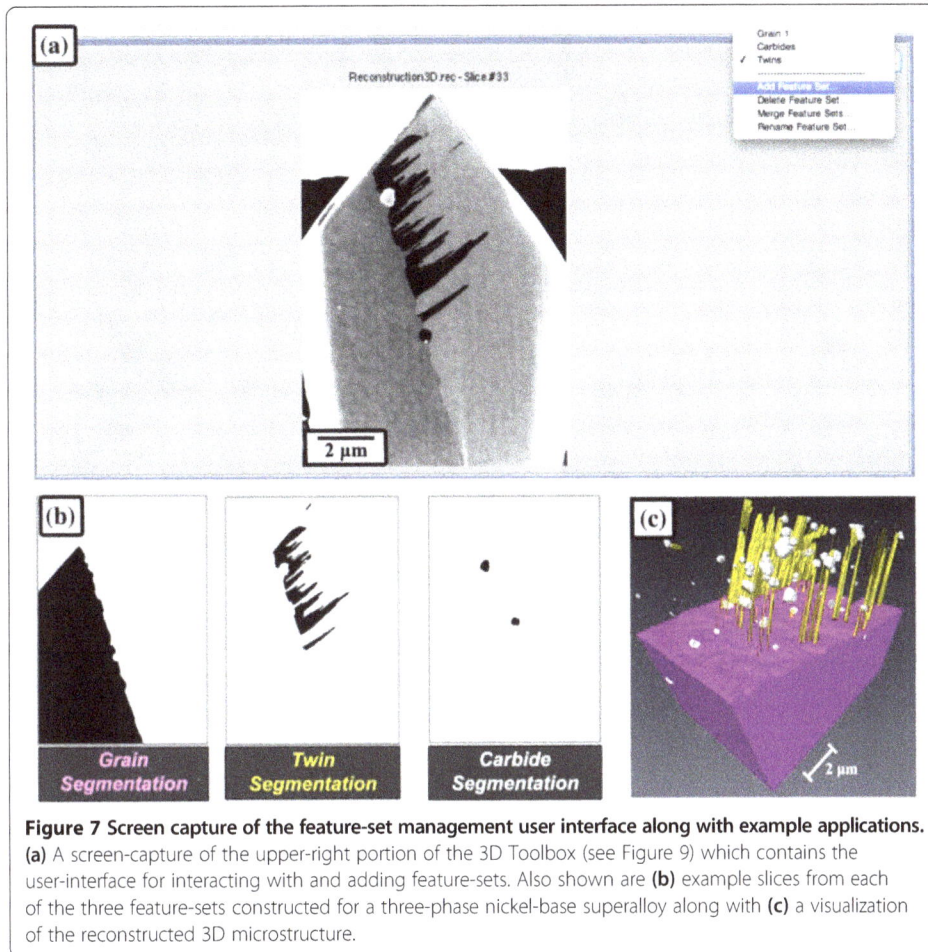

Figure 7 Screen capture of the feature-set management user interface along with example applications. **(a)** A screen-capture of the upper-right portion of the 3D Toolbox (see Figure 9) which contains the user-interface for interacting with and adding feature-sets. Also shown are **(b)** example slices from each of the three feature-sets constructed for a three-phase nickel-base superalloy along with **(c)** a visualization of the reconstructed 3D microstructure.

datasets (i.e. datasets whose voxels are comprised of multiple discrete variables and/or spectra).

MIPAR's multi-dimensional data structure offers a promising solution to multi-modal data storage and processing. For example, tools for 3D EBSD data import, cleanup, and visualization (see Figure 8(a)) have been developed. Figure 8(b, c) presents two 3D EBSD reconstructions that were imported as raw data into MIPAR™, cleaned, and visualized using inverse pole figure colormaps. These tools will be extended to handle the input of compositional measurements from techniques such as EDS and WDS. Upon adding these tools to MIPAR's comprehensive 3D characterization package, and by storing crystallographic data along side BSE/SE/optical intensity, data fusion tools remained as the final requisite for handling multi-modal data.

The primary challenge related to data fusion involves the registration of voxels of different dimensions. Separate datasets must first be resampled such that their voxel dimensions match. Second, the data must be registered. Inter-modal registration presents greater challenges than conventional serial section alignment and often requires non-rigid volume transformations. MIPAR™ has already incorporated the open-source Medical Image Registration Toolbox (MIRT) [26] for the free-form registration of 2D images. At present, MIPAR™ offers both manual and automated tools for assigning crystallographic orientations and phase identifications to BSE/SE/optical datasets.

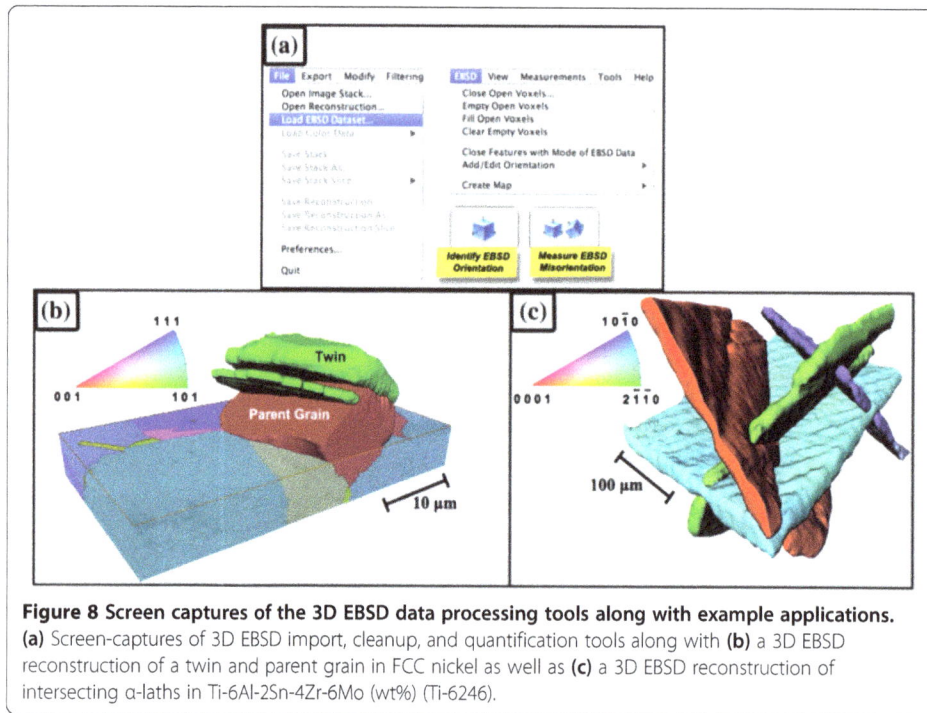

Figure 8 Screen captures of the 3D EBSD data processing tools along with example applications.
(a) Screen-captures of 3D EBSD import, cleanup, and quantification tools along with (b) a 3D EBSD reconstruction of a twin and parent grain in FCC nickel as well as (c) a 3D EBSD reconstruction of intersecting α-laths in Ti-6Al-2Sn-4Zr-6Mo (wt%) (Ti-6246).

Quantification

As stated earlier, quantification has become the desired result of most 3D characterization experiments. Therefore, MIPAR™ was equipped with a host of tools in the areas of global, feature-by-feature, and localized quantification. The following sub-sections will present several examples of the included tools.

Global quantification

The most basic 3D quantification metrics are classified as global. These are defined as any which extract a single value from a reconstructed volume. In the case of binary segmentations, they tend to operate on those voxels which have been assigned to the features or phase of interest. Common global metrics include total volume, volume fraction, total surface area, surface area density, surface area per volume, and mean linear intercept. Each of these and several others are available in MIPAR's 3D Toolbox from a dropdown menu (see Figure 3).

Feature-by-feature quantification

Since 3D reconstructions tend to capture multiple microstructural features in their entirety, individual measurements of each discrete feature are often desired. Such metrics are of the class termed "feature-by-feature". Feature labeling is performed automatically within MIPAR™ using a conventional 6-voxel-connectivity scheme. Discretization can be aided using a variety of algorithms including watersheding and iterative erosion/dilation techniques. Once the segmented features are sufficiently discretized, a graphical interface allows users to extract specific metrics from each labeled feature. Figure 9 displays this interface and reveals the list of available metrics.

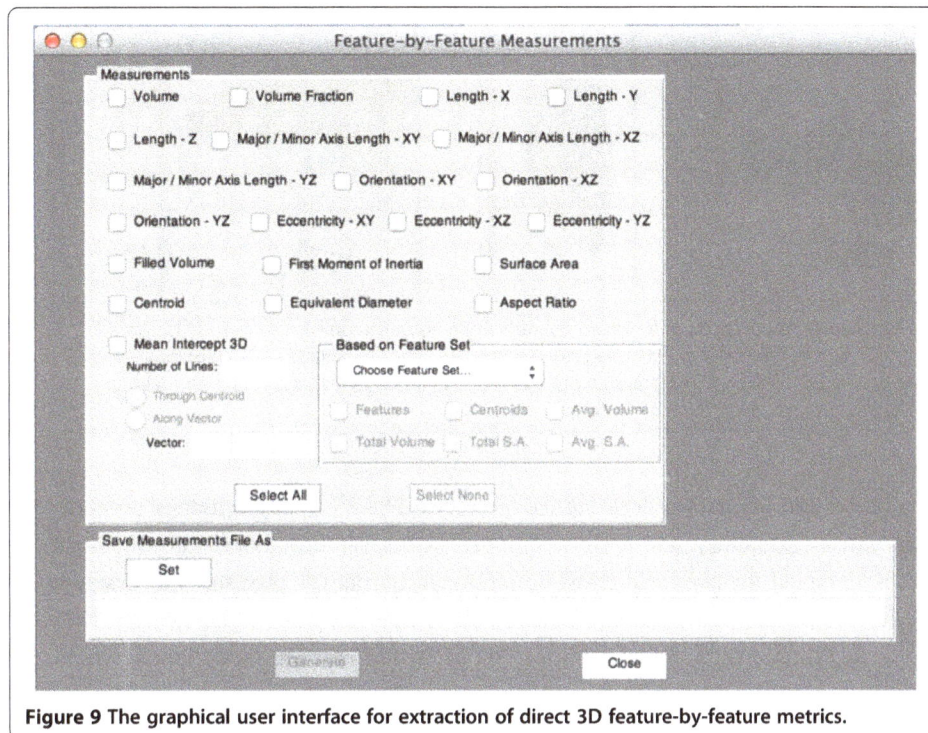

Figure 9 The graphical user interface for extraction of direct 3D feature-by-feature metrics.

Localized quantification

Additionally, 3D quantification can be performed locally about each point on the surface of a reconstructed feature. These points are known as vertices, and are the Cartesian coordinates in the sample reference frame of all points which constitute a reconstructed surface. Incorporating codes from the MATLAB™ file exchange, MIPAR™ can report fundamental parameters from each vertex such as Cartesian position, local normal, and local curvature. More complex algorithms have also been developed including thickness and roughness mapping. The former measures the local thickness at each vertex by measuring the distances between opposing vertices which lie along a vector parallel to either the local surface normal or a user-defined global normal. Roughness mapping is accomplished by fitting a plane or curved surface to the quantified vertex positions and subsequently coloring the surface according to each vertices' elevation relative to the fit plane or surface.

Regardless of the quantities measured at each vertex, all of which are exported as comma-separated text files, the reconstructed surface can subsequently be colored according to each vertices' measurement. A simple dropdown menu allows users to select this visualization scheme and choose the data on which to base the localized coloring. Therefore, additional custom algorithms can be subsequently performed on MIPAR's exported vertex measurements in software such as Microsoft Excel®, the results of which can be color-mapped on the associated surface. Figure 10 reveals example applications of visualized local quantification such as local curvature, topography, and thickness.

Advantages of MATLAB™ Development

Although some have questioned the utility of MATLAB™ as a professional software development platform, it offers several advantages over other computing languages in

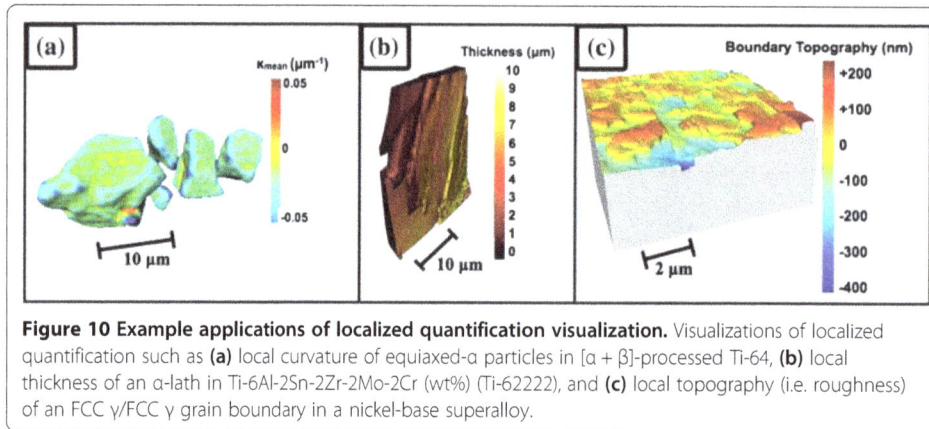

Figure 10 Example applications of localized quantification visualization. Visualizations of localized quantification such as **(a)** local curvature of equiaxed-α particles in [α + β]-processed Ti-64, **(b)** local thickness of an α-lath in Ti-6Al-2Sn-2Zr-2Mo-2Cr (wt%) (Ti-62222), and **(c)** local topography (i.e. roughness) of an FCC γ/FCC γ grain boundary in a nickel-base superalloy.

the area of 2D and 3D data processing. For one, MATLAB™ is appropriately an acronym for *Matrix Lab*oratory, and is thus equipped with several toolboxes and functions specifically designed for two- and three-dimensional matrices – precisely the raw format of 2D images and 3D volumes. These built-in libraries offer rapid development and powerful functionality to any 2D or 3D analytical software developed within MATLAB™. In MIPAR's case, they comprise roughly 55% of the constitutive functions.

Second, included in this built-in functionality are a number of 2D and 3D visualization tools. As discussed in the introduction, even the most robust analytical software cannot reach its full potential if it fails to provide users immediate visual feedback on the result of a 2D or 3D filter. MATLAB's built-in visualization library has aided in MIPAR's evolution as both a processing and visualization platform.

Third, MATLAB's broad user-base has resulted in extensive public code repositories, the most popular of which is the MATLAB™ File Exchange. The numerous codes contained within such repositories are provided open source, under the stipulation that the code's license file be included with the deployed application. In MIPAR's case, several algorithms were harvested from these repositories and make up nearly 20% of its function count. The corresponding authors have been listed as contributors to the software and their licenses included. This contribution has accelerated MIPAR's growth and underlined the fact that a single developer cannot produce the most efficient form of every algorithm, nor is it effective to "reinvent the wheel" when it comes to well-accepted data processing algorithms.

Finally, MATLAB's popularity has led to the development of third-party commercial software written specifically for MATLAB™ (e.g. Jacket™ by Accelereyes®). This particular software offers a collection of MATLAB™ codes written to run on the graphics processor unit (GPU). Such codes are often orders of magnitudes faster than corresponding CPU-based codes. Functions such as GPU-based parallel for-loops have dramatically accelerated complex image-processing algorithms offered in MIPAR™.

Conclusions

The paradigm of 3D data processing has motivated the development of a multi-faceted 3D processing/analytical software package named MIPAR™ (Materials Image Processing and Automated Reconstruction). Through development in MATLAB™, MIPAR™ takes advantage of MATLAB's powerful 2D and 3D data processing libraries, public code repository, and third-party GPU-acceleration software, while being deployed as a standalone cross-platform application.

In addition to offering many conventional pre-processing filters, MIPAR™ includes a unique graphical interface for the development and application of noise-reducing FFT filters which are effective at reducing the influence of data acquisition artifacts such as ion-milling curtains and polishing scratches. Furthermore, in the area of segmentation, MIPAR™ offers an interactive framework for the automated optimization of segmentation parameters through the use of mutual information. Once optimized, MIPAR's numerous binary segmentation and cleanup algorithms can be employed for the effective segmentation of even complex, multi-phase 3D microstructures. Such is possible via a multi-dimensional data structure which works in conjunction with a simple user interface to permit the multi-dimensional layering of multiple binary segmentations of the same 3D dataset. Although MIPAR™ offers nowhere near as comprehensive a 3D EBSD toolset as other programs, its employed 3D data structure has enabled the import, basic processing, quantification and visualization of 3D EBSD data, as well as import/export functionality to and from additional programs. Future integration of such programs with MIPAR™ would offer a powerful 3D EBSD characterization platform. Additionally, MIPAR's multi-dimensional data structure has permitted the development of tools for the fusion of multi-modal datasets. In the near feature, the medical image registration toolbox (MIRT) 3D free-form registration tools will be incorporated into MIPAR's 3D Toolbox for the automated registration of 3D EBSD with BSE/SE/optical datasets.

In the area of quantification, MIPAR™ includes common global quantification algorithms for metrics such as volume fraction, surface area density, and mean linear intercept. In addition, a variety of direct 3D feature-by-feature metrics such as individual volume, surface area, shape factor, and aspect ratio may be extracted. Finally, MIPAR™ offers a simple yet powerful means of locally quantifying a variety of metrics at each point on a reconstructed surface and subsequently visualizing the spatial distribution of such measurements on various color scales. The available metrics range from basic, such as position, local normal, and local curvature, to complex, such as local thickness and surface roughness.

Competing interests
The authors declare that they have no competing interests.

Authors' contributions
JMS carried out the software development, dataset post-processing, and drafted the manuscript. DEH carried out data acquisition and provided suggestions for software development. BW carried out data acquisition and provided suggestions for software development. HLF provided oversight to the work, managed the affiliated authors, and reviewed the manuscript. All authors read and approved the final manuscript.

Acknowledgements
The authors gratefully acknowledge Nicholas Hutchinson, Dr. Santhosh Koduri, Dr. Paul Shade, Dr. Jennifer Carter, and Samuel Kuhr for their contributions in the area of data acquisition, as well as all additional supporters whose data collection efforts and software-related feedback have greatly contributed to this work.

References
1. Spowart JE (2006) Automated Serial Sectioning for 3D analysis of Microstructures. Script Materialia 55(1):5–10
2. Uchic MD, Holzer L, Inkson BJ, Principe EL, Munroe P (2007) Three-Dimensional Beam Tomography 32(May):408–416
3. Elliott JC, Dover SD (1982) X-ray microtomography. J of Microscopy 126(2):211–213

4. Dierksen K, Typke D, Hegerl R, Koster AJ, Baumeister W (1992) Towards automatic electron tomography. Ultramicroscopy 40(1):71–87
5. Bronnikov AV (2006) Phase-contrast {CT}: {Fundamental} theorem and fast image reconstruction algorithms. Developments in X-Ray Tomography V 6318:Q3180
6. Cloetens P, Ludwig W, Baruchel J, Dyck D, Van Landuyt J, Van Guigay JP, Schlenker M (1999) Holotomography: {Quantitative} phase tomography with micrometer resolution using hard synchrotron radiation {X}-rays. Appl Phys Lett 75(19):2912–1914
7. Brown LG (1992) A survey of image registration techniques. ACM Comput Surveys (CSUR) 24(4):325–376
8. Roshni V, Revathy K (2008) Using mutual information and cross correlation as metrics for registration of images. J of Theoretical & Applied Information 4:474–481
9. Maes F, Collignon A, Vandermeulen D, Marchal G, Suetens P (1997) Multimodality image registration by maximization of mutual information. Medical Imaging, IEEE Transactions on 16(2):187–198
10. Gulsoy E, Simmons J, De Graef M (2009) Application of joint histogram and mutual information to registration and data fusion problems in serial sectioning microstructure studies. Scripta Materialia 60(6):381–384
11. Russ JC ((1986) Practical Stereology. In: Practical Stereology. Kluwer Academic Pub, The Netherlands, pp 1–381
12. Pal NR, Pal SK (1993) A review on image segmentation techniques. Pattern Recognition 26(9):1277–1294
13. Haralick RM, Shapiro LG (1985) Image segmentation techniques. Computer Vision, Graphics, and Image Processing 29(1):100–132
14. Westenberger P, Estrade P, Lichau D (2012) Fiber Orientation Visualization with Avizo Fire®. Conference of Industrial Computed Tomography (ICT) 2012, Wels, Austria
15. Rasband WS ImageJ, U.S. National Institutes of Health. U.S. National Institutes of Health, Bethesda, Maryland, USA, imagej.nih.gov/ij/, 1997—2012
16. Groeber M, Jackson M (2011–2013) DREAM.3D: Digital Representation Environment for Analyzing Microstructure in 3D. BlueQuartz Software. BlueQuartz Software
17. Henderson A (2007) ParaView Guide. Kitware Inc., A Parallel Visualization Application
18. Volkert CA, Minor AM (2007) Focused Ion beam microscopy and micromachining. MRS Bulletin 32(05):389–399
19. Venkatakrishnan SV, Drummy LF, Jackson MD, Graef M, Simmons J, Bouman C (2013) A model based iterative reconstruction algorithm for high angle annular dark field-scanning transmission electron microscope (HAADF-STEM) tomography. IEEE transactions on image processing : a publication of the IEEE Signal Processing Society 22(11):4532–44
20. Suter RM, Hennessy D, Xiao C, Lienert U (2006) Forward modeling method for microstructure reconstruction using x-ray diffraction microscopy: Single-crystal verification. Review of Scientific Instruments 77(12):123905
21. MacKay DJC (2003) Information theory, inference and learning algorithms. Cambridge university press, Cambridge, United Kingdom
22. Rigau J, Feixas M, Sbert M, Bardera A, Boada I (2004) Medical image segmentation based on mutual information maximization. In: Medical Image Computing and Computer-Assisted Intervention - Miccai 2004, Pt 1, Proceedings, 3216th edn., pp 135–142
23. Gonzalez RC, Woods RE (2008) Digital Image Processing. Pearson/Prentice Hall, Upper Saddle River, New Jersey
24. Shi DJ, Liu ZQ, He J (2010) Genetic algorithm combined with mutual information for image segmentation. Advanced Materials Res 108–111:1193–1198
25. Carter JLW, Zhou N, Sosa JM, Shade PA, Pilchak AL, Kuper MW, Mills MJ (2012) In: Huron ES, Reed RC, Hardy MC, Mills MJ, Montero RE, Portella PD, Telesman (eds) Characterization of strain accumulation at grain boundaries of nickel-based superalloys. Superalloys, Seven Springs, Pennsylvania, pp 43–52
26. Myronenko A (2007) Medical Image Registration Toolbox [Computer software]. Oregon Health & Science University, Portland, OR

4

Crystallographic texture evolution in 1008 steel sheet during multi-axial tensile strain paths

Adam Creuziger[1]*, Lin Hu[2,3], Thomas Gnäupel-Herold[1] and Anthony D Rollett[2]

*Correspondence:
adam.creuziger@nist.gov
[1] National Institute of Standards and Technology, 100 Bureau Dr., 20899, Gaithersburg, MD, USA
Full list of author information is available at the end of the article

Abstract

This paper considers the crystallographic texture evolution in a 1008 low carbon steel. The texture evolution along uniaxial, plane strain and balanced biaxial strain states were measured. For uniaxial testing, grains tend to rotate such that the $\{111\}\langle 1\bar{1}0 \rangle$ slip directions are aligned with the loading axis. For plane strain and balanced biaxial strain states, the majority of grains are distributed with the $\{111\}$ plane parallel to the sample normal direction. Accompanying visco-plastic self consistent (VPSC) predictions of the texture evolution were made along same strain paths and strain increments. Comparing between the measured texture evolution and computational texture evolution indicate that the VPSC model qualitatively predicts the measured texture evolution, but the rate at which the texture evolution occurs is over predicted.

Keywords: Crystallographic texture; Steel; Neutron diffraction; Metal forming

Introduction

Steel has been in use for nearly 4000 years, and remains one of the main materials (along with concrete, glass and asphalt) used around the world. A large majority of the steels created are carbon steels, with a large proportion of these carbon steels strained after production to create a required shape. Accurate prediction of the final shape and mechanical properties after straining continues to be an important industrial goal. One of the methods applied to this problem is measurement of the crystallographic texture (or preferred orientation) as a function of strain path and strain increment. Accurate prediction of the crystallographic texture and how the texture evolves will allow better estimates of the flow stress and plastic contraction, which in turn provides a more accurate prediction of the final shape and failure during forming. The particular case of deformation of sheet stock material is discussed in this paper. The majority of the literature discusses the high strain increments relevant for rolling sheet material. Unfortunately existing [1-6] experimental data on the texture evolution along multiaxial strain states at intermediate strain increments ($\Delta\epsilon$) relevant for forming are incomplete, so assessment of model accuracy is difficult in this regime.

This paper focuses on the measurement of the crystallographic texture evolution in 1008 AISI/SAE low carbon steel sheets after strain increments typical in forming structures from the rolled sheet. The texture is measured by neutron diffraction to interrogate a large number of grains. Five different strain states were investigated: uniaxial strain along the rolling direction (URD), strain along the rolling direction with the transverse direction fixed near zero strain (a plane strain condition, or PSRD), balanced biaxial strain (BB) in the rolling and transverse directions, strain along the transverse direction with the rolling direction fixed near zero strain (PSTD), and uniaxial strain along the transverse direction (UTD). In the plane of the sheet (normal direction or ND) the sample is assumed to be in a plane stress condition. In addition, the texture evolution is predicted from the as received texture using the visco-plastic self consistent (VPSC) model and compared with the experimental data. Data on the texture evolution of low carbon steels will also provide a well characterized reference for comparison to more complicated materials, such as high strength steels (HSS) like high strength low alloy (HSLA) steel or advanced high strength steels (AHSS) such as dual phase (DP) or transformation induced plasticity (TRIP) steels.

The data described in this paper has been submitted to the NIST/MatDL DSpace repository as part of the IMMI General Articles Collection [7]. The repository contains files that include a description of the data contained in the repository, a schematic of the the analysis and computational processes, some of the programs used to analyze the data, a collection of the texture data files at various points in the analysis, and input files to the VPSC model.

Background

Using the American Iron and Steel Institute (AISI) and Society for Automotive Engineers (SAE) grade designation, 1008 steel is defined as an iron alloy possessing a mass fraction of carbon $\leq$ 0.0010, a mass fraction of manganese between 0.0030 and 0.0050, a mass fraction of phosphorus $\leq$ 0.0004, and a mass fraction of sulfur $\leq$ 0.0005 [8]. Owing to the low carbon content and absence of austenitic stabilizers, the structure is almost entirely ferritic (body centered cubic crystal structure) with some carbides. This allows for high ductility (approximately 40% elongation), but low strength (tensile strength approximately 350 MPa) [9].

The texture evolution of steels has been widely studied in the past, particularly with respect to the complete processing path from the melt to hot and cold rolling. These topics are of interest to create desirable textures for deep drawing. Some articles on these topics include [2,3,10,11], which discuss steel textures and their origin from deformation and recrystalization of the austenite phase and subsequent transformation from austenite to ferrite. The complex details of the austenitic transformation are beyond the scope of this paper, but it is important to note that deformation and recrystallization in the austenite phase are one of the primary causes of crystallographic texture in the ferritic phase. The review by Inagaki [12] summarizes work on the inter-relationship between texture, grain boundaries and dislocation structures in low carbon steels. Hutchinson [13] discusses texture control in low carbon steels as a function of processing parameters. Bodin et al. [2] shows the strong effects of processing path and temperature on the texture.

The predicted plastic strain ratio (R value) has been used to correlate the effect of a particular texture component to the macroscopic properties. For deep drawing applications, sheet material that has a large volume of grains aligned with the γ-fiber ({111} plane parallel to the ND) are preferred as they have a high average R value ($\bar{R}$) and a planar anisotropy (ΔR) value near 0 [10,13]. Orientations with the {100} plane parallel to the ND direction such as 'cube' and 'shear' orientations are not desirable, since they have low average R values [10,13]. This is due to the orientation of slip systems for body centered cubic (BCC) materials. Conversely, for high strength materials where high ductility is not required, orientations along the α-fiber ($\langle 110 \rangle$ crystal direction parallel to rolling direction) are preferred for high strength [14] but may produce anisotropy and brittleness, particularly for the $\{113\}\langle 110 \rangle$ orientation [10].

There is also a significant amount of investigation into the textures caused by cold rolling in the ferrite phase. This data is particularly relevant to this paper as cold rolling, apart from a hydrostatic term, produces a stress state similar to plane strain along the rolling direction (PSRD). As shown in a series of review papers, cold rolling ferrite strengthens the α and γ fibers, particularly the α fiber at higher reduction ratios [3,10,11]. Figure thirteen in Kestens and Jonas [11] shows relative amounts of texture intensity as a function of thickness reduction. Inagaki [12] discusses some of the microstructural reasons for this, and theorizes that grain pairs with $\{111\}[uvw]$ type grain boundaries that allow the grain to rotate easily, keeping orientations along the γ fiber stable.

Kestens and Jonas [11] proposes that $\{211\}\langle 011 \rangle$ along the α fiber is the stable final orientation in plane strain compression, while $\{111\}\langle 110 \rangle$ and $\{554\}\langle 225 \rangle$ orientations are caused by ferritic recrystallization after plane strain compression. Inagaki [12] proposes that the $\{223\}\langle 011 \rangle$ orientation is the final stable orientation in plane strain compression. There is additional literature on cross-rolling of low carbon steel sheets [5], a strain state similar to plane strain along the transverse direction (PSTD). Results from Wronski et al. [5] show that the shear orientation and γ fiber are stable orientations after near 80% thickness reduction.

There has been limited investigation into the texture evolution along other strain paths relevant for sheet metal forming. Yerra et al. [6] is one of the most complete, following five different strain paths from equal biaxial to uniaxial tension, all with the largest strain values in the rolling direction. The orientation distribution functions (ODFs) are measured at a single level of strain for each strain path. In Yerra et al. [6], the experimental textures were not found to change greatly between the different strain modes nor the volume fractions of specific orientations. However a calculation using the Lamel model by the same authors show a significant difference between the strain paths. Other notable works include Luzin et al. [15], which measured texture after balanced biaxial straining. The γ fiber in a balanced biaxial strain mode is intensified, in agreement with Mirshams et al. [16] and theoretical calculations that predict the θ ($\langle 100 \rangle$ parallel to ND direction) and γ fibers to be the stable orientations in uniaxial compression along the ND. Other strain modes, such as torsion [17] and simple shear [18] are outside the scope of this paper.

As steel is such a relevant industrial material, there have been attempts to predict the texture evolution. Savoie and Jonas [19] used a rate sensitive analysis to model the texture along a flange during deep drawing. These simulations include a rotation rate map that shows the stable end orientations for different constraints imposed on the models. They found that all initial orientations tend to rotate toward the $\{223\}\langle 110 \rangle$ orientation. Toth et al. [20] used a rate dependent slip theory that contains models for pancake, lath and Taylor models to calculate texture evolution in low carbon steels during rolling. Starting with the 'RGoss' orientation, $\{111\}\langle 110 \rangle$, and shear orientations; they found the $\{111\}\langle 112 \rangle$ orientation also forms, but becomes unstable at higher strains. Many more stable orientations were found using this method, but most were observed along the α fiber and γ fiber as well as $\{554\}\langle 225 \rangle$. Raabe [21] used a Taylor type model to predict the rolling texture in low carbon deep drawing steel as a function of reduction ratio. The Taylor model showed increases in the intensity of the α fiber and γ fiber in good agreement with experiment [21].

Methods

Experimental

Chemical analysis

Chemical analysis was performed following ASTM E415 [22] for all elements except sulfur which followed ASTM E1019 [23]. The resulting elemental mass fractions are given in Table 1. These are within the required parameters for AISI/SAE 1008 steel with the exception of manganese which is slightly lower than specified.

Grain size measurements

The average grain size was measured according to ASTM E112 [24]. Linear intercept methods along the rolling direction (RD), transverse direction (TD) and normal direction (ND) were performed along five lines in three images over two different samples for all six cross-sectional views (N = 60 per direction). These averages were summed together to find the overall average grain size of 18.7 μm and aspect ratios of 1.0:0.78:0.46 for RD:TD:ND, normalized to RD. Complete information on the grain size is shown in Table 2.

Uniaxial sample testing

Samples were rough cut from 1.17 mm (0.046 in, 18 gauge) thick sheet stock into sample blanks approximately 19 mm (0.75 in) × 200 mm (8 in) using a hydraulic shear. A TENSILKUT[a] machine was used to create dogbone samples that conform to ASTM E8 [25] standards for sheet specimens with a 50.8 mm (2 inch) gauge length and 12.7 mm (0.5 inch) width. After machining, samples were loaded into a screw driven tensile frame moving at a crosshead speed of 0.1 $\frac{mm}{sec}$ resulting in a nominal strain rate of 0.002 $\frac{strain}{sec}$. Axial and transverse strains were measured using extensometers. The strains imparted to the samples are listed in Table 3. The von Mises effective strain was calculated assuming volume conservation and neglecting shear strains:

$$\epsilon_{\text{eff}} = \frac{\sqrt{2}}{3}\left[(\epsilon_{11} - \epsilon_{22})^2 + (\epsilon_{22} - \epsilon_{33})^2 + (\epsilon_{33} - \epsilon_{11})^2\right]^{\frac{1}{2}} \tag{1}$$

Multi-axial sample testing

Samples for non-uniaxial stain paths were loaded in a custom testing machine [26] using a sample geometry developed by Raghavan [27]. A clamp hold down force of -450 kN (force control) and a ram rate of 0.1 $\frac{mm}{sec}$ was applied resulting in a nominal strain rate of 0.2 $\frac{strain}{sec}$. Digital Image Correlation (DIC) was used to measure the plastic strains from a circular area with an undeformed diameter of 32 mm (1.25 inches). Strains are listed in Table 3.

Table 1 Mass fraction analysis performed by an independent lab per [22] and [23] for sulfur

Element	Measured	1008 AISI-SAE carbon steel grade [8]
Carbon	0.0007	0.0010 max
Manganese	0.0029	0.0030 to 0.0050
Phosphorus	0.00010	0.00040 max
Sulfur	0.00014	0.00040 max
Iron	Balance	

Table 2 Average grain sizes for 1008 steel using the linear intercept method

Plane	Direction	Average grain size [μm]	Standard deviation [μm]	Confidence interval [μm]	Relative accuracy [%]
ND	RD	24.15	2.94	1.10	4.55
TD	RD	28.72	3.25	1.21	4.22
ND	TD	17.85	1.50	0.56	3.14
RD	TD	23.40	2.27	0.85	3.62
RD	ND	11.79	0.93	0.35	2.96
TD	ND	12.59	0.78	0.29	2.32

Images were recorded after etching with Nital (2%) for 45 seconds. Images were recorded at 200 × magnification on all three planes on two different samples each. n = 30 for each row.

Texture measurements

Crystallographic texture measurements were performed using neutron diffraction at the BT-8 Residual Stress Diffractometer at NIST [28]. Strained and as received samples were cut into 8 mm squares and stacked and glued together until a cube was formed. For the average grain sizes calculated in Table 2, there are on the order of 2×10^7 grains in each sample. Three complete pole figures from the (200), (220), and (211) reflections were measured. A 5° hexagonal grid [29] of the pole figure space was used, resulting in 469 separate measurements, with the exception of the URD and UTD samples strained to 10% and 20%. In these four samples a hexagonal grid of 6° spacing was used, resulting in 331 separate measurements. The raw data was then processed in the program PF [30] and converted to pole figures in the popLA format [31]. The as received samples exhibited orthotropic

Table 3 Applied strains, measured after unloading the samples, expressed as true strain

Name	Nominal strain (%)	ϵ_{RD}	ϵ_{TD}	ϵ_{ND}	ϵ_{eff}
URD10	10%	0.100	-0.063	-0.037	0.101
URD20	20%	0.200	-0.127	-0.072	0.202
URD30	30%	0.302	-0.186	-0.116	0.305
PSRD10	10%	0.10	-0.01	-0.09	0.11
PSRD10	10%	0.10	-0.01	-0.09	0.11
PSRD20	20%	0.20	-0.02	-0.17	0.21
PSRD20	20%	0.20	-0.02	-0.17	0.21
PSRD30	30%	0.33	-0.03	-0.29	0.36
PSRD30	30%	0.33	-0.03	-0.30	0.37
BB10	10%	0.10	0.10	-0.19	0.19
BB20	20%	0.20	0.20	-0.40	0.41
BB30	30%	0.32	0.32	-0.63	0.63
PSTD10	10%	-0.01	0.10	-0.09	0.11
PSTD10	10%	-0.01	0.10	-0.09	0.11
PSTD20	20%	-0.02	0.22	-0.20	0.24
PSTD20	20%	-0.02	0.22	-0.20	0.24
PSTD30	30%	-0.03	0.34	-0.31	0.37
PSTD30	30%	-0.03	0.33	-0.30	0.37
UTD10	10%	-0.063	0.100	-0.037	0.101
UTD20	20%	-0.125	0.201	-0.076	0.203
UTD30	30%	-0.167	0.297	-0.130	0.297

For uniaxial testing, strains are measured along the RD and TD by extensometers, for BB and PS testing RD and TD strains are measured by digital image correlation. ND strains are calculated from volume conservation. The von Mises effective strain is calculated as shown in equation 1.

sample symmetry from the rolling process. All sample strains occurred aligned with the initial orthotropic symmetry resulting in orthotropic symmetry in the strained samples. For these reasons, orthotropic sample symmetry was applied to the pole figures and the orientation distribution functions (ODFs).

Orientation Distribution Function (ODF) calculation

From the series of pole figures, orientation distribution functions (ODFs) were calculated to determine the distribution of grains as a function of crystallographic orientation in the bulk material. Two different ODF solvers were investigated, the WIMV method [32] in popLA and mTex [33]. The calculation of the ODF is an inverse problem, which are known to not necessarily have a unique solution. For each solver a range of different ODFs can be calculated that match the pole figures depending on the, number of iterations of the ODF solving loop, resolution of the ODF, and other input parameters. To facilitate comparison with the VPSC ODF calculations, the WIMV method was used. Uncertainty in ODF measurements is an ongoing measurement challenge currently under investigation [34]. For the purposes of this paper, an uncertainty of 10% in the ODF values is assumed [35].

Good agreement (2% calculation error) with the pole figures was found after 12 iterations. Experimental and recalculated pole figures were qualitatively inspected to check agreement as well. The ODF from the as received sample was used to calculate an input file for the VPSC analysis. The ODF is expressed in a crystal orientation distribution file that has a regular 5° grid in the Bunge Euler angles ϕ_1, Φ, ϕ_2. Due to the crystal and sample symmetries, the ODF can be expressed in a subspace with the range $0° \leq \phi_1 \leq 90°$, $0° \leq \Phi \leq 90°$, $0° \leq \phi_2 \leq 90°$, albeit with an additional three fold symmetry in this subspace. A value for intensity is listed at each point in this grid, expressed in units of 'multiples of a uniform distribution' or MUD (often referred to as 'multiples of a random distribution' MRD).

Excerpts from the ODF were also extracted along fiber textures. The fibers of interest for this paper are the α fiber, γ fiber and θ fiber. The α fiber runs along the Φ axis, with $\phi_1 = 0°$ and $\phi_2 = 45°$. The γ fiber runs along the ϕ_1 axis, with $\Phi = 55.4°$ and $\phi_2 = 45°$. The γ_1, γ_3 and γ_2, γ_4 pairs are equivalent points by symmetry. The θ fiber runs along the ϕ_1 axis, with $\Phi = 0°$ and $\phi_2 = 45°$. Specific orientations of interest are listed in Table 4, with a shorthand name, Miller indices, and Bunge Euler angles. Additionally, difference ODF plots were created between adjacent strain steps. These difference ODFs show the locations that grains are rotating to or away from. However, as a consequence of the non-linear nature of the intensity scale of the ODF, caution should be applied to ascribing quantitative significance to the values in the difference plots. To aid in the idenification of the orientations listed in Table 4, a key showing the location of each orientation located in the $\phi_2 = 45°$ cross section is shown in Figure 1.

Computational

The viscoplastic self-consistent (VPSC) model developed by Lebensohn and Tomé [36] was used to simulate the texture development. The VPSC model treats each grain as a viscoplastic ellipsoidal inclusion embedded in and interacting with a homogeneous effective matrix (HEM). Under the HEM assumption, individual grain response varies as a function of its orientation and the stiffness of the grain-to-matrix interaction, depending on

Table 4 Relevant orientations investigated

Name	{hkl}	⟨uvw⟩	ϕ_1	Φ	ϕ_2
Cube	{001}	⟨100⟩	45°	0°	45°
Shear	{001}	⟨1$\bar{1}$0⟩	0°	0°	45°
Goss	{011}	⟨100⟩	90°	90°	45°
RGoss	{011}	⟨0$\bar{1}$1⟩	0°	90°	45°
Copper	{112}	⟨$\bar{1}\bar{1}$1⟩	90°	35.26°	45°
Brass	{011}	⟨2$\bar{1}$1⟩	54.74°	90°	45°
S	{213}	⟨$\bar{3}\bar{6}$4⟩	58.98°	36.70°	63.44°
{554}	{554}	⟨$\bar{2}\bar{2}$5⟩	90°	60.50°	45°
γ_1	{111}	⟨1$\bar{1}$0⟩	0°	54.74°	45°
γ_2	{111}	⟨1$\bar{2}$1⟩	30°	54.74°	45°
γ_3	{111}	⟨0$\bar{1}$1⟩	60°	54.74°	45°
γ_4	{111}	⟨$\bar{1}\bar{1}$2⟩	90°	54.74°	45°
α_1	{115}	⟨1$\bar{1}$0⟩	0°	15.79°	45°
α_2	{113}	⟨1$\bar{1}$0⟩	0°	25.24°	45°
α_3	{112}	⟨1$\bar{1}$0⟩	0°	35.26°	45°
α_4	{223}	⟨1$\bar{1}$0⟩	0°	43.31°	45°

All orientations are described by Miller Indices and Bunge Euler angles.

the choice of homogenization methods at grain level. Details of the model can be found in [36].

A file containing 24 938 discrete orientations was created to approximate the calculated ODF from the as received sample. This list of orientations was used as an input file to the VPSC algorithm. Three other input files were also required: a file containing the applied velocity gradients and applied Cauchy stresses; a file containing the unit cell parameters, elastic moduli of the single crystal, slip planes, and hardening behavior of slip families; and an input file containing the path and name of the subfiles, grain aspect ratio, hardening model, and interaction model.

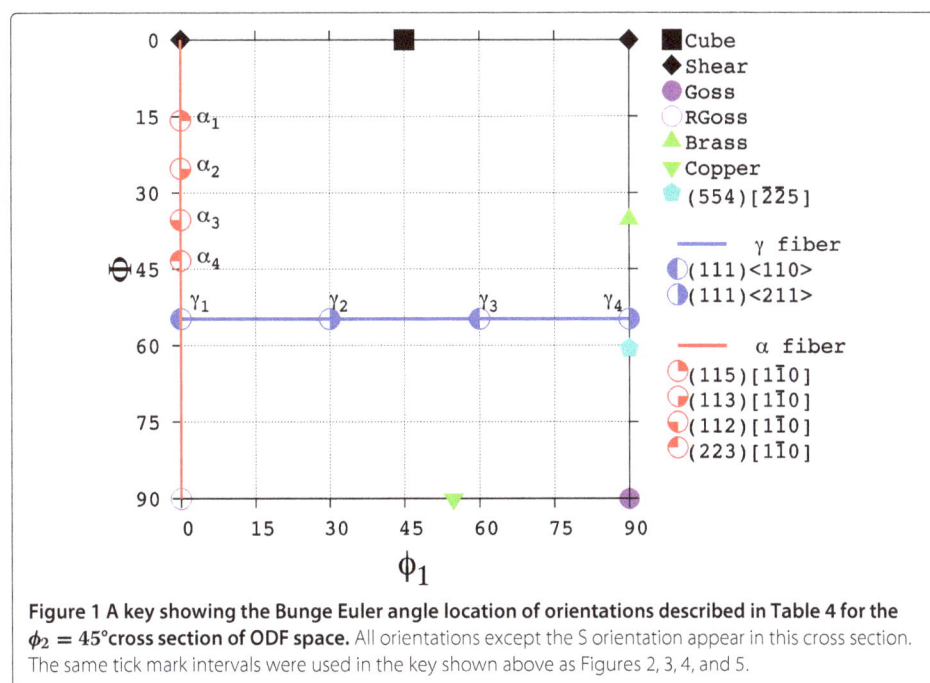

Figure 1 A key showing the Bunge Euler angle location of orientations described in Table 4 for the $\phi_2 = 45°$ cross section of ODF space. All orientations except the S orientation appear in this cross section. The same tick mark intervals were used in the key shown above as Figures 2, 3, 4, and 5.

For all simulations a velocity gradient is imposed on all components of the velocity gradient tensor matching the macroscopic strains in Table 3 in the following manner. Imposed velocity tensors were the ϵ_{RD}, ϵ_{TD} and ϵ_{ND} strains normalized such that the highest positive strain was set equal to 1. The strain increment was the value of strain along the highest positive strain direction. All stresses calculated by the VPSC model are therefore deviatoric stresses. Only the $\{110\}\langle 111\rangle$ and $\{112\}\langle 111\rangle$ families of slip planes and directions were considered in this model as initial experimentation including the $\{123\}\langle 111\rangle$ family of planes did not show a significant effect on the texture. The ellipsoid ratios used were based on the experimental data listed in Table 2 normalized along the rolling direction and having values of 1.0, 0.78 and 0.46 for the RD, TD and ND respectively. Secant linearization and first nearest neighbor interaction models were chosen. The outputs used here from the VPSC algorithm include a list of discrete orientations that have been rotated to new orientations due to the imposed velocity gradient. These discrete orientations were binned into a regular grid to create an ODF.

Results

Experimental textures

Three different techniques have been used to show how the experimentally measured texture evolves as a function of applied strain. Figure 2 shows the intensities of the ODF

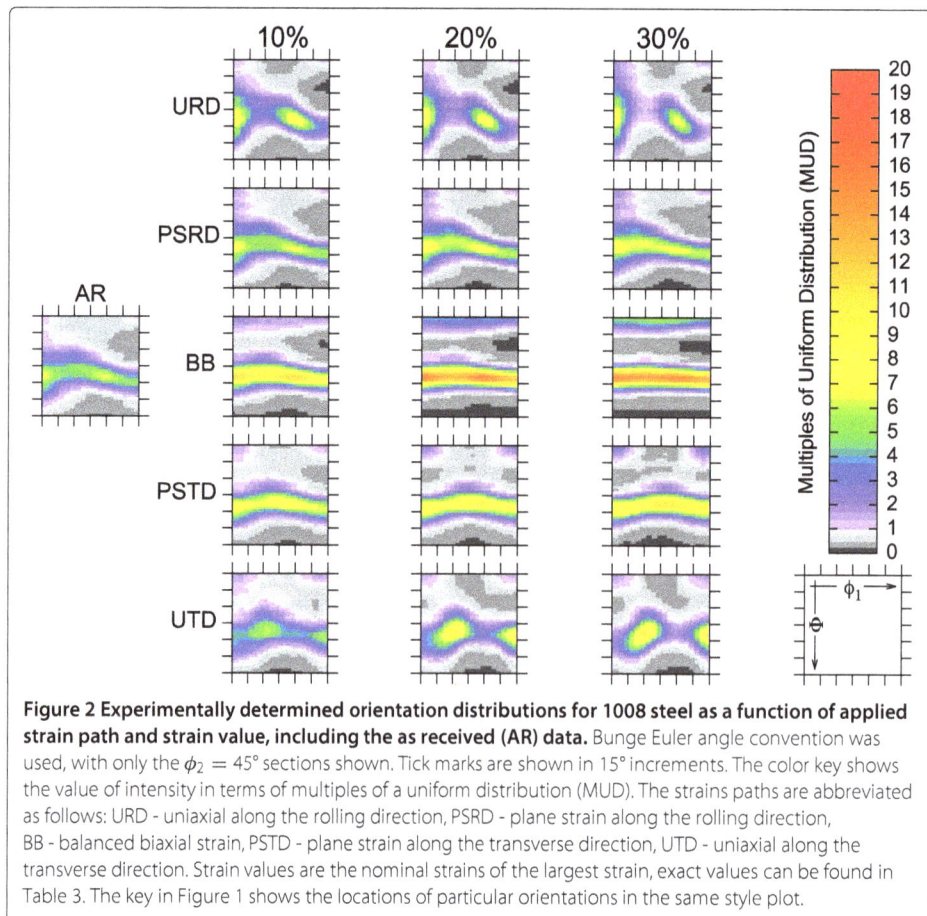

Figure 2 Experimentally determined orientation distributions for 1008 steel as a function of applied strain path and strain value, including the as received (AR) data. Bunge Euler angle convention was used, with only the $\phi_2 = 45°$ sections shown. Tick marks are shown in 15° increments. The color key shows the value of intensity in terms of multiples of a uniform distribution (MUD). The strains paths are abbreviated as follows: URD - uniaxial along the rolling direction, PSRD - plane strain along the rolling direction, BB - balanced biaxial strain, PSTD - plane strain along the transverse direction, UTD - uniaxial along the transverse direction. Strain values are the nominal strains of the largest strain, exact values can be found in Table 3. The key in Figure 1 shows the locations of particular orientations in the same style plot.

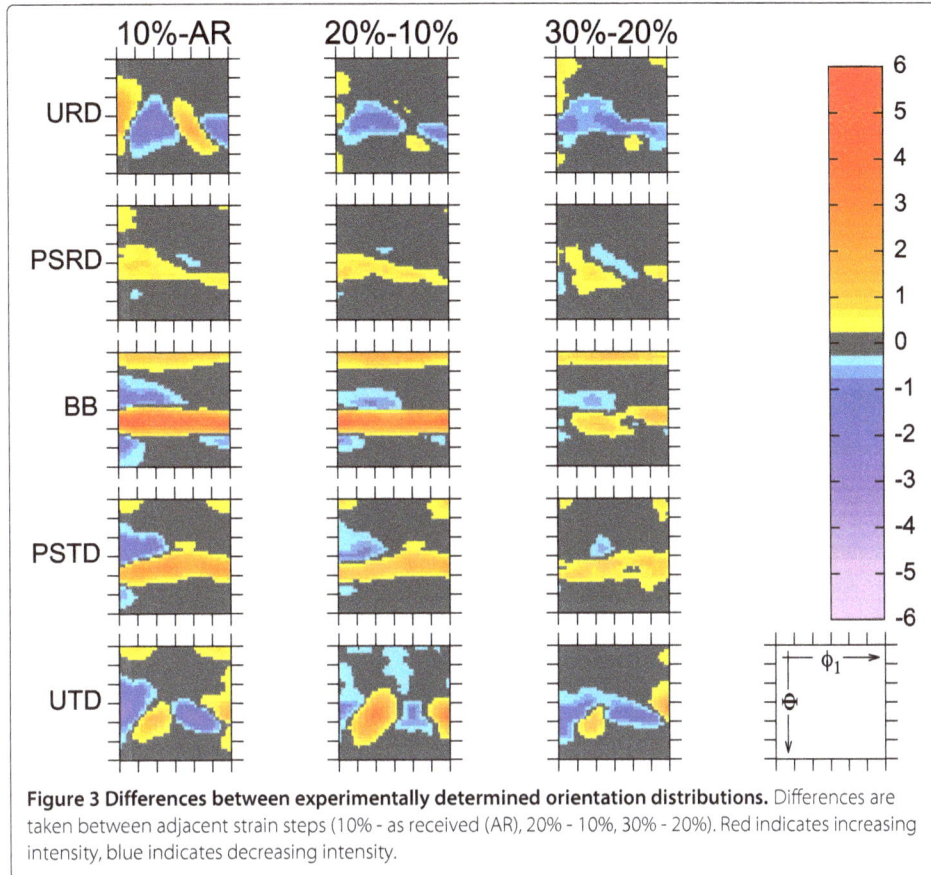

Figure 3 Differences between experimentally determined orientation distributions. Differences are taken between adjacent strain steps (10% - as received (AR), 20% - 10%, 30% - 20%). Red indicates increasing intensity, blue indicates decreasing intensity.

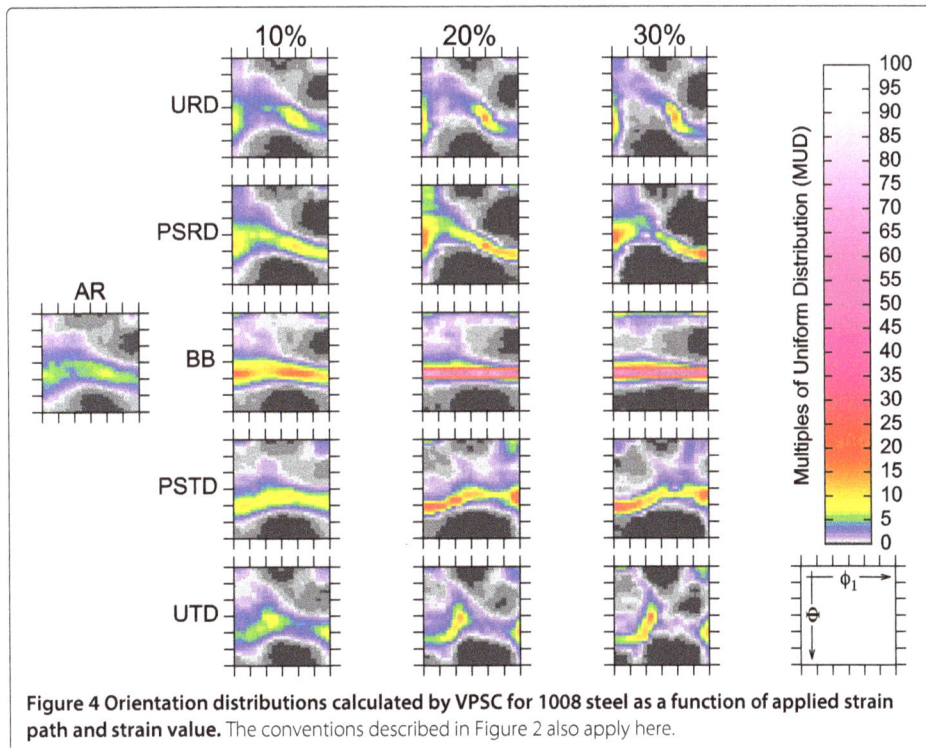

Figure 4 Orientation distributions calculated by VPSC for 1008 steel as a function of applied strain path and strain value. The conventions described in Figure 2 also apply here.

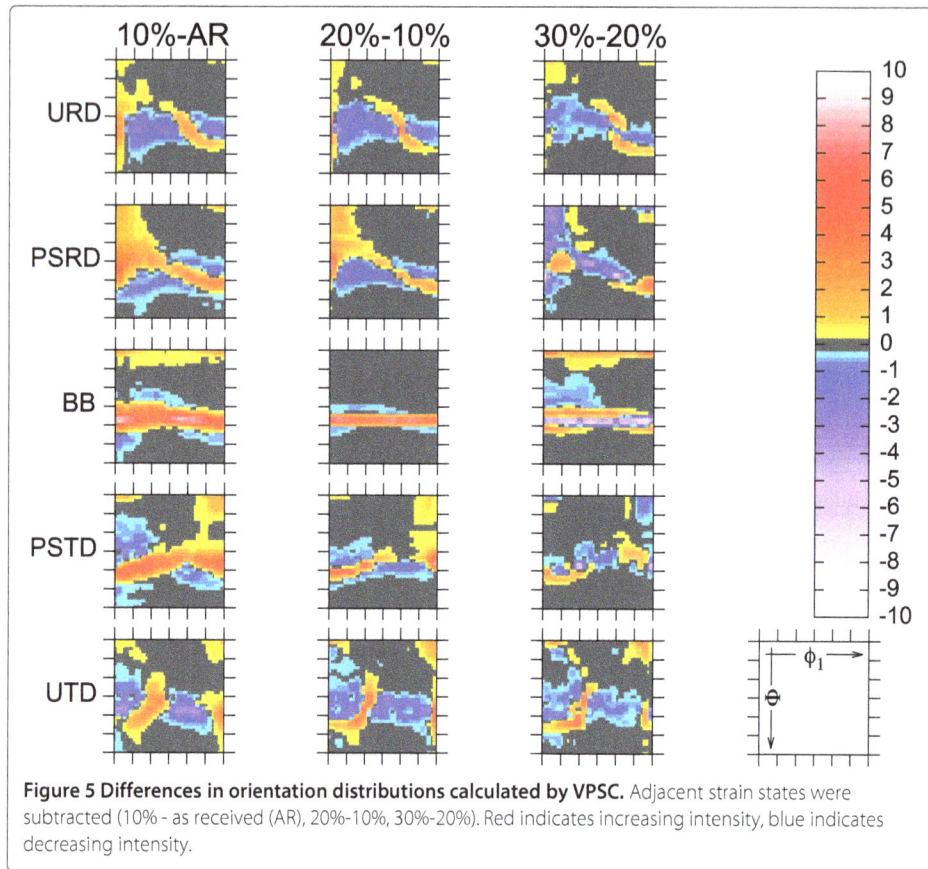

Figure 5 Differences in orientation distributions calculated by VPSC. Adjacent strain states were subtracted (10% - as received (AR), 20%-10%, 30%-20%). Red indicates increasing intensity, blue indicates decreasing intensity.

on $\phi_2 = 45°$ cross sections. Figure 3 is a plot of the differences between adjacent strain steps (10% - as received, 20% - 10%, 30% - 20%) on the same $\phi_2 = 45°$ cross sections shown in Figure 2. The smaller strain step is subtracted from the larger strain step, so positive values indicate an increase in intensity at the given orientation. The third technique extracts line profiles along a path with two fixed Euler angles. These line profiles correspond to the α fiber (Figure 6 top), γ fiber (Figure 7 top), and θ fiber (Figure 8 top). For brevity, the discussion of the URD deformation contains the relevant figure numbers for discussion of the fiber components, they will not be repeated for each strain mode.

Figure 2 includes the as received texture, as well as the three nominal strain increments along the five strain paths explored. The as received texture of the 1008 steel is typical for a rolled body centered cubic (BCC) material for deep drawing applications, with strong intensity along the γ fiber, and peaks along the $\langle 110 \rangle$ RD directions of the (111) plane. The intensity values for α fiber orientations and the shear orientation are also larger than in a uniform texture distribution. The intensities of cube, shear, Goss, RGoss, copper, brass, and S orientations are low, as expected in a BCC material.

When samples are strained in uniaxial tension along the rolling direction (URD), the ODF intensity along the γ_1 and γ_3 orientations increases, as shown in the first row in Figure 2, Figure 3, and Figure 7. The ODF intensity along the $\{554\}\langle \bar{2}\bar{2}5 \rangle$, γ_2, and γ_4 decrease. The largest change in the ODF intensities are along these orientations occurs while straining from 0% to 10%. The ODF intensity along all of the α fiber orientations

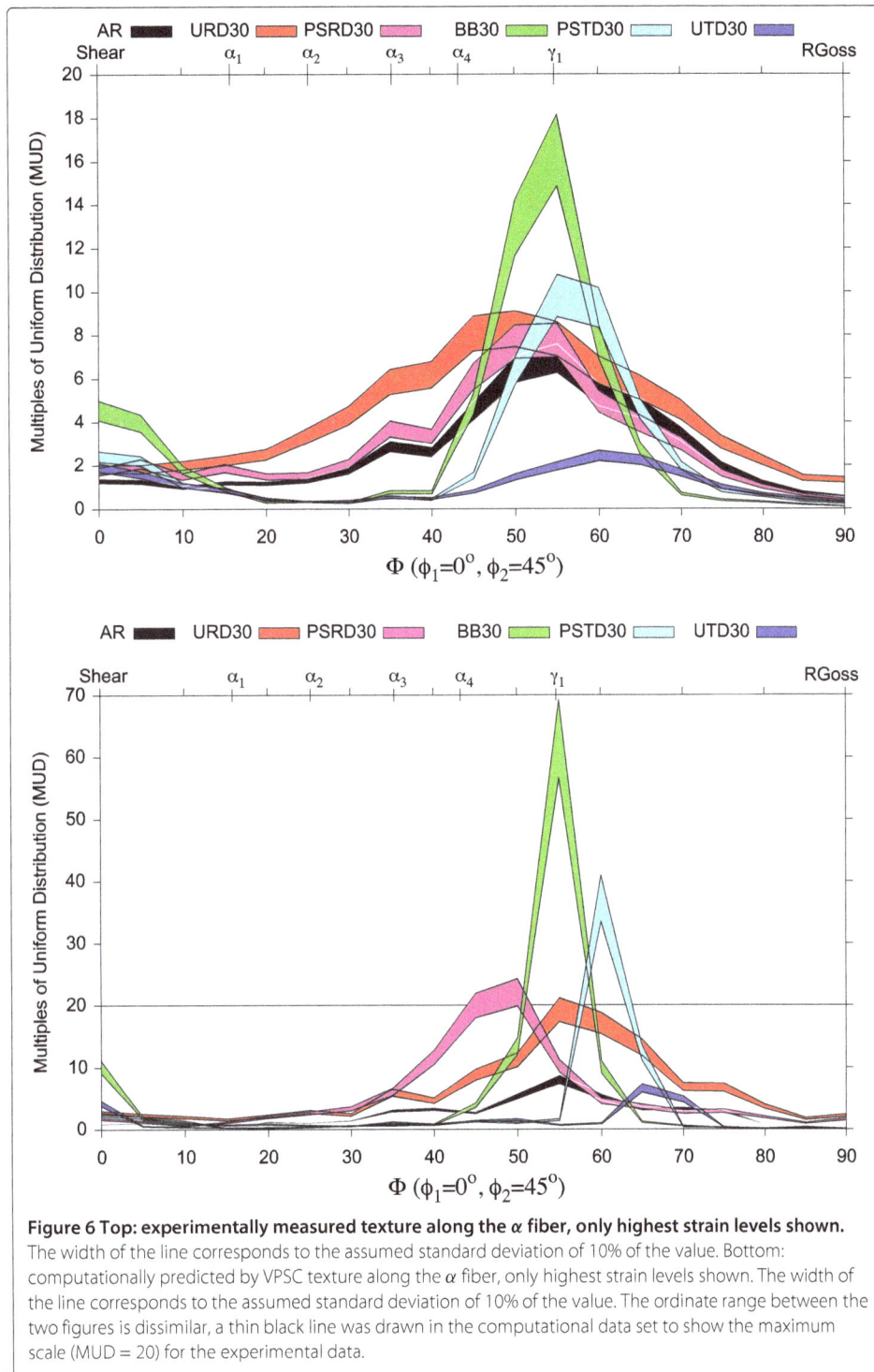

Figure 6 Top: experimentally measured texture along the α fiber, only highest strain levels shown.
The width of the line corresponds to the assumed standard deviation of 10% of the value. Bottom:
computationally predicted by VPSC texture along the α fiber, only highest strain levels shown. The width of
the line corresponds to the assumed standard deviation of 10% of the value. The ordinate range between the
two figures is dissimilar, a thin black line was drawn in the computational data set to show the maximum
scale (MUD = 20) for the experimental data.

increase, shown in Figure 2, Figure 3, and Figure 6. Along the θ fiber the grains along
the shear orientation increase, shown in Figure 2, Figure 3, and Figure 8, but the ODF
intensity along the cube orientation remains the same.

The PSRD strain state is notable in that there are very few changes from the as received
structure, depicted in the second row in Figures 2 and 3. For all of the orientations

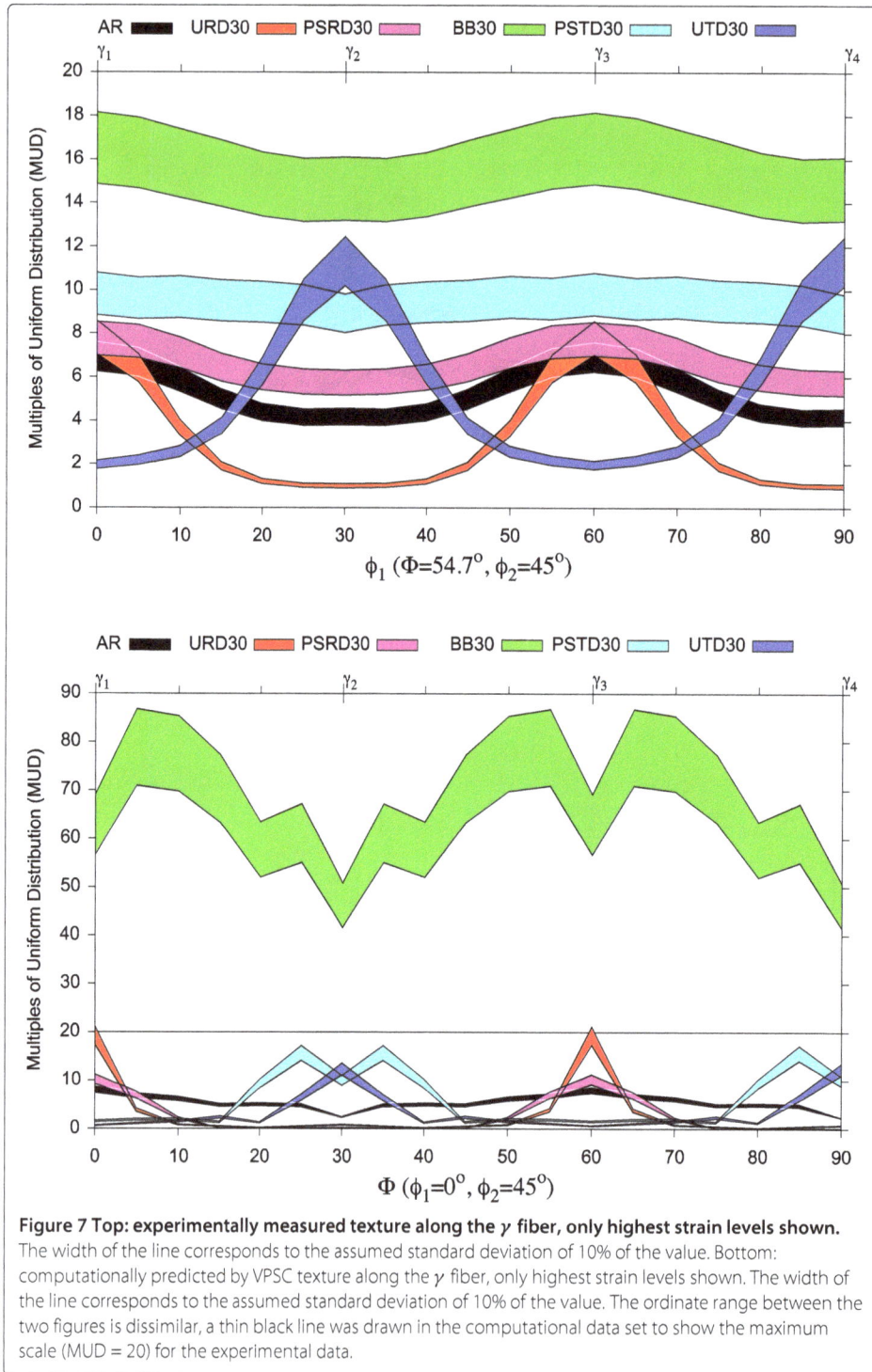

Figure 7 Top: experimentally measured texture along the γ fiber, only highest strain levels shown.
The width of the line corresponds to the assumed standard deviation of 10% of the value. Bottom: computationally predicted by VPSC texture along the γ fiber, only highest strain levels shown. The width of the line corresponds to the assumed standard deviation of 10% of the value. The ordinate range between the two figures is dissimilar, a thin black line was drawn in the computational data set to show the maximum scale (MUD = 20) for the experimental data.

investigated, the ODF intensity changes only very slightly at each strain level. The ODF intensity of orientations along the α fiber are basically unchanged. There are some modest increases in the intensity along the γ fiber, particularly filling in the γ_2 and γ_4 orientations to make the intensity along the γ fiber slightly higher and more uniform in intensity. The θ fiber shows the same trend as the samples deformed with uniaxial tension

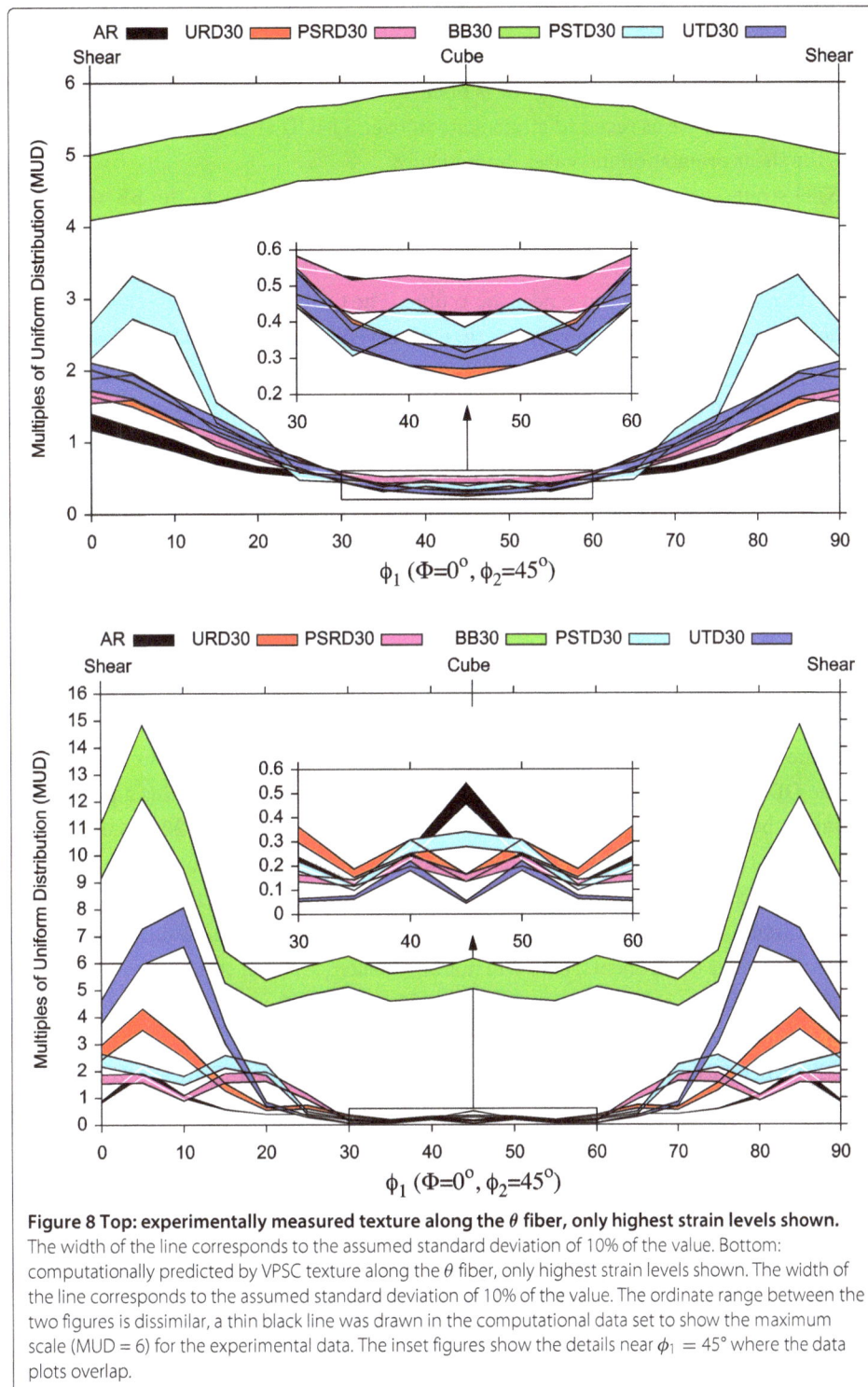

Figure 8 Top: experimentally measured texture along the θ fiber, only highest strain levels shown. The width of the line corresponds to the assumed standard deviation of 10% of the value. Bottom: computationally predicted by VPSC texture along the θ fiber, only highest strain levels shown. The width of the line corresponds to the assumed standard deviation of 10% of the value. The ordinate range between the two figures is dissimilar, a thin black line was drawn in the computational data set to show the maximum scale (MUD = 6) for the experimental data. The inset figures show the details near $\phi_1 = 45°$ where the data plots overlap.

along the rolling direction with the shear orientation increasing and the cube orientation unchanged.

The dominant texture evolution in the BB strain state is intensification of the γ fiber. The 20% sample still shows the highest intensity along the $\langle 110 \rangle$ RD directions, but by 30% strain all orientations along the γ fiber are approximately equal and have an intensity of 15 to 16 times higher than a uniform distribution. Grains oriented along the α fiber have decreased sharply outside the range of the γ fiber and θ fiber. The BB strain state also shows the strongest changes along the θ fiber. The cube orientation increases from a value of 0.5 MUD in the as received strain state to over 5 MUD after 30% BB straining. Similarly, the shear orientation increases dramatically.

The PSTD strain state results in a texture evolution that is similar to the BB strain state. Grains oriented along the γ fiber increase, particularly the γ_2 and γ_4 orientations, to become nearly constant in intensity at 10 MUD by 30% strain. Grains oriented along the α fiber decrease, except for those near the γ fiber. The ODF intensity along the lower part of the α fiber ($\phi_1 \geq 60°$) increases more than observed in the RD, PSRD or BB strain states. There is an increase in the number of grains oriented along the shear orientation, while the cube orientation remains unchanged.

Due to the symmetries in this projection of Euler space, the texture evolution of UTD samples look like the corresponding URD ODF mirrored about the $\phi_1 = 90°$ axis. Grains oriented along the γ_2 and γ_4 orientations increase, while γ_1 and γ_3 intensity decreases. Orientations along the α fiber decrease. There is some increase in the copper orientation, which can be related by symmetry operations to an orientation near the α_3 orientation. Along the θ fiber the fraction of grains oriented along these directions are similar to measurements of the URD and PSRD samples.

Computational textures

Similar to the experimental data shown in Figure 2, Figure 4 shows the calculated intensities of the ODF in $\phi_2 = 45°$ cross sections from VPSC simulations. Figure 5 is a plot of the differences between adjacent strain steps (10% - as received, 20% - 10%, 30% - 20%) in the ODFs calculated with VPSC. Line profiles along the α fiber (Figure 6 bottom), γ fiber (Figure 7 bottom) and θ fiber (Figure 8 bottom) are also shown. Similar to the experimental data, the discussion of the URD deformation contains the relevant figure numbers for discussion of the fiber components, they will not be repeated for each strain mode.

The intensity scale for each of the computational figures was adjusted to accommodate the higher intensity values calculated in the VPSC code. However the color scale in the range of the experimental intensities are unchanged for Figure 4 and Figure 5 to allow for comparison, additional colors were added for intensities outside of the experimental range. For Figures 6, 7, and 8 a thin black line is used to depict the maximum in the experimental line profile figures.

The VPSC as received texture shown in Figure 4 is similar to, but less smooth than the experimental as received texture shown in Figure 2. Regions around γ_2 and γ_4 are slightly depleted in intensity compared to the experimental data. Similarly, areas around the copper and brass orientations are also lower than observed experimentally.

The main intensity changes for URD occur along γ_1 and γ_3. There are corresponding decreases in the γ_2 and γ_4 orientations. As the amount of strain increases, the intensity around the γ_3 orientation increases in a very narrow band in the ϕ_1 axis while broadening in the Φ axis at higher strains. The α fiber increases from the as received sample, but with larger increases at $\Phi > 55°$ than at $\Phi < 55°$ observed in the experimental data. Along the θ fiber, the shear orientation shows a sharp drop, observed in all VPSC data sets. These observations are likely computational artifacts due to the low number of grains (2.5×10^5) compared to the experimental data ($\approx 2 \times 10^7$). The θ fiber orientations show modest increases in the shear texture, but the cube texture is essentially unchanged.

When the strain state changes to plane strain along the rolling direction (PSRD), the texture evolution is similar to the VPSC URD sample. The γ_1 through γ_3 orientations increase but the γ_2 and γ_4 orientations decrease. However, the {554} orientation, which is close to the γ_4 orientation, increases like the γ_1 and γ_3 orientations. Unlike experimental data, orientations near the α_3 and α_4 orientations show large increases, particularly at 20% strain, but the other α fiber orientations only show slight increases in ODF intensity. Similar to the α_4 orientations, the shear orientation shows a large increase at 20% strain, and then decreases back to the as received intensity at 30%. The cube and shear orientations in the θ fiber are also largely unchanged.

Balanced biaxial (BB) strains show the sharpest texture evolution. The texture sharpening is primarily along the γ fiber. All orientations along γ fiber increase by similar amounts for each strain step, such that by 30% strain, intensities of 80 MUD are observed. The α fiber shows a sharp peak where it intersects with the γ fiber, but decreases in all other orientations. The θ fiber shows a large increase as well, predominantly along the shear orientation.

Plane strain along the transverse direction (PSTD) shows increases in the ODF intensity at all orientations along the γ fiber after 10% strain, and then increases in the γ_2 and γ_4 thereafter. Similar to the URD sample, the α fiber increases more at $\Phi > 55°$ than at $\Phi < 55°$. The shear orientation increases in intensity, but the cube orientation remains nearly constant.

Finally, the UTD samples show intensity increases along γ_2 and γ_4, the $\langle 110 \rangle$ TD directions of the (111) plane, but of lower magnitude than the PSTD strain state. The intensity of orientations γ_1 and γ_3 decrease. Like the URD case, as strain increases the highest intensity bands tend to align themselves with the Φ axis. The intensity along the α fiber orientations generally decrease, with only a slight increase at $\Phi = 65°$ to 70°. Orientations near the shear texture also show large increases.

Discussion

Qualitatively there is good agreement between the experimental and computational texture evolution models. Most changes from the texture evolution have the correct 'sense' to them, experimental increases are matched to computational increases and vice versa. Both the computational and experimental textures show significant changes even at relatively low (10%) values of strain. There are some subtle differences between the Experimental and VPSC as received textures. As discussed in the results and methods sections, the VPSC as received texture is less smooth, likely due to the three order of magnitude reduction in the number of grains explored in the VPSC model (24 938 orientations) than in the experimental data (estimated at $\approx 2 \times 10^7$). For orientations with $\phi_2 = 45°$ and $\Phi < 20°$ there are less than 10 grains for each of the orientations in the 5° bin and at $\Phi = 0$ most of the orientations only have 1 grain associated with each bin. This may also be the cause of oddities around the shear orientation in Figure 8, where large peaks are observed 5° from the orientation that are not observed experimentally in Figure 8.

There are a few places where there is qualitative disagreement between the trends in the VPSC predictions and experimental data. In the experimental URD and PSRD strain states there are broad increases in the α fiber, particularly at $\Phi < 55°$, shown in Figure 6.

The VPSC data in Figure 6 shows some increase in the intensity at $\Phi < 55°$, but a larger increase in intensity at $\Phi > 55°$. Similarly the increase in cube orientation over the shear orientation observed experimentally in Figure 8 under BB straining was not observed in the VPSC predictions in Figure 8. At higher strain levels the VPSC model seemed to rotate grains away from the γ fiber to orientations $\pm 5°$ to $\pm 10°$ away from $\Phi = 55°$ (γ fiber), as shown by the decreases in intensity along the γ fiber in the difference map in Figure 5. This may also affect the PSTD results data, where the intensity pattern observed experimentally in Figure 7 does not match the VPSC predictions shown in Figure 7.

Quantitatively, the VPSC model tends to over predict the rate of texture evolution leading to much higher intensities and sharper orientation distributions. Overall, the intensity values for experimental data at 30% strain better match the VPSC data at 10% strain. The BB 30% strain state is a particularly good example of the difference in rate of texture evolution, with the intensity predicted by the VPSC over-predicted by a factor of 5. Another example is the PSRD strain state, where the α_4 and {554} orientations are predicted to have increases larger than are observed experimentally.

Comparing this data with prior computational and analytical work, the uniaxial samples show the $\langle 110 \rangle$ slip directions of the {111} plane as stable orientations, and instability of slip directions $\langle 211 \rangle$ in the {111} plane. The PSRD strain state is indeed similar to the predictions for cold rolling the ferrite phase. The experimental data indicates that while the α_4 ({223}$\langle 1\bar{1}0 \rangle$) orientation is the most stable, there seems to be a broad distribution of orientations that are increasing at the strain levels investigated. The stability of the α fiber changes with more transverse direction strain, such that in the BB and PSTD strain states this fiber is not as stable. This result could be used in production, for applications where α fiber is not desirable, as more lubrication on the rollers or cross rolling could be used to remove some of the α fiber and increase the number of grains with orientations aligned with the γ fiber; albeit with some increase in the cube and shear components as well. The increase in γ fiber agrees well with observations by [15] and [16] but in our work more cube orientated grains are observed than by Luzin et al. [15]. The experimental and VPSC data are more consistent with the Lamel predictions than experimental data from Yerra et al. [6].

A similar study 5754 aluminum compared Los Alamos polycrystal plasticity (LApp), VPSC and visco-plastic fast Fourier transform (VPFFT) models for multiaxial strain states [37]. Hu et al. [37] showed similar trends to those observed here, with the rate of texture evolution over predicted in the VPSC model, particularly for the BB strain state. The VPFFT model calculated a rate of texture evolution closer to those observed experimentally. In the VPFFT model, the inter-grain and intra-grain interactions are included in the model, likely improving the accuracy. However, the VPFFT model described by Hu et al. [37] is discretized as voxel elements. Approximately 500 voxels were required for each grain, as opposed to the single ellipsoid per grain in the VPSC model, resulting in the VPFFT model containing one fifth as many grains as the VPSC model for a similar computational time. The reduction in number of grains resulted in a lower fidelity to the as received texture. The texture in the 5754 aluminum studied was much closer to a uniform distribution than the steel texture measured here, so a larger number of grains would be required to accurately model the steel microstructure. Following work by Dawson et al. [38], the addition of the elastic component to the viscoplastic models (EVPSC or EVPFFT approaches) is not expected to improve the model as the strains in this study are relatively large strains along monotonically increasing strain paths. Additionally, the construction of an accurate microstructure with respect to the spatial correlation of adjacent grains required as an input to a VPFFT or crystal plasticity based finite element analysis (CP-FEA) models is a topic of current research.

However, there is some concern that even accurate microstructure may be insufficient to model the deformation of polycrystalline materials. Pokharel et al. [39] compare a modeled microstructure in VPFFT that is taken from experimental data from near field high-energy X-ray diffraction microscopy (nf-HEDM) [40]. Even with a microstructure model that closely matches the experimental data, the agreement between experiment and model are not as close as desired. Similar to the results shown in this paper, there is reasonable qualitative agreement, but there are significant qualitative and quantitative variations. Pokharel et al. [39] conclude that the "empirically derived constitutive relations used in continuum scale simulations" (i. e. the Voce model in VPFFT and VPSC) as the likely cause of the disagreement and recommend further work on the development of a constitutive relation based on dislocation interactions. Pokharel et al. [39] also touches on CP-FEA, but states that the empirically derived continuum level constitutive relationships used in FEA may suffer similar problems as observed in the VPFFT model.

Conclusions

To summarize, the crystallographic texture evolution in a 1008 low carbon steel along five different strain states have been experimentally measured and computationally predicted using VPSC. Comparisons between measured texture evolution and computational texture evolution indicates that the VPSC model is generally qualitatively accurate in predicting the orientations to which grains rotate. Quantitatively, the VPSC model over predicts the rate at which grains will rotate, particularly in multi-axial strain states such as balanced biaxial.

Endnote

[a] Certain commercial equipment, instruments, or materials are identified in this paper in order to specify the experimental procedure adequately. Such identification is not intended to imply recommendation or endorsement by the National Institute of Standards and Technology, nor is it intended to imply that the materials or equipment identified are necessarily the best available for the purpose.

Competing interests
The authors declare that they have no competing interests.

Authors' contributions
AC calculated the ODFs, compared the data and lead writing of the manuscript; LH conducted the VPSC simulations; TGH measured and analyzed the experimental pole figures; ADR advised LH in VPSC, reviewed our work and assisted in interpretation. All authors read and approved the final manuscript.

Acknowledgements
The authors thank Steve Banovic, Tim Foecke, Mark Iadicola and Mark Stoudt at the NIST Center for Metal Forming for advice and assistance, and the Research Associate Program at the National Academies for supporting this work.

Author details
[1] National Institute of Standards and Technology, 100 Bureau Dr., 20899, Gaithersburg, MD, USA. [2] Carnegie Mellon University, 5000 Forbes Avenue, 15213, Pittsburgh, PA, USA. [3] IBM Semiconductor Research and Development Center, 2070 Route 52, 12533, Hopewell Junction, NY, USA.

References

1. Bleck W, Grossterlinden R, Lotter U, Reip CP (1991) Textures in steel sheets. Steel Res 62(12): 580–586. ISSN 0177-4832
2. Bodin A, Sietsma J, van der Zwaag S (2002) Texture and microstructure development during intercritical rolling of low-carbon steels. Metallurg Mater Trans A 33(6): 1589–1603. ISSN 1073-5623
3. Hutchinson B (1999) Deformation microstructures and textures in steels. Philos Trans R Soc Lond A 357(1756): 1471–1485. ISSN 1364-503X
4. Kestens L, Jonas JJ (1999) Deep drawing textures in low carbon steels. Metals And Mater-Korea 5(5): 419–427. ISSN 1225-9438
5. Wronski S, Wrobel M, Baczmanski A, Wierzbanowski K (2013) Effects of cross-rolling on residual stress, texture and plastic anisotropy in f.c.c. and b.c.c. metals. Mater Characterization 77: 116–126. ISSN 1044-5803
6. Yerra SK, Vankudre HV, Date PP, Samajdar I (2004) Effect of strain path and the magnitude of prestrain on the formability of a low carbon steel: On the textural and microtextural developments. J Eng Mater Technol-Trans ASME 126(1): 53–61. ISSN 0094-4289
7. Creuziger A, Hu L, Gnäupel-Herold T, Rollett AD (2013) Data citation: crystallographic texture evolution in 1008 steel sheet during multi-axial tensile strain paths. http://hdl.handle.net/11115/231 Accepted 11 Dec 2013
8. Oberg E, Jones FD, Horton HL, Ryffel H H (1996) Machinery's handbook, 25th edition. Industrial Press Inc., 32 Haviland Street, Unit 2C | South Norwalk, CT 06854
9. (2013) MatWeb. AISI 1008 steel, CQ, DQ, and DQSK sheet. http://www.matweb.com/search/datasheet.aspx?MatGUID=145867c159894de286d4803a0fc0fe0f
10. Ray RK, Jonas JJ (1990) Transformation textures in steels. Int Mater Rev 35(1): 1–36. ISSN 0950-6608
11. Kestens L, Jonas JJ (2006) ASM handbook, volume 14A:685–700. ASM, ASM International Materials Park, OH: 44073-0002
12. Inagaki H (1994) Fundamental aspect of texture formation in low-carbon steel. ISIJ Int 34: 313–321
13. Hutchinson B (1994) Practical aspects of texture control in low carbon steels. Mater Sci Forum 157–162: 1917–1928
14. Inagaki H, Kurihara K, Kozasu I (1977) Influence of crystallographic texture on strength and toughness of control-rolled high-tensile strength steel. Trans Iron And Steel Institute of Japan 17(2): 75–81. ISSN 0021-1583
15. Luzin V, Banovic S, Gnaupel-Herold T, Prask HJ, Ricker RE (2005) Measurement and calculation of elastic properties in low carbon steel sheet. In: Houtte PV, Kestens L (eds) ICOTOM 14: Textures Of Materials, volume 495-497 of Materials Science Forum, pp 1591–1596
16. Mirshams RA, Crosby KE, Mohamadian HP, Burris CL (1993) The influence of biaxial stretching on the texture of ultra-low carbon (ulc) steel sheets. Scripta Metallurgica Et Materialia 29(4): 433–438. ISSN 0956-716X
17. Baczynski J, Jonas JJ (1996) Texture development during the torsion testing of alpha-iron and two IF steels. Acta Mater 44(11): 4273–4288. ISSN 1359-6454
18. Kang JY, Bacroix B, Regle H, Oh KH, Lee HC (2007) Effect of deformation mode and grain orientation on misorientation development in a body-centered cubic steel. Acta Mater 55 15: 4935–4946. ISSN 1359-6454
19. Savoie J, Jonas JJ (1994) Simulation of the deformation textures induced by deep-drawing in extra low-carbon steel sheets. Acta Metallurgica Et Materialia 42(12): 4101–4116. ISSN 0956-7151
20. Toth LS, Jonas JJ, Daniel D, Ray RK (1990) Development of ferrite rolling textures in low-carbon and extra low-carbon steels. Metallurgical Trans A 21(11): 2985–3000. ISSN 0360-2133
21. Raabe D (1995) Simulation of rolling textures of b.c.c. matals considering grain interactions and crystalline slip on {100}, {112} and {123} planes. Mater Sci Eng A 197: 31–37
22. ASTM International (2008a) Standard test method for atomic emission vacuum spectrometric analysis of carbon and low-alloy steel. ASTM Standards E 415
23. ASTM International (2011) Standard test methods for determination of carbon, sulfur, nitrogen, and oxygen in steel, iron, nickel, and cobalt alloys by various combustion and fusion techniques. ASTM Standards E 1019
24. ASTM International (2004) Standard test methods for determining average grain size E 112. ASTM Standards E 112
25. ASTM International (2008b) Standard test methods for tension testing of metallic materials E8/ E8M. ASTM Standards E 8
26. Foecke T, Iadicola MA, Lin A, Banovic SW (2007) A method for direct measurement of multiaxial stress-strain curves in sheet metal. Metallurgical and Materials Transactions A 38A: 306–313
27. Raghavan KS (1995) A simple technique to generate in-plane forming limit curves and selected applications. Metallurgical and Mater Trans A 26A: 2075–2084
28. Brand PC, Prask HJ, Gneaupel-Herold T (1997) Residual stress measurements at the NIST reactor. Physica B 241–243: 1244–1245
29. Matthies S, Wenk HR (1992) Optimization of texture measurements by pole figure coverage with hexagonal grids. Physica Status Solidi (a) 133: 253–257
30. Gnaeupel-Herold T (2012) Bt8 data analysis. http://www.ncnr.nist.gov/instruments/bt8/BT8DataAnalysis.htm. Accessed 29 Jan 2013
31. Kallend UF, John S, Kocks, Rollett AD, Wenk HR (1991) popLA - an integrated software system for texture analysis. Textures and Microstructures 14: 1203–1208
32. Matthies S, Vinel GW (1982) On the reproduction of the orientation distribution function of texturized samples from reduced pole figures using the conception of a conditional ghost correction. Physica Status Solidi B 112(2): K111–K114. ISSN 0370-1972
33. Hielscher R, Schaeben H (2008) A novel pole figure inversion method: specification of the MTEX algorithm. J Appl Crystallography 41: 1024–103
34. Creuziger A, Syed K, Gnaeupel-Herold T (2014) Measurement of uncertainty in orientation distribution function (odf) calculations. Scripta Mat 72-73: 55–58
35. Bunge HJ, Haessner F (1968) 3-dimensional orientation distribution function of crystals in cold-rolled copper. J Appl Phys 39(12): 5503–5515
36. Lebensohn RA, Tomé CN (1993) A self-consistent anisotropic approach for the simulation of plastic deformation and texture development of polycrystals: application to zirconium alloys. Acta Metallurgica et Materialia 41: 2611–2624

37. Hu L, Rollett AD, Iadicola M, Foecke T, Banovic S (2012) Constitutive relations for AA 5754 based on crystal plasticity. Metallurgical and Mater Trans A 43A: 854–869
38. Dawson PR, MacEwen SR, Wu PD (2003) Advances in sheet metal forming analyses: dealing with mechanical anisotropy from crystallographic texture. Int Mater Rev 48: 86–122
39. Pokharel R, Lind J, Kanjarala AK, Lebensohn RA, Li SF, Kenesei P, Suter RM, Rollett AD (2013) Polycrystal plasticity: comparison between grain scale observations of deformation and simulations. submitted to Annual Reviews Condensed Matter Physics
40. Li SF, Lind J, Hefferan CM, Pokharel R, Lienert U, Rollett AD, Suter RM (2012) Three-dimensional plastic response in polycrystalline copper via near-field high-energy X-ray diffraction microscopy. J Appl Crystallography 45: 1098–1108

5

Prediction of microstructure evolution during multi-stand shape rolling of nickel-base superalloys

Kannan Subramanian[1]* and Harish P Cherukuri[2]

*Correspondence:
ksubramanian@stress.com
[1] Stress Engineering Services, Inc.,
3314 Richland Ave, 70002 Metairie,
LA, USA
Full list of author information is
available at the end of the article

Abstract

In this paper, a comprehensive numerical approach to predict the microstructure of nickel-base superalloys during multi-stand shape rolling is presented. This approach takes into account the severe deformation that occurs during each pass and also the possible reheating between passes. In predicting the grain size at the end of the rolling process, microstructural events such as dynamic recrystallization (DRX), metadynamic recrystallization (MDRX), and static grain growth are captured at every deformation step for superalloys. Empirical relationships between the average grain size from various microstructural processes and the macroscopic variables such as temperature (T) and effective strain ($\bar{\varepsilon}$) and strain rate ($\dot{\bar{\varepsilon}}$) form the basis for the current work. These empirical relationships are based on Avrami equations. The macroscopic variables are calculated using a finite element analysis package wherein the material being rolled is modeled as a non-Newtonian fluid with viscosity that depends on the effective strain rate, strain, and temperature. A two-dimensional transient thermal analysis is carried out between passes that can capture the MDRX and/or static grain growth during the microstructural evolution. The presented microstructure prediction algorithm continuously updates two families of grains, namely, the recrystallized family and strained family at the start of deformation in any given pass. In addition, the algorithm calculates various subgroups within these two families at every deformation step within a pass. As the material undergoes deformation between the rolls, recrystallization equations are invoked depending on critical strain and strain rate conditions that are characteristics of superalloys. This approach predicts the microstructural evolution based on recrystallization kinetics and static grain growth only. The methodology was successfully applied to predict the microstructure evolution during the multi-pass rolling of nickel-base superalloys. The predicted results for Alloy 718 for a 4-stand rolling followed by air cooling and for a 16-stand rolling followed by a combination of air and water cooling are also compared with experimental observations.

Keywords: Multi-stand; Multi-pass; Shape rolling; Microstructure; Modeling; Nickel-base; Superalloys

Background

Superalloys are used for high temperature (>650°C) applications such as those encountered in the aircraft, petrochemical, and nuclear utility industries and where resistance to deformation is a primary requirement. Nickel-base alloys such as Waspaloy and Alloy 718 (IN 718) are examples of superalloys that resist deformation at elevated temperatures and are therefore difficult to hot work. Hot working is the term often used to describe the plastic deformation at temperatures high enough to overcome strain hardening. The major hot-working operations are open-die press forging, radial forging, extrusion, and rolling. In the case of rolling and forging, there may be many passes and some reheats involved. In this paper, the hot-working operation under consideration is the continuous shape rolling process. In the considered shape rolling process, billets transform from round-to-oval and oval-to-round until the desired shape and size are obtained in multiple stands or passes.

At a given alloy composition, the high temperature flow stress is largely influenced by the grain size of the microstructure. During shape rolling, the correct working forces, which relate to gauge and shape control as well as to power requirements, can be estimated accurately only if the microstructure relevant to the specific pass of rolling is known. The microstructure present at the end of the rolling and cooling operations also controls the product properties. Coarser grain size (ASTM 4 to 8) favors creep strength and crack-growth resistance while a fine grain (ASTM 10 to 14) structure favors improved low-cycle fatigue life and tensile yield strength. In addition, control of grain size is an important characteristic in any hot-working due to the stringent ultrasonic inspectability requirements.

Modeling the microstructural changes due to deformation during a multi-pass continuous rolling is the main focus of this paper. Due to the severe economical and practical aspects of industrial trials and laboratory experiments, numerical methods are often used to study the influence of the vast number of variables present in a typical industrial multi-stand rolling on the mictrostructure. Currently, numerical techniques based on the finite element (FE) method are extensively used in solving plastic deformation problems. Process variables such as strain and temperature are predicted from FE analysis of the deformation process. In general, microstructural modeling relates those process variables to microstructural evolution. Typical microstructural modeling involves two major steps. In the first step, constitutive equations describing the microstructural evolutions are developed using experiments. In the second step, the microstructural constitutive equations are implemented in commercial FE packages or a custom-built software.

In this paper, a comprehensive microstructure prediction methodology specifically suited for nickel-base superalloys is presented and validated in the context of multi-stand rolling involving flow formulation. Information on microstructure prediction can be found in the open literature for non-superalloy materials such as steel and aluminum that do not have a clear distinction between some of the microstructural events found in superalloys. In addition, microstructure prediction techniques do not appear to be currently available for multi-stand shape rolling of nickel-base superalloys where the number of stands may be as high as 16. This is attributed to the fact that conducting experiments and gathering data for multi-stand rolling result in significant down-time of the equipment. Furthermore, numerical simulations to model multi-stand rolling are also expensive to carry out due to the large deformations, contact conditions, and thermomechanical couplings involved. Therefore, this paper focuses primarily on a flow-formulation based on the primary goal of predicting microstructure expeditiously so that real-time control of multi-stand rolling can be achieved. The approach is validated using experimental data for four stands [1] and then for 16 stands [2]. In contrast, a vast majority of current FE methods simulate shape rolling using rigid plastic or elastoplastic formulations with microstructure as internal variables which is computationally intensive and real-time predictions are not feasible.

This paper is organized as follows. In the 'Methods' section, a discussion of the finite element formulation of the multi-stand shape rolling process is provided. An introduction to the microstructural processes, models, and microstructure evolution algorithm is also discussed in the 'Methods' section. Alloy 718, the alloy for which results are presented in this work, is governed by the fcc lattice structure of the γ matrix [3] and a number of characteristic precipitates such as γ'' (Ni3(Nb,Ti)), γ', δ, and carbides (MC and M6C). The high temperature strength of Alloy 718 is derived essentially from the coherent γ'' and to a smaller extent from γ'. The presence of the other precipitates improves hot-working to produce very fine-grained billet structures. However, modeling the complex precipitation processes requires a more detailed understanding of precipitation kinetics than is presently available. A mechanism-based model considering the δ phase precipitate effects has been discussed in literature by Thomas, et al. [4]. The proposed phenomenological model in this work results in good agreement with the observed microstructure as long as it is applied to materials where the initial microstructure does not contain excessive δ phase since these precipitates retard metadynamic recrystallization (MDRX). The proposed model is recommended when the rolling occurs at temperatures above δ solvus as observed in this work. Due to this, the current work aims to find the average grain sizes from recrystallization processes alone and does not take into account the precipitation of phases. The results are discussed and conclusions and future direction are provided in the last two sections of this paper.

Methods

Since the microstructure prediction methodology is closely tied to the FE formulation of the continuous shape rolling process, a brief discussion of the FE approach is presented initially. This is followed by a discussion on the microstructure prediction methodology.

FE formulation of multi-stand rolling

The multi-stand shape rolling process is simulated using an FE code in which the material behavior is modeled as that of a non-Newtonian fluid with a viscosity that is dependent on the temperature, strain rate, and effective strain. In this flow formulation, the governing momentum and energy balance equations are solved simultaneously to find the steady-state velocity components and the temperature at each node while satisfying the incompressibility constraint. A transient heat transfer analysis is carried out between the passes to take into account the heating or cooling between the passes.

The three-dimensional FE mesh corresponding to the control volume considered in the analysis consists of a number of equi-sized slices. Initially, a two-dimensional mesh using quadrilaterals (see Figure 1a) is generated. With appropriate connectivity between the nodes of cross-sectional elements in any given slice, the three-dimensional mesh is created (see Figure 1b). The three-dimensional element considered for the analysis is

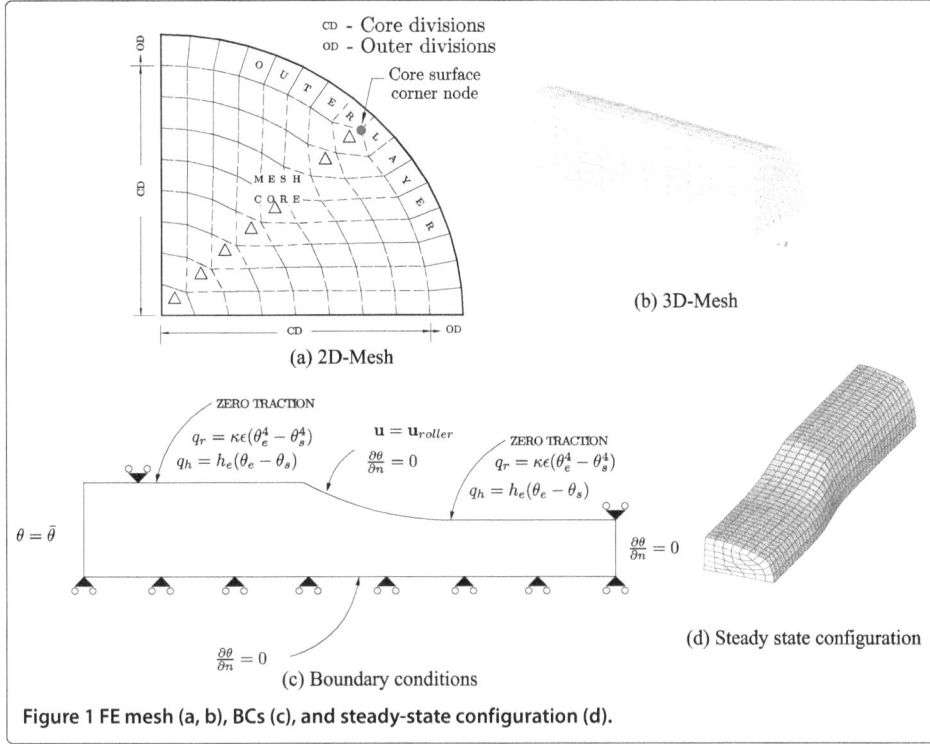

(a) 2D-Mesh

(b) 3D-Mesh

(c) Boundary conditions

(d) Steady state configuration

Figure 1 FE mesh (a, b), BCs (c), and steady-state configuration (d).

the commonly used brick element (tri-linear hexahedron) with *velocity-pressure* (*u-p*) formulation capabilities.

A brief description of the flow formulation is given in the following. The augmented potential energy functional for the flow under consideration is

$$J^* = \int_\Omega \sigma'_{ij} \dot{\varepsilon}_{ij} d\Omega - \int_{\Gamma_t} \bar{T}_i u_i d\Gamma + \int_\Omega p \dot{\varepsilon}_{ii} d\Omega. \qquad (1)$$

In the absence of body forces, the first and second terms in Equation 1 represent the rates of internal and external works, respectively, while the last term represents the addition of incompressibility constraint in the flow potential. Upon substituting the constitutive equation, $\sigma'_{ij} = 2\mu \dot{\varepsilon}_{ij}$, and applying Gauss theorem, the first variation of this functional simplifies to

$$\delta J^* = \int_\Omega \dot{\varepsilon}_{ij} 2\mu \delta \dot{\varepsilon}_{ij} d\Omega - \int_{\Gamma_t} \bar{T}_i \delta u_i d\Gamma + \int_\Omega p \delta \dot{\varepsilon}_{ii} d\Omega + \int_\Omega \delta p \dot{\varepsilon}_{ii} d\Omega. \qquad (2)$$

If $\mathbf{K}_D$ and $\mathbf{Q}$ refer to the stiffness contribution for velocity and pressure, respectively, the FE formulation of Equation 2 leads to

$$\begin{bmatrix} \mathbf{K}_D & \mathbf{Q} \\ \mathbf{Q}^T & 0 \end{bmatrix} \begin{Bmatrix} \mathbf{u} \\ \mathbf{p} \end{Bmatrix} = \begin{Bmatrix} \mathbf{f} \\ 0 \end{Bmatrix}.$$

The above system is augmented with a penalty function [5], typically the Lagrange multiplier γ:

$$\begin{bmatrix} \mathbf{K}_D & \mathbf{Q} \\ \mathbf{Q}^T & -\frac{1}{\gamma}\mathbf{I} \end{bmatrix} \begin{Bmatrix} \mathbf{u} \\ \mathbf{p} \end{Bmatrix} = \begin{Bmatrix} \mathbf{f} \\ -\frac{1}{\gamma}\mathbf{p} \end{Bmatrix}.$$

This can be iteratively solved to determine the pressure and velocity as follows:

$$\mathbf{p}^{i+1} = \mathbf{p}^i + \gamma\left(\mathbf{Q}^T\mathbf{u}^i\right)$$

$$\left[\mathbf{K}_D + \gamma\left(\mathbf{Q}\mathbf{Q}^T\right)\right]\mathbf{u}^{i+1} = \mathbf{f} - \mathbf{Q}\mathbf{p}^{i+1}.$$

In a similar manner, a finite element formulation using Galerkin's weighted residual method is applied for the coupled heat transfer equations. Appropriate boundary conditions (see Figure 1c) are applied and the steady-state configuration (see Figure 1d) is determined. In this shape rolling approach, the locations of the nodes at the steady state are obtained by an iterative method using a modified Euler integration of the current velocity field. This FE formulation was (previously) successfully implemented for a four-stand multi-stand rolling [1]. In the present work, this FE formulation is adopted to analyze 16 or more stands [2].

Microstructure evolution and algorithm

Microstructure theory

Work hardening and dynamic softening coexist during hot deformation of nickel-base superalloys. Various deformation parameters such as strain, strain rate, and temperature influence the microstructure. The strain rate accelerates the accumulation of dislocations that results in strain-rate hardening. Temperature is related to the softening process through the resulting decrease or rearrangement of dislocations. In general, the thermomechanical processing encompasses recovery, recrystallization, and grain growth. Recovery and/or recrystallization may occur during deformation at high temperatures which are the common softening or restoration processes. In addition, the rates of cooling of the material are generally very low in large-scale metal forming operations, allowing recovery, recrystallization, and grain growth to occur immediately after hot deformation. These dynamic restoration processes are different from the static annealing processes which occur during post-deformation heat treatment. These processes are of special importance to the metal industry as they lower the flow stress of the material and enable the material to deform more easily. In addition, they also have an influence on the texture and the grain size of the worked material. In the case of metals of low or medium stacking fault energy (copper, nickel, and austenitic iron), the recovery processes are slow, and dynamic recrystallization dominates after a critical deformation condition is reached [6].

Microstructure evolution

During high strain and strain-rate thermomechanical processes such as multi-stand rolling, recrystallization is the major restoration process followed by static grain growth between passes that influences the development of microstructure. Therefore, further discussion will focus only on the details of recrystallization and grain growth listed below:

- Dynamic recrystallization (DRX)
- Metadynamic recrystallization (MDRX)
- Static grain growth (SGG)

DRX occurs during deformation when the strain exceeds a certain critical strain $\bar{\varepsilon}_c$. Use of critical strain can be found in a pioneering work by Sellars [7]. This occurs somewhat before the peak of the stress-strain curve. In this process, the nuclei for recrystallization are formed. MDRX occurs after deformation because the strains required to complete the

DRX are not continuously achieved. The strains are still greater than the critical strain $\bar{\varepsilon}_c$. During MDRX no new nuclei are formed but the dislocation density is reduced. Even though the straining is stopped, annealing continues and the existing nuclei will grow.

The progress of recrystallization during isothermal annealing is commonly represented by the volume fraction of material recrystallized (X) using a sigmoidal curve given as follows,

$$X = 1 - \exp\left(-\beta x^n\right). \tag{3}$$

In Equation 3, x can be time or strain, depending on the phase of recrystallization process discussed later. This is commonly known as the Johnson-Mehl-Avrami-Kolmogorov (JMAK) model. β is typically a function of the rate at which the nuclei are formed and the rate at which the grains grow. The exponent n is usually defined as JMAK or Avrami exponent. A significant feature of the JMAK approach is that the nucleation sites are assumed to be randomly distributed. However, it is too simple to quantitatively model a process as complex as recrystallization. The strain rate and deformation temperature are often incorporated into a single parameter, the 'Zener-Hollomon parameter' (Z) also known as the 'temperature compensated strain rate', defined as:

$$Z = \dot{\varepsilon} \exp\left(\frac{Q}{RT}\right), \tag{4}$$

where Q is the activation energy of the process. In the present work, the flow stress is incorporated in the FE formulation as a function of temperature, strain, and strain rate in a tabular form.

When the material is fully recrystallized, further grain growth may occur. Static grain growth occurs after deformation. The strains are less than the critical strain $\bar{\varepsilon}_c$.

Empirical modeling

Empirical laws describing the various processes mentioned above establish the relationships between grain size and process parameters such as tool and workpiece geometry, temperature, deformation speed, and amount of deformation through regression analysis of experimental data as developed by Huang et al. [8] and Shen [9]. Sellars [7] considered the relationship between the grain size obtained after each process (DRX and MDRX) and the stress for steel. The evolution of microstructure was studied as a function of temperatures, pass reductions, speeds, and times in rolling schedules by Sellars and Whiteman [10] during the plate rolling of low-carbon manganese steel. The constitutive equations were written in terms of the temperature during the deformation, strains, and strain rate for the various recrystallization and recovery processes. Davenport et al. [11] standardized the constitutive equations in terms of Z and included the flow stress behavior for steel during hot deformation. Shen [9,12] and Shen et al. [13] developed constitutive equations involving the Z and carefully studied the effect of DRX, MDRX, and SRX on the microstructure evolution during the forging of Waspaloy turbine discs. In the present work, a generalized form of empirical laws similar to those proposed by Shen [9] are taken as the basis and the development of the proposed microstructural algorithm incorporates appropriate modifications to predict the behavior of various superalloys including Alloy 718.

Computations

Anderson et al. [14] were some of the pioneers in numerically simulating the grain growth in materials during the early 1980s. Beynon and Sellars [15] developed an approach that can calculate rolling loads and torques with an accurate prediction of mean flow stress. Modeling the dynamic microstructural events is important for determining flow stress levels, and hence rolling loads. Lin et al. [16] treated the microstructure evolution in the context of dislocation densities using viscoplastic equations for C-Mn steel. Mirza et al. [17] incorporated microstructure predicting algorithms in an FE package to determine the microstructure in aluminum alloys. Similar attempts can be found using the commercial FE packages such as FORGE [18-20], ABAQUS [21], LARSTRAN/SHAPE [22], and DEFORM [23,24]. Goerdeler et al. [22] and Hirch et al. [25] developed simulation procedures that can predict the grain orientation or texture in addition to the usual grain size prediction during the multi-stand rolling of aluminum alloys. Davenport et al. [11] suggest the incorporation of constitutive equations into first-stage equations, describing the stress at a given strain as a function of Z, and second-stage equations, resulting in a continuous flow stress curve. Serajzadeh [26,27] discusses an approach involving the basic balance laws coupled with the microstructural behaviors. The current work employs the deformation variables calculated from the FE formulation discussed in the previous section and uses the formulation discussed in the next section to model the microstructure as a separate microstructure analysis package.

Formulations

Some of the published works take specific recrystallization processes into consideration. The effects of DRX and MRX are discussed by Zhou et al. [28] on wrought Alloy 718 in hot deformation, and the kinetics of DRX is predicted by Serajzadeh [29]. Semiatin et al. [30] divide the DRX into discontinuous DRX (DDRX) and continuous DRX (CDRX) for low stacking fault-energy materials such as nickel-base alloys. Karhausen et al. [31] developed a procedure in which the effective strain used in the calculations is assumed to be a function of the volume recrystallized during the rolling process in predicting the microstructure during a five-pass rolling of Cr-V Steel. Pauskar and Shivpuri [32] introduced an averaging procedure for various families of grains as the deformation proceeds in various passes during the rolling of TMS-80R steel. Thomas et al. [4] developed a microstructure model for forging of Alloy 718. The formulation presented in the current work is applied on multi-stand rolling of various nickel-base superalloys.

Microstructure algorithm

The presented algorithm is applied in the context of rolling processes that are modeled with flow formulations described in the 'FE formulation of multi-stand rolling' section. However, minor modifications can be applied to the procedure presented in this paper in developing a generic approach to include other FE formulations used for simulating the rolling process. In the proposed approach, the development of microstructure variables is predicted along a streamline from the flow formulation of the rolling process during a pass. Examples of such streamlines in a pass from the FE analysis are shown in Figure 2.

For the rolling problem characterized by flow formulation, all nodes on the free surface coincide with points on the flow streamlines, and each node in the whole mesh is essentially an integral part of a flow streamline. In the case of the steady-state rolling process

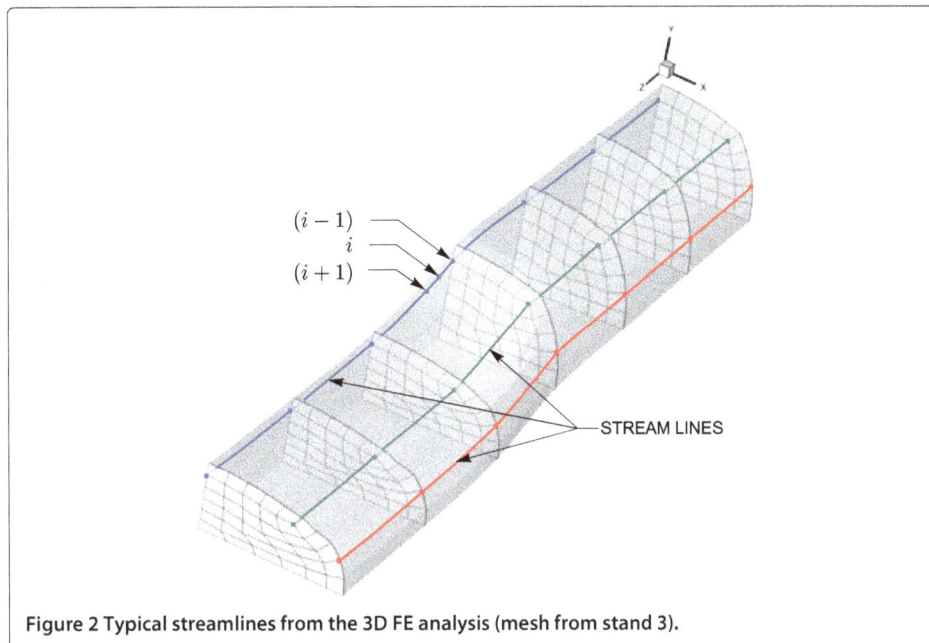

Figure 2 Typical streamlines from the 3D FE analysis (mesh from stand 3).

as described in this flow formulation, the deformation step is assumed to be character-ized by the resident time at a particular node. The resident time is the time that elapses between a material particle's entry into the control volume and the time at its current location in space. Therefore, the microstructure evolution algorithm can be implemented in a dynamic fashion along the flow streamline. In addition, the microstructure is allowed to develop during interpass periods as well.

The FE code discussed earlier, predicts only the temperature distribution during the interpass by simulating a two-dimensional transient heat-transfer analysis. Therefore, the continuation of microstructure development is implemented along a hypothetical straight streamline whose nodes correspond to the points in time during the interpass. Typically, the points in time correspond to the locations in space since the rolled material moves during the interpass without undergoing any change in shape. The application of stream-line outputs from commercial FE codes to calculate the evolution of material properties based on elementary rolling was implemented previously by Goerdeler et al. [22] in the context of aluminum alloys. Shen [9] employs an element-based approach in contrast to the presented node-based streamline approach for Waspaloy, in the context of forging analysis.

The proposed procedure considers two grain families: strained and recrystallized, at any given location in a streamline. A similar approach was developed by Thomas et al. [4]. Both the families of grains undergo recrystallization based on a deformation crite-rion discussed later. The schematic shown in Figure 3 shows that an initial uniform grain size, described by D_{st}, develops two grain families primarily characterized by the average volume recrystallized (F) and the average grain size (D).

The subscripts 'st' and 'rex' denote respectively the strained and recrystallized family of grains. These families undergo further recrystallization based on the achievement of cer-tain critical parameters and develop into four subgroups. The strained family develops recrystallized grains characterized by (d_{strex}, X_{st}) that represent the instantaneous grain

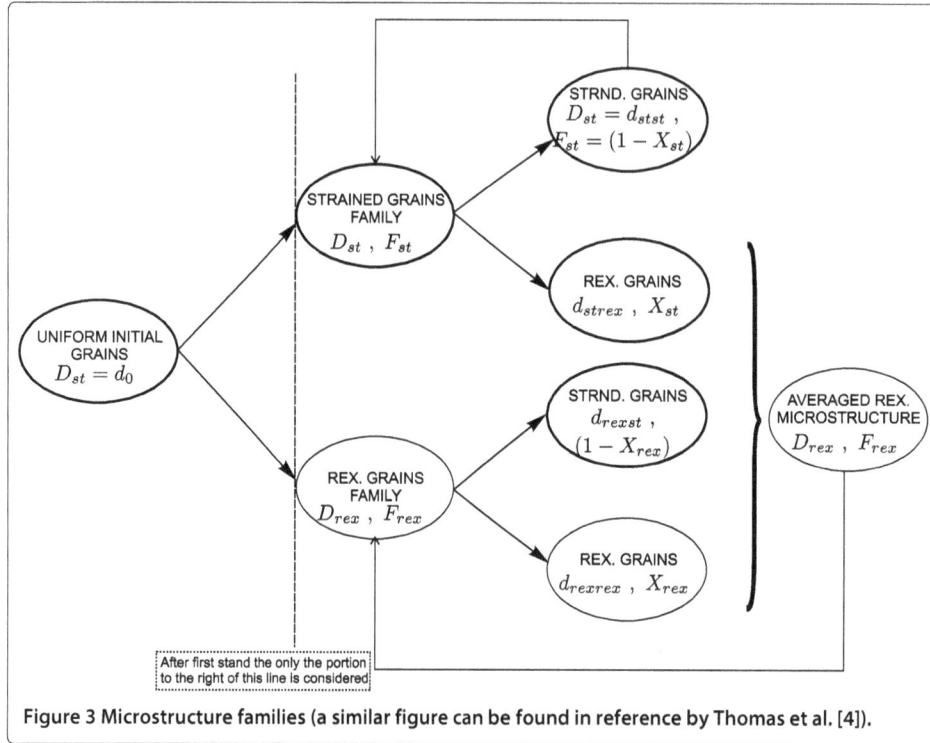

Figure 3 Microstructure families (a similar figure can be found in reference by Thomas et al. [4]).

size due to recrystallization laws and the recrystallization fraction, respectively. The non-recrystallized portion is the strained portion that is characterized by $(d_{\text{stst}}, (1-X_{\text{st}}))$ which are functions of the recrystallized subgroup characteristics. A similar analogy is applicable to the recrystallized grain family subgroups and are characterized by $(d_{\text{rexrex}}, X_{\text{rex}})$ for the recrystallized portion and $(d_{\text{rexst}}, (1 - X_{\text{rex}}))$ for the strained portions. These families are expected to evolve during the deformation in a pass and during interpass. Then, a weighted averaging algorithm is applied to calculate D_{rex} and F_{rex} prior to the achievement of critical deformation parameters during the next pass. D_{st} and F_{st} are calculated as follows:

$$D_{\text{st}} = D_{\text{st}} \left(1 - X_{\text{st}}\right)^{\frac{1}{n\alpha}} \tag{5}$$

$$F_{\text{st}} = (1 - X_{\text{st}}) F_{\text{st}} \text{ or } = 1 - F_{\text{rex}}. \tag{6}$$

A detailed description of the microstructure evolution based on the deformation criterion can be explained with the schematic in Figure 4. The deformation during a pass is depicted with sawtooth lines at the bottom of the schematic. The figure is drawn for any arbitrary pair of stands with an interpass during continuous multi-stand rolling. The strain used for the microstructure calculation is given by Equation 7.

$$\bar{\varepsilon}_x^i = \bar{\varepsilon}^i + \nu(T)\bar{\varepsilon}_x^{(i-1)}, \tag{7}$$

where the subscript x can be st or rex to represent the strained and recrystallized families, $\bar{\varepsilon}^i$ is the instantaneous effective strain due to deformation, $\nu(T)$ is the temperature dependent factor used as a fraction for the previously stored strain $\bar{\varepsilon}_x^{(i-1)}$. The factor $\nu(T)$ varies between 0 and 1. The second term in Equation 7 denotes the retained strain from a previous pass. When this equivalent strain reaches a critical strain $\bar{\varepsilon}_c$, and if the effective strain rate $\dot{\bar{\varepsilon}}$ reaches a material dependent value greater than equal to $\dot{\bar{\varepsilon}}_c$, DRX is initiated.

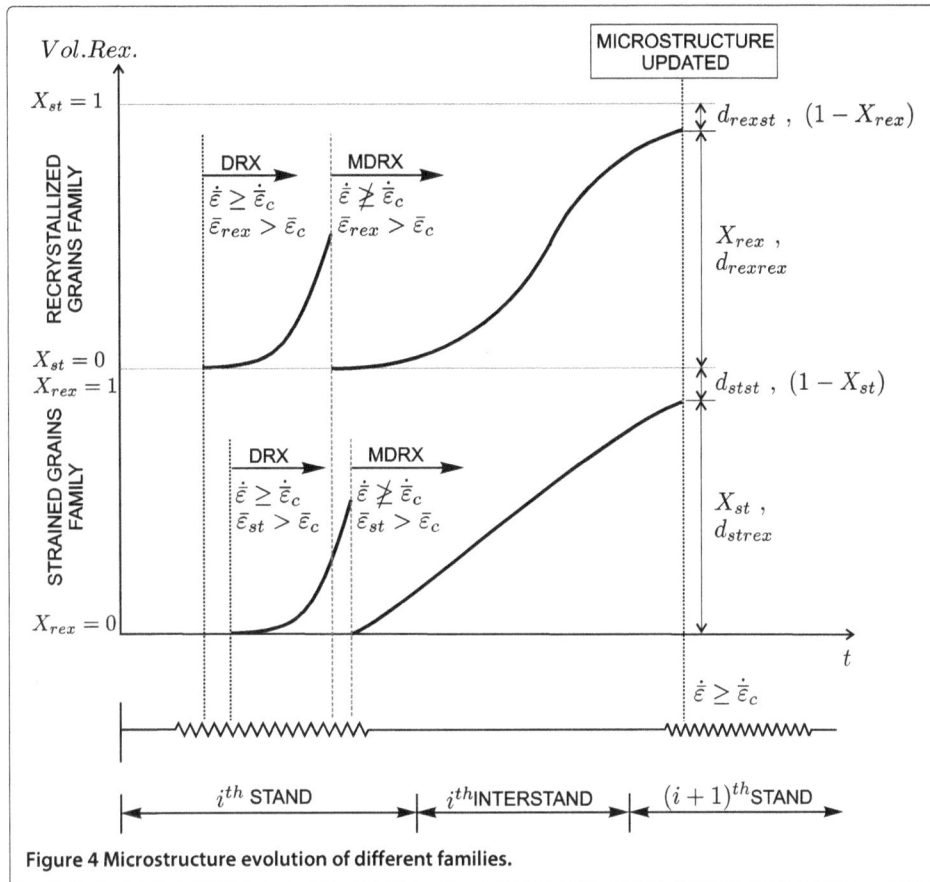

Figure 4 Microstructure evolution of different families.

Consider a single streamline shown in the Figure 5, extracted from Figure 2. Based on Figure 5, the strain rate is calculated as,

$$\dot{\bar{\varepsilon}} = \frac{\delta\bar{\varepsilon}}{\delta t} = \frac{\bar{\varepsilon}^i - \bar{\varepsilon}^{i-1}}{t^i - t^{i-1}}. \tag{8}$$

MDRX follows DRX and does not require the strain rate to be greater than $\dot{\bar{\varepsilon}}_c$. However, the need for critical strain to be achieved still holds. MDRX assumes that the recrystallization initiates from the beginning; even though, some of the grains may be partially recrystallized due to DRX. The initial grain size used for MDRX calculations are the original initial grain sizes used in calculating the DRX grain size. That is, DRX is considered

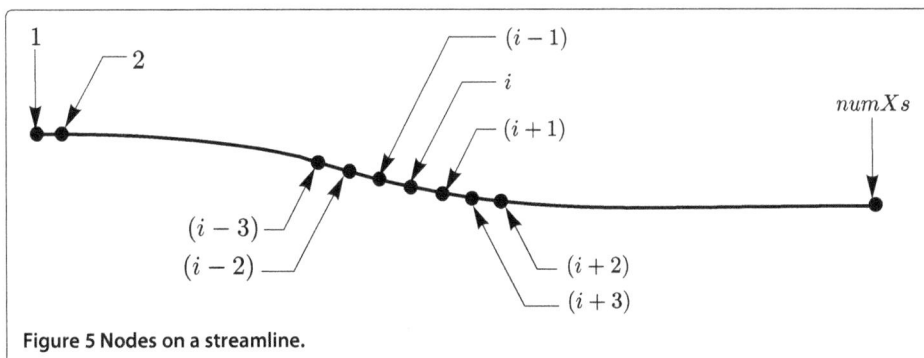

Figure 5 Nodes on a streamline.

only as a recrystallization initiation process and the contribution to the microstructure is primarily due to MDRX. When the recrystallization fraction reaches a set value greater than 90%, the grains at the location corresponding to the applicable node are assumed to be fully recrystallized and static grain growth is initiated until further deformation changes the microstructure evolution process. In this work, the nodes in the FE mesh are considered to follow the streamline, and the nodal deformation variables are used in calculating the microstructure variables.

Implementation

Empirical laws describing the various processes mentioned above establish the relationships between grain size and process parameters such as tool and workpiece geometry, temperature, deformation speed, and amount of deformation through regression analysis of experimental data as developed by Huang et al. [8] and Shen [9]. Recrystallization is a continuously evolving process, and an instantaneous application of the empirical laws may not capture the continuous nature of the process under consideration. In addition, the actual deformation is not an isothermal process. The approach described in this section calculates the evolution based on time integration detailed in the following.

Dynamic recrystallization

The rate of DRX increases with an increase in temperature and strain and decreases with an increase in the strain rate. For a particular location in a streamline, this can be observed from the expression for fraction recrystallized expressed as,

$$X = 1 - \exp\left(-\ln(2)\left[\frac{\bar{\varepsilon}}{\bar{\varepsilon}_{0.5}}\right]^a\right),$$ (9)

where $\bar{\varepsilon}_{0.5}$ refers to the strain at which the grains are 50% recrystallized. In other words, $\bar{\varepsilon}_{0.5}$ refers to the strain at which $X = 0.5$ and is expressed as follows:

$$\bar{\varepsilon}_{0.5} = d \, (d_o)^c (\dot{\bar{\varepsilon}})^{d_1} \, \exp\left(d_2 \frac{Q_h}{R\bar{T}}\right).$$ (10)

In order to capture the continuous evolution of the fraction recrystallized, a time integration is necessary that incorporates the increment in strain in each increment of time. Accordingly, the recrystallization fraction for DRX is a function of the equivalent strain named as the virtual strain in this work. That is,

$$X = f(\bar{\varepsilon}_v).$$ (11)

Rewriting Equation 9 in line with Equation 11,

$$X = 1 - \exp\left[-\ln(2)\left(\frac{\bar{\varepsilon}_v}{\bar{\varepsilon}_{0.5}}\right)^a\right]$$

we find that the virtual strain is given by,

$$\bar{\varepsilon}_v = \bar{\varepsilon}_{0.5} \left(\frac{\ln(1-X)}{-\ln(2)}\right)^{\frac{1}{a}}.$$ (12)

During rolling, the material undergoes deformation continuously under the rolls and the strain continues to increase and so does the recrystallization fraction. In general, the recrystallization fraction for DRX during rolling is the sum of the recrystallization fraction from previous deformation and an incremental recrystallization fraction from the current deformation. That is,

$$X^i = X^{i-1} + \delta X^{i-1}. \tag{13}$$

In terms of virtual strain, the right-hand side of Equation 13 can be expressed as,

$$X^i = 1 - \exp\left[-\ln(2)\left(\frac{\bar{\varepsilon}_v^{i-1} + \delta\bar{\varepsilon}^{i-1}}{\bar{\varepsilon}_{0.5}^{i-1}}\right)^a\right]. \tag{14}$$

From Figure 5, initially $i = 1$, $X^0 = 0$, and therefore,

$$\bar{\varepsilon}_v^0 = \bar{\varepsilon}_{0.5}^0\left[\frac{\ln(1 - X^0)}{-\ln(2)}\right]^{\frac{1}{a}}$$
$$= 0.$$

This algorithm is incorporated into an interactive package entitled GRANARY (the microstructure prediction package developed for the current work) for DRX based on the achievement of certain critical strain. The critical strain is expressed as a fraction of the peak strain. A peak strain is a material and process-dependent variable. The peak strain, critical strain, and the strain corresponding to 50% recrystallization for any location are given by,

$$\bar{\varepsilon}_c = n_{\text{dynctop}}\,\bar{\varepsilon}_p, \tag{15}$$

$$\bar{\varepsilon}_p = f\,d_o^g\,\dot{\bar{\varepsilon}}^{h_1}\,\exp\left(h_2\frac{Q_h}{R\overline{T}}\right), \tag{16}$$

and

$$\bar{\varepsilon}_{0.5}^{(i)} = d\,(d_o)^c(\dot{\bar{\varepsilon}})^{d_l}\,\exp\left(d_2\frac{Q_h}{R\overline{T}}\right). \tag{17}$$

Equations 14 through 17 are evaluated by substituting the strain rate calculated using Equation 8 and average temperature calculated using the following expression,

$$\overline{T}^i = \frac{T^i + T^{i+1}}{2}.$$

The grain size expression developed by empirical methods gives only the grain size at the steady state, that is when the recrystallization is 100% complete. However, the grain size also evolves continuously and the instantaneous grain size need to be incorporated in the recrystallization fraction. Equation 18 is the commonly found expression for the steady-state grain size due to recrystallization. Equation 19 is the instantaneous recrystallized grain size as a function of the fraction recrystallized. The third expression (given by Equation 20) denotes the grain size of the strained grains in the current family of grains.

$$d_{\text{drx}}^{(ss)} = p(\dot{\bar{\varepsilon}})^{q_1}\,\exp\left(q_2\frac{Q_d}{R\overline{T}}\right), \tag{18}$$

$$d_{\text{drx}}^{(i)} = d_{\text{drx}}^{(ss)}\left(X^{(i)}\right)^{n_{\text{xdrx}}}, \tag{19}$$

and

$$d_{\text{str}}^{(i)} = d_0\left(1 - X^{(i)}\right)^{n_{\text{xdrxst}}}. \tag{20}$$

In these expressions, the following parameters,

$$n_{\text{xrex}},\ n_{\text{xrexst}},\ Q_h,\ Q_d,\ n,\ Q_p,\ a*,\ b*,\ c*,$$
$$d1*,\ p*,\ q1*,\ f*,\ g*,\ h1*,\ h2*,\ d2*,\ q2*,$$

are pertinent to the specific material and are described in the nomenclature. For those parameters with the symbol $*$, they are further defined specifically

$$\text{for sub-solvus } \left(\overline{T} < 1,010°\text{C} \right),$$
$$\text{for solvus } \left(\overline{T} = 1,010°\text{C} \right),$$
$$\text{for super-solvus } \left(\overline{T} > 1,010°\text{C} \right).$$

As a first step to describing similar laws for Alloy 718, the empirical laws that describe the behavior of Waspaloy found in the literature by Shen [9] were incorporated into the microstructure algorithm as one of the many models that may be utilized by the user.

Metadynamic recrystallization

When DRX is not 100% complete, further recrystallization occurs without the addition of any strain which is characterized as MDRX. During MDRX, the recrystallization fraction is primarily a function of the time (t) and a time constant ($t_{0.5}$) at which the recrystallization is 50% complete. The general expression is similar to Equation 9, and the curve is a similar sigmoidal curve, except that the independent strain variable $\overline{\varepsilon}$ is replaced by time variable t indicating the fact that MDRX evolves with time. The time at which 50% recrystallization occurs can be expressed in general by the following equation:

$$t_{0.5} = b \, d_o^c \, \overline{\varepsilon}^f \, \dot{\overline{\varepsilon}}^{d_1} \, \exp\left(d_2 \frac{Q_h}{R\overline{T}} \right). \tag{21}$$

In the present work, the average strain rate during deformation has been incorporated to calculate this variable. To capture the continuous evolution of MDRX, an approach similar to that developed in the previous section is applied which includes a virtual time similar to the virtual strain shown in Equation 14. The following equations for the grain sizes due to MDRX are similar to Equations 18, 19, and 20 for calculating the grain sizes due to DRX,

$$d^{ss}_{\text{mdrx}} = p \, d_0^s \, \overline{\varepsilon}^r \, \dot{\overline{\varepsilon}}^{q_1} \, \exp\left(q_2 \frac{Q_d}{R\overline{T}} \right),$$
$$d^{(i)}_{\text{mdrx}} = d^{(ss)}_{\text{mdrx}} \left(X^{(i)} \right)^{n_{\text{xmdrx}}},$$

and

$$d^{(i)}_{\text{str}} = (1 - \left(X^{(i)} \right)^{n_{\text{xmdrxst}}}.$$

Similar empirical parameters in the aforementioned expressions can be found from the literature by Huang et al. [8], Shen [9], Yeom et al. [33] for various superalloy materials.

Static grain growth

When the MDRX process is 100% complete, and the material does not undergo any additional strain, annealing occurs. Any extended hold at elevated temperatures causes the grains to grow statically. There are quadratic and cubic laws that describe the static grain growth. A general expression that describes the static grain growth is expressed by Equation 22,

$$d_{\text{ggr}} = \left[d_{\text{ini}}^{n_{\text{ggr}}} + t_{\text{ggr}} \delta t \exp\left(\frac{-Q_{\text{ggr}}}{R\overline{T}} \right) \right]^{\frac{1}{n_{\text{ggr}}}}, \tag{22}$$

where the parameters n_{ggr}, t_{ggr}, and Q_{ggr} can be found from literature. The grain growth typically occurs during the long interpasses and during hold times at the end of the rolling process.

Results and discussion

The presented microstructure algorithm can predict the microstructure evolution during the multi-stand rolling of nickel-base superalloys such as Waspaloy and Alloy 718. The algorithm was applied on three nickel-base superalloys: i) Waspaloy, ii) Alloy 718, and iii) a proprietary alloy. In this paper, for the verification of the developed algorithm, only the results from Alloy 718 are presented based on a proprietary rolling pass schedule that was used to carry out the FE analysis of rolling. This paper presents predicted results in comparison with the actual observed microstructure for a four-stand rolling followed by air cooling and for a 16-stand rolling followed by a combination of air and water cooling. The mesh considered for the rolling and microstructure analysis was developed with five core divisions and one outer division (see Figure 6a). The details of the core and outer divisions are given in the 'FE formulation of multi-stand rolling' section in line with Figure 1. Figure 6a shows some locations with symbols ●, ◆, ▼, ■, and ▲ chosen to study the history of microstructure evolution over many stands. These locations characterize center, mid radius, sub surface, and surface locations chosen to permit a comparison of predicted microstructure results with the experimental observations.

Based on a separate mesh sensitivity study, it was determined that the chosen mesh, that is, five core and one outer division, gives accurate enough results when compared with the measured temperature and shape at intermediate and final stands. It is to be noted, that the calculations are carried out at each node in a cross section and each node constitutes a point on a streamline.

Since the deformation variables are extracted to the nodes from the integration points, nodal variables are chosen to predict the microstructure. In the current work, the parameters used for Waspaloy and Alloy 718 are similar to those found in the literatures by Shen [9] and Huang et al. [8], respectively, for DRX. Parameters similar to those used in

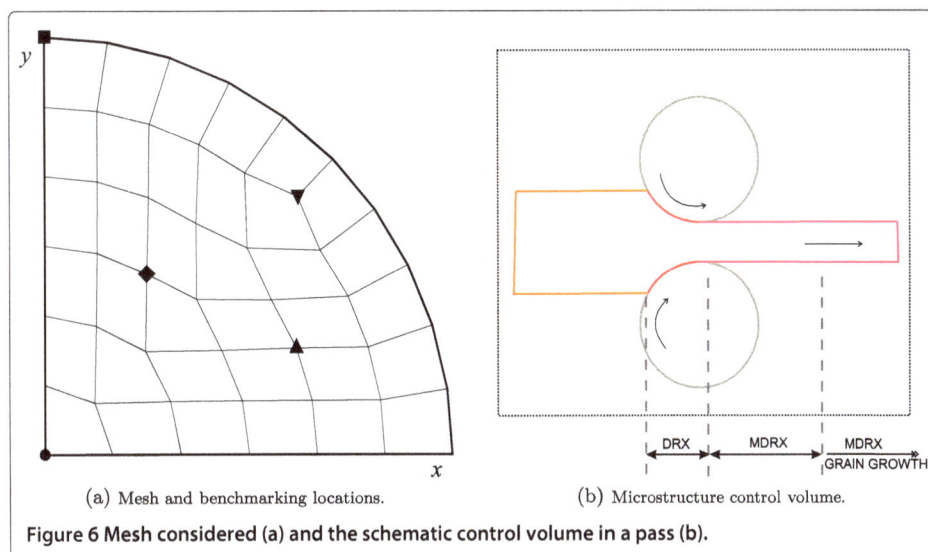

(a) Mesh and benchmarking locations.   (b) Microstructure control volume.

Figure 6 Mesh considered (a) and the schematic control volume in a pass (b).

the current work for the MDRX process and grain size calculation during MDRX can be found in literature by Huang et al. [8] and by Yeom et al. [33]. The static grain growth was calculated based on cubic laws similar to those found in the reference by Shen [9]. An initial uniform grain size of 90 μm (ASTM rating 4) is used as input to the analysis.

Cooling

Two scenarios are considered. In the first case that corresponds to four-stand rolling, at the end of rolling, the bars are assumed to be air cooled. At high strain rates, quench/cooling time becomes a critical parameter for MDRX. In addition, the micrographs to compare the predicted results were captured at the end of 4th stand with 5 s of air cooling. In the case of a second scenario that corresponds to the 16-stand rolling process, the cooling is primarily due to water. In the analysis, at the end of the 16th stand, a combination of air and water quenching is applied. Firstly, an air cooling for 2 s captures the time for transferring the billet, and an additional 2 s captures the evolution during immersion water quenching. This type of combined cooling is considered to mimic the process for which the micrograph results are available.

As a brief review of the discussion made in the previous section, Figure 6 gives a larger picture of the microstructure evolution process. That is, DRX process under the rolls as deformation continues, MDRX in the region close to the rolls and a continuation of the MDRX and a static grain growth during the interpass or cooling depending on the achievement of 100% recrystallization.

Four-stand analysis with air cooling at the end of the 4th stand

The temperature and equivalent strain histories at the chosen locations are shown in Figures 7 and 8, respectively. From these figures, it can be observed that the strains are low in the initial two stands, specifically very low during the 2nd pass with a maximum value of 0.25. In Figure 8, the independent axis is not to scale. That is, the interpass times are much larger compared to the pass times, and hence, the figure was created with actual strains and not-to-scale times.

Sixteen-stand analysis with air and water cooling at the end of the 16th stand

The temperature and equivalent strain history at the chosen locations (see Figure 6a) are shown in Figures 9 and 10, respectively. In Figure 10, the plot was created with actual strains and not-to-scale times since the interpass times are much larger compared to the pass times. From these figures, it can be observed that the instantaneous equivalent strains reach values as high as 1.45 during the 13th stand. Similarly, due to severe deformations imposed during the later stands, the temperatures also experience a significant increase.

Streamline results

Figures illustrating the effectiveness of the presented microstructure algorithm were developed with normalized variables such as temperature, strain, and other microstructure variables over time for the recrystallized and strained families of grains on streamlines passing through the considered locations. For simplicity, only the discussion of the observations are presented here in the context of the four-stand analysis, and the reader is encouraged to see reference [34] for details.

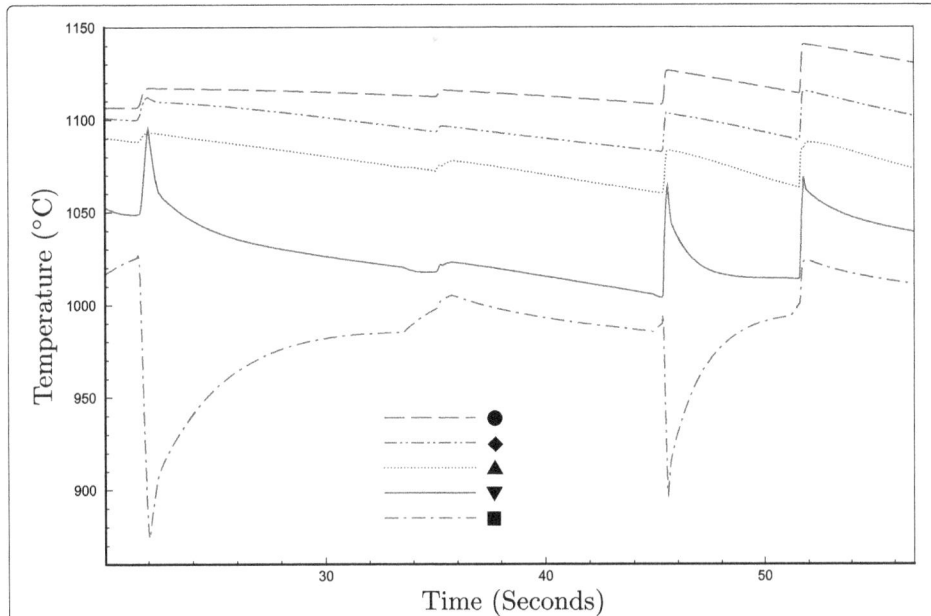

Figure 7 Temperature history on various streamlines (four-stand analysis). (At time ranges 20 to 23 s, 34 to 36 s, 45 to 46 s, and 51 to 52 s, the bar is under the rolls in stands 1, 2, 3, and 4, respectively, undergoing deformation).

At the central location represented by ●, recrystallization initiates with DRX and continues with MDRX as soon as the addition of the strain stops and progresses until the recrystallization is complete during the interpass. The strains are reset to zero for the strained family, and the grains start to grow statically during the interpass. Prior to achieving the strain rate and critical strain condition in the second pass, the

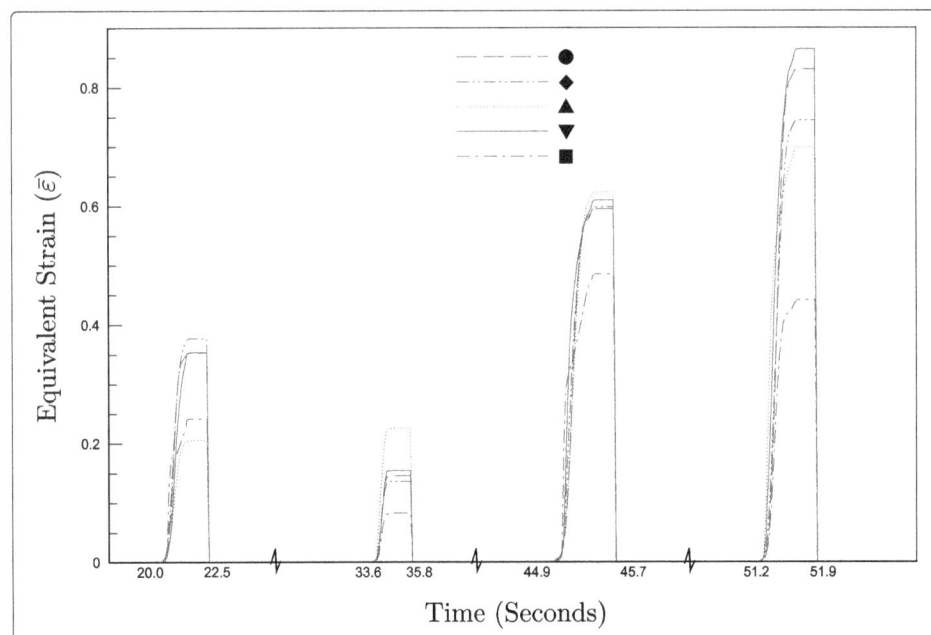

Figure 8 Strain history on various streamlines (four-stand analysis).

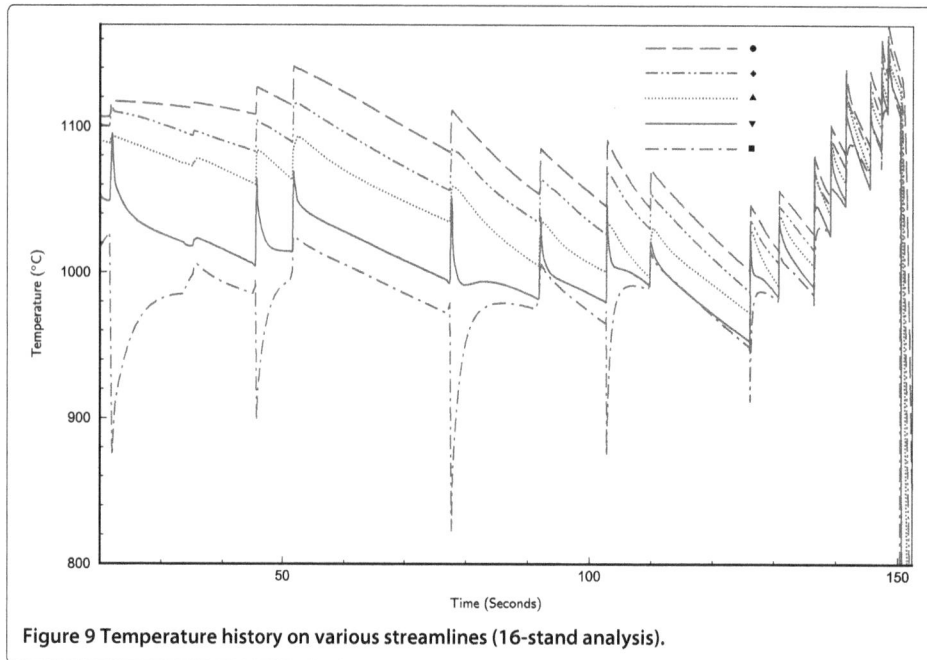

Figure 9 Temperature history on various streamlines (16-stand analysis).

microstructure averaging algorithm updates the microstructure, and F_{st} is set to zero if all the grains at this location are fully recrystallized.

During the deformation in the 2nd stand, the recrystallized family undergoes recrystallization and the MDRX does not recrystallize the REX family of grains fully. However, due to the high strains and temperatures in stand 3, the MDRX completes and a similar phenomenon is observed in stand 4 as well. Therefore, at this location, at the end of 4th stand, only the recrystallized family exists and the grain size is approximately 70 μm.

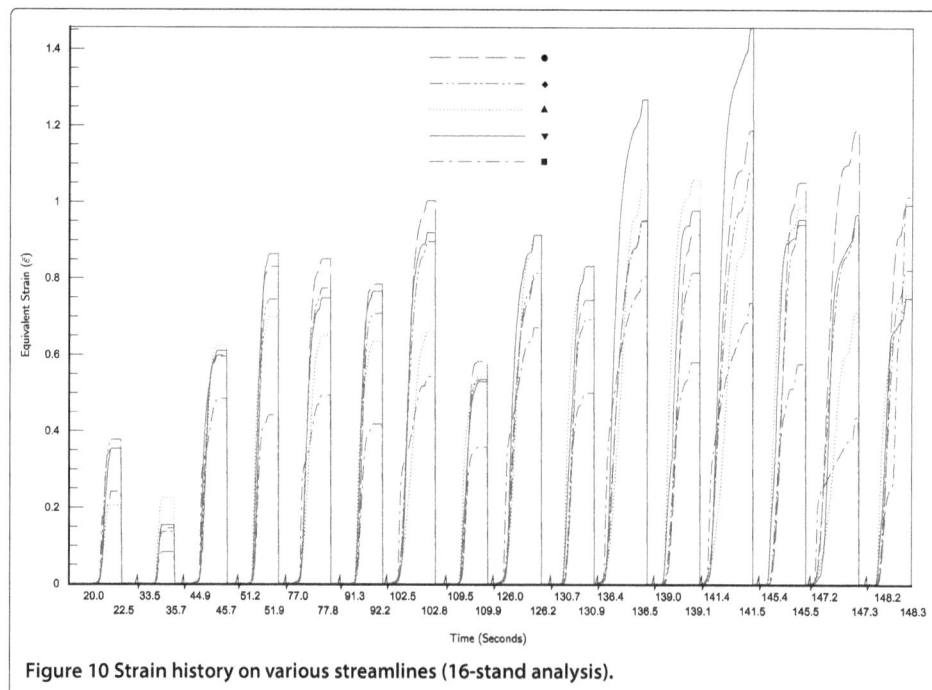

Figure 10 Strain history on various streamlines (16-stand analysis).

Similar characteristics are observed at the interior locations represented by ◆ and ▼; however, a rapid grain growth is observed at the location represented by ▼. For the location indicated by the symbol ▲, the strained family of grains vanish only at the end of 2nd stand. On the other hand, it takes four stands for the surface location indicated by the symbol ■ to recrystallize completely.

If the air cooling is not considered at the end of 4th stand but an interpass is assumed to continue the analysis, an interesting phenomenon is observed in the case of the subsurface location (indicated by ▼) where an insignificant portion of strained grain family experienced full recrystallization during the 9th interpass. However, this does not contribute much to the global microstructure characteristics. At the surface location represented by ■, contains both strained and recrystallized grain families till the 6th stand and then the strained grain family vanishes.

More details on the microstructure evolution for all the locations studied can be found in the literature by the first author [34].

Final observations for the four-stand analysis

A contour plot of the recrystallized fraction (F_{rex}) at the beginning of the deformation during the 4th stand and the recrystallized fraction (X_{rex}) at the end of the air cooling analysis after the 4th stand are depicted in Figure 11a,b, respectively. F_{rex} is calculated based on the averaging algorithm proposed in the 'Formulations' subsection. It can be observed that there are very few portions in the cross section near the surface that are not fully recrystallized.

An observation on the grain sizes due to Figure 12a,b indicates that the region near the core experienced a significant grain growth due to the high temperatures while the surface regions show smaller grains. Figure 13 indicates the observed microstructure at various locations and the grain sizes at those locations. It is clear that the grains show signs of complete recrystallization at all locations and groups of recrystallized families as seen in Figure 13c. Also, at the center (Figure 13a) the grains are largest and smaller at the midradius location (Figure 13b) while they are smallest at the surface (Figure 13c).

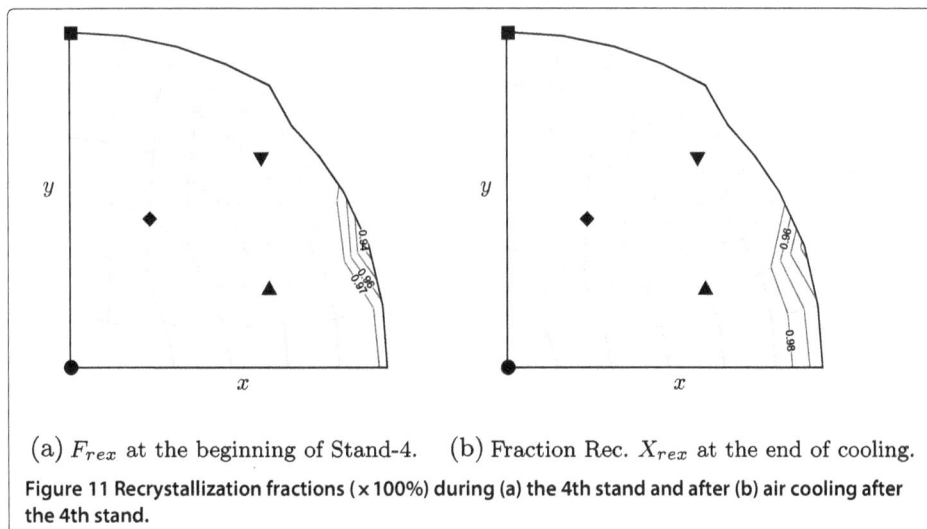

(a) F_{rex} at the beginning of Stand-4.   (b) Fraction Rec. X_{rex} at the end of cooling.

Figure 11 Recrystallization fractions (×100%) during (a) the 4th stand and after (b) air cooling after the 4th stand.

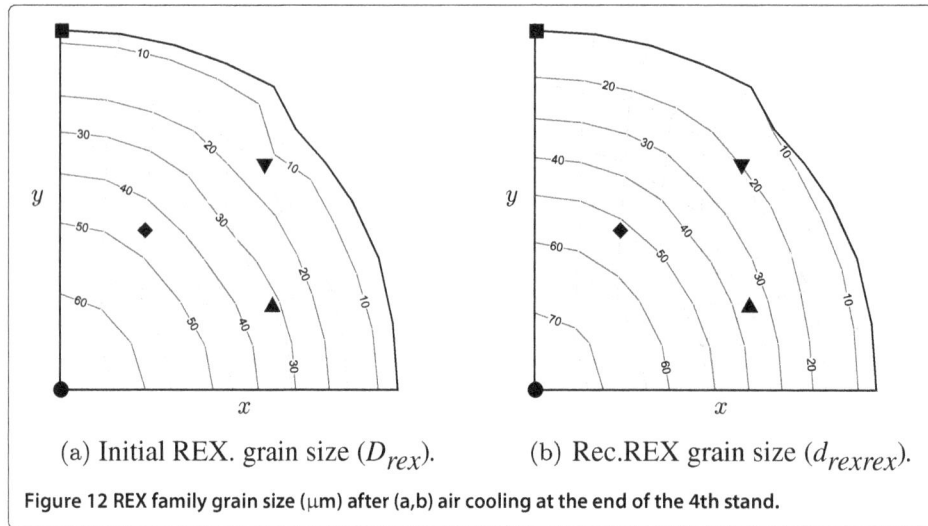

(a) Initial REX. grain size (D_{rex}).   (b) Rec.REX grain size (d_{rexrex}).

Figure 12 REX family grain size (μm) after (a,b) air cooling at the end of the 4th stand.

Final observations for the 16-stand analysis

This section describes the results at the end of the 16th stand if the rolling continued without any cooling at the end of 4th stand but with interpass conditions.

A contour plot of the recrystallized grain sizes characterized by the variables D_{rex} and d_{rexrex} are shown in Figure 14. The grain size distribution is almost uniform around 20 μm (see Figure 14a) for the D_{rex} which characterizes the overall grain size distribution while the recrystallized grains (represented by d_{rexrex} in Figure 14b) show slightly larger grains close to the center, since the center does not cool quickly. The actual microstructure observed at the end of cooling after 16-stand rolling of the considered material is shown in Figure 15. The calculated microstructure results are in good agreement with the observed microstructure.

Discussion

The predicted microstructure from the presented approach is in good agreement with the observed microstructure for the four-stand analysis and for the 16-stand analysis.

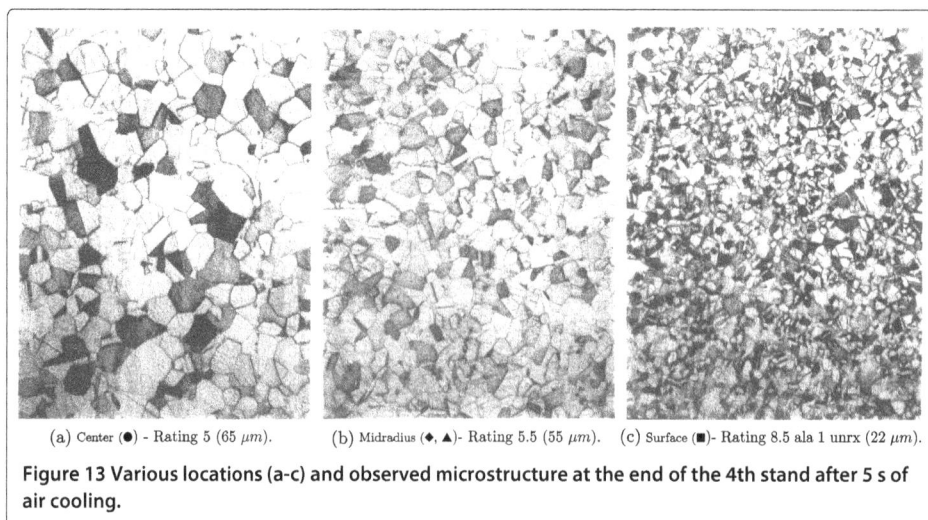

(a) Center (●) - Rating 5 (65 μm).   (b) Midradius (◆, ▲)- Rating 5.5 (55 μm).   (c) Surface (■)- Rating 8.5 ala 1 unrx (22 μm).

Figure 13 Various locations (a-c) and observed microstructure at the end of the 4th stand after 5 s of air cooling.

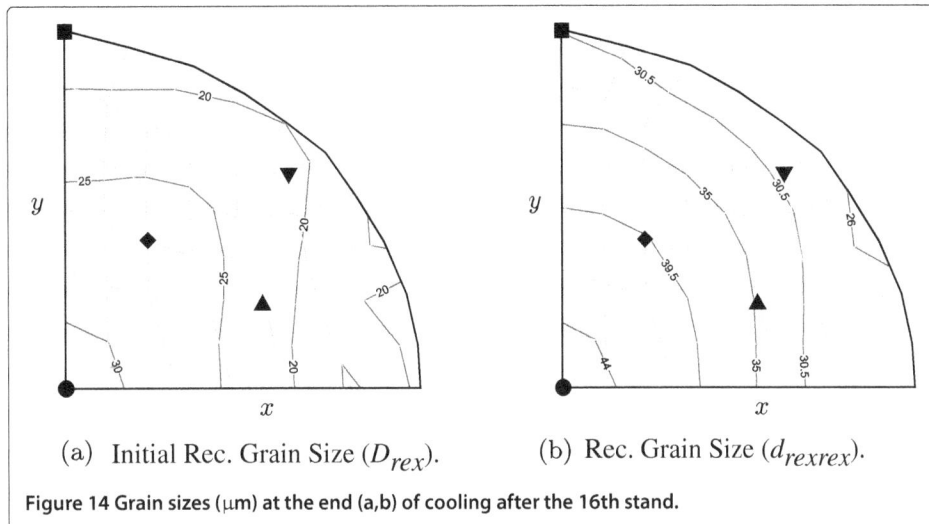

(a) Initial Rec. Grain Size (D_{rex}).   (b) Rec. Grain Size (d_{rexrex}).

Figure 14 Grain sizes (μm) at the end (a,b) of cooling after the 16th stand.

When microstructural processes such as DRX occur, the increases in strain and temperature favors dynamic softening. Grain size is a function of the deformation variables such as temperature, strain, and strain rate. Temperature enhances dynamic softening and, as observed during the later stands where a significant increase in temperature occurs, the grain sizes tend to increase. Strain influences the microstructure significantly as observed by other authors [4,35,36]. An increased rate in MDRX after deformation at high strain rates as a result of adiabatic heating is explained by Brand et al. [3]. The streamline figures confirm this conclusion. The fraction recrystallized due to DRX increases due to increase in the addition of strain, and the accumulated strain at the end of deformation influences the MDRX since DRX does not completely recrystallize the grains in any case for Alloy 718 material.

The variation of grain size is correlative to the behavior of work hardening and dynamic softening existing in Alloy 718 during hot deformation. When the temperature increases, dynamic softening occurs and the grain size increases. When the strain rate increases, work hardening occurs and the grain size decreases. Increasing the holding

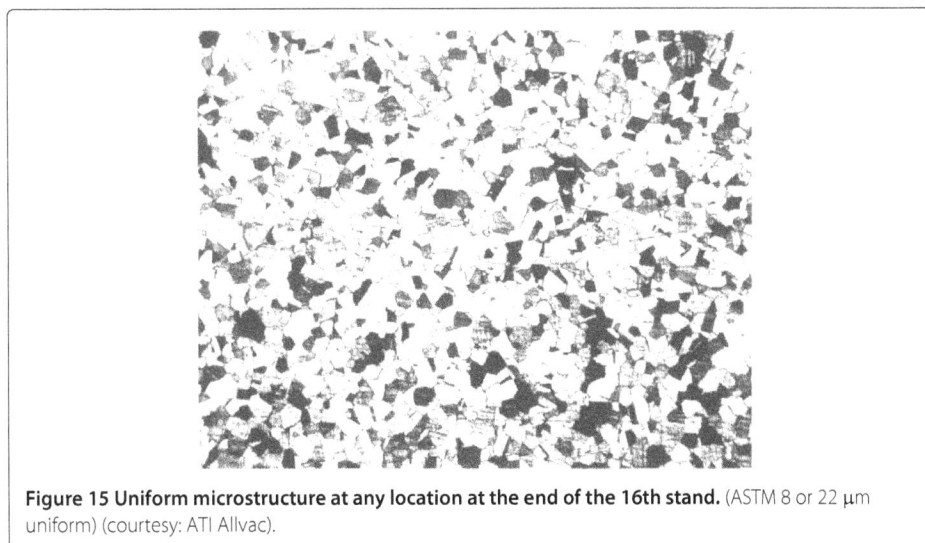

Figure 15 Uniform microstructure at any location at the end of the 16th stand. (ASTM 8 or 22 μm uniform) (courtesy: ATI Allvac).

time increases the volume fraction recrystallized. Long holding times after deformation allows complete recrystallization of microstructure. This is witnessed during the interpasses, where the MDRX is a significant contributor.

Conclusions

The prediction of microstructure by employing appropriate empirical models available in the literature requires a rigorous approach that captures the evolution of the microstructure. The current work encompassed the development of such a comprehensive procedure and its validation in the context of three superalloys, with results presented specifically for Alloy 718 in this paper. The predicted numerical results are in good agreement with the observed microstructure for a four-stand and a 16-stand multi-pass rolling involving different cooling conditions at the end of the rolling process by incorporating suitable empirical models found in literature for various microstructural processes.

In summary, the presented procedure incorporates the approach by which a total recrystallization fraction is calculated by taking into account the residual strain and also incorporates a total recrystallized fraction that incorporates fractions of previously recrystallized grain families. The developed procedure assumes that the grain matrix does not contain excessive δ phase. However, δ phase disappears at temperatures above 1,000°C, close to the solvus temperature of $\approx$1,010°C. In addition, the observed temperatures are mostly above the solvus region, and the averaging procedure is considered applicable in predicting the microstructure dominated by γ matrix. The predicted microstructure results are in agreement with the observed microstructures within 1 ASTM number of the grain size. In the current work, the microstructure is considered isotropic and equiaxed. In the future, this algorithm will be tested for other superalloy materials and stands beyond 16 in designing a generic forming process.

Nomenclature

J^*	Potential Energy Functional
Ω	Domain
Γ_t	Traction-specified boundary
$\bar{T}_i$	Components of the traction vector
$\dot{u}_i$	Velocity component
$\dot{\varepsilon}_{ij}$	Components of the strain rate tensor
σ'_{ij}	Components of the deviatoric stress tensor
δa	Small increment or variation in a quantity a
a_{ii}	Summation implied over the index i implied on any quantity a
I	Second-order identity tensor
p	Hydrostatic pressure
a^T	Transpose of the quantity a
μ	Viscosity
$\boldsymbol{u}$	Velocity vector
$\boldsymbol{f}$	Force vector
$\boldsymbol{p}$	Pressure
$\mathbf{K}_D$	Global stiffness matrix for velocity
$\mathbf{Q}$	Global stiffness matrix for pressure
γ	Penalty parameter

θ	Temperature variable
κ	Boltzmann's constant
ϵ	Emissivity
$\bar{\theta}$	Defined temperature
q_r	Heat flux due to radiation
q_h	Heat flux due to convection
DRX	Dynamic recrystallization
MDRX	Metadynamic recrystallization
u_i	ith component of velocity
D	Average grain size
F	Average volume recrystallized
R	Universal gas constant
$\bar{\varepsilon}_c$	Critical strain
F_{rex}	Fraction of recrystallized grains due to averaging
D_{rex}	Grain size of recrystallized grains due to averaging
D_{st}	Grain size of the strained grains due to averaging
F_{st}	Fraction of strained grains due to averaging
d_{strex}	Instantaneous recrystallized grain size evaluated using the equations for the strained grain family
d_{stst}	Instantaneous strained grain size evaluated using the equations for the strained grain family
d_{rexrex}	Instantaneous recrystallized grain size evaluated using the equations for the recrystallized grain family
d_{rexst}	Instantaneous strained grain size evaluated using the equations for the recrystallized grain family
n_α	Exponent for calculating the strained grain size
ν	Fraction to find the retained strain
$\bar{\varepsilon}_{0.5}$	Strain at which the grains are 50% recrystallized during DRX
a	Temperature-specific DRX or MDRX parameter to evaluate X and $\bar{\varepsilon}_{0.5}$ for DRX and $t_{0.5}$ for MDRX
b	MDRX parameter to evaluate X and $t_{0.5}$
c	Temperature-specific DRX or MDRX parameter to evaluate X and $\bar{\varepsilon}_{0.5}$ for DRX and $t_{0.5}$ for MDRX
d	Temperature-specific DRX parameter to evaluate X and $\bar{\varepsilon}_{0.5}$
d_1	Temperature-specific DRX or MDRX parameter to evaluate X and $\bar{\varepsilon}_{0.5}$ for DRX and $t_{0.5}$ for MDRX
d_2	Temperature-specific DRX por MDRX parameter to evaluate X and $\bar{\varepsilon}_{0.5}$ for DRX and $t_{0.5}$ for MDRX
Q_h	Activation energy for DRX and MDRX in evaluating respectively the $\bar{\varepsilon}_{0.5}$ and $t_{0.5}$
Q_d	Activation energy for DRX and MDRX in evaluating the steady state grain size
$\bar{\varepsilon}_p$	Peak strain
f	Temperature-specific parameter to evaluate $\bar{\varepsilon}_p$ during DRX and a parameter to evaluate X and $t_{0.5}$ during MDRX
g	Temperature-specific DRX parameter to evaluate $\bar{\varepsilon}_p$
h_1	Temperature-specific DRX parameter to evaluate $\bar{\varepsilon}_p$

h_2	Temperature-specific DRX parameter to evaluate $\bar{\varepsilon}_p$
n_{dynctop}	Fraction to calculate the critical strain $\bar{\varepsilon}_c$
d_0	Initial grain size used in evaluating the microstructure variables
$\bar{\varepsilon}_v$	Virtual strain
d_a^{ss}	Steady state grain size when recrystallization is 100% for any grain family represented by a
n_{xdrx}	Exponent to calculate the instantaneous grain size due to DRX
n_{xdrxst}	Exponent to calculate the instantaneous strained grain size due to DRX
$t_{0.5}$	Time required to achieve 50% recrystallization during MDRX
t_v	Virtual time
p	MDRX parameter to evaluate the steady-state grain size
s	MDRX parameter to evaluate the steady-state grain size
r	MDRX parameter to evaluate the steady-state grain size
q_1	MDRX parameter to evaluate the steady-state grain size
q_2	MDRX parameter to evaluate the steady-state grain size
n_{xmdrx}	Exponent to calculate the instantaneous grain size due to MDRX
n_{xmdrxst}	Exponent to calculate the instantaneous strained grain size due to MDRX
d_{ggr}	Grain size during static grain growth
n_{ggr}	Exponent denoting the type of grain growth law (quadratic or cubic)
t_{ggr}	Grain growth parameter which is material specific
Q_{ggr}	Activation energy for grain growth
$\overline{T}$	Average temperature used in microstructure calculations

Competing interests
The authors declare that they have no competing interests.

Authors' contributions
KS did all the formulations and prepared the article. HPC discussed the simulation results and corrected the article. Both authors read and approved the final manuscript.

Acknowledgements
Financial support for this research work is provided by ATI Allvac, Monroe, NC. The authors express gratitude towards Drs. Minisandram and Thomas from ATI Allvac for their valuable inputs on Alloy 718 and multi-stand rolling simulation.

Author details
[1] Stress Engineering Services, Inc., 3314 Richland Ave, 70002 Metairie, LA, USA. [2] Department of Mechanical Engineering and Engineering Science, University of North Carolina at Charlotte, 9201 University City Blvd, 28223 Charlotte, NC, USA.

References
1. Minisandram RS, Thompson EG, Forbes Jones RM, Stedje-Larsen R (2001) Numerical simulation of a multi-stand rolling mill. In: Mori K (ed). Proceedings of the 7th international conference on numerical methods in industrial forming processes - NUMIFORM 2001, Toyohashi, Japan, 18–21 June 2001. Swets & Zeitlinger B.V., Lisse
2. Subramanian K, Minisandram RS, Cherukuri HP (2007) Mesh re-zoning in multi-stand rolling. In: César de Sá JMA (ed). Proceedings of the 9th international conference on numerical methods in industrial forming processes - NUMIFORM 2007, Porto, Portugal, 17–21 June 2007. American Institute of Physics (API), College Park
3. Brand AJ, Karhausen K, Kopp R (1996) Microstructural simulation of nickel based inconel 718 in production of turbine discs. Mater Sci Tech 12(11):963–969
4. Thomas JP, Bauchet E, Dumont C, Montheillet F (2004) EBSD Investigation and modelling of the microstructural evolutions of superalloy 718 during hot deformation. In: Green KA (ed). Proceedings of Superalloys 2004. The Minerals, Metals & Materials Society (TMS), Warrendale. pp 959–968
5. Zienkiewicz O (1984) Flow formulation for numerical solution of forming processes. In: Numerical analysis of forming processes. Wiley, New York. pp 1–69
6. Humphreys FJ, Hatherly M (2004) Recrystallization and related annealing phenomena. Elsevier, Oxford
7. Sellars CM (1978) Recrystallization of metals during hot deformation. Philos T Roy Soc A 288(1350):147–158
8. Huang D, Wu WT, Lambert D, Semiatin SL (2001) Computer simulation of microstructure evolution during hot forging of waspaloy and nickel alloy 718. In: Srinivasan R, Semiatin SL, Beaudoin A, Fox S, Jin Z (eds). Proceedings of

symposium: microstructure modeling and prediction during thermomechanical processing, Indianapolis, November 4–8 2001. TMS, Warrendale. pp 137–147

9. Shen G (1994) Modeling microstructural development in the forging of waspaloy turbine engine disks. Dissertation, Ohio State University

10. Sellars CM, Whiteman JA (1979) Recrystallization and grain growth in hot rolling. Met Sci 13:187–194

11. Davenport SB, Silk NJ, Sparks CN, Sellars CM (2000) Development of constitutive equations for modelling of hot rolling. Mat Sci Tech 16(5):539–546

12. Shen G (2005) Microstructure modeling in superalloy forging. In: Altan T (ed). Cold and hot forging: fundamentals. ASM International, Novelty. pp 247–255

13. Shen G, Semiatin SL, Shivpuri R (1995) Modeling microstructural development during the forging of waspaloy. Metall Mater Trans A 26(7):1795–1803

14. Anderson MP, Srolovitz DJ, Grest GS, Sahni PS (1983) Computer simulation of grain growth – I. Kinetics. Acta Metall 32(5):783–791

15. Beynon JH, Sellars CM (1992) Modeling microstructure and its effects during multipass hot rolling. ISIJ Intl 32(3):359–367

16. Lin J, Liu Y, Farrugia DCJ, Zhou M (2005) Development of dislocation-based unified material model for simulating microstructure evolution in multipass hot rolling. Phil Mag 85(18):1967–1987

17. Mirza MS, Sellars CM, Karhausen K, Evans P (2001) Multiphase rolling of aluminium alloys: finite element simulation and microstructural evolution. Mat Sci Tech 17(7):874–879

18. Kusiak J, Kuziak R, Wajda W, Kowalski B (1999) Finite-element modeling of forging of nickel based superalloys. In: Kanagy DL (ed). Proceedings of 41st Mechanical Working and Steel Processing Conference, Baltimore, Maryland, October 24–27 1999. Iron and Steel Society, London, vol. XXXVII. pp 683–688

19. Dandre CA, Walsh CA, Evans RW, Reed RC, Roberts SM (2000) Microstructural evolution of Inconel 718 during ingot breakdown: process modelling and validation. Mat Sci Tech 16(1):14–26

20. Duan X, Sheppard T (2001) Prediction of temperature evolution by FEM during multi-pass hot flat rolling of aluminum alloys. Model Sim Mater Sci Engng 9(6):525–538

21. Mukhopadhyay A, Howard IC, Sellars CM (2004) Development and validation of a finite element model for hot rolling using ABAQUS/STANDARD. Mat Sci Tech 20(9):1123–1133

22. Goerdeler M, Crumbach M, Gottstein G, Neumann L, Luce R, Kopp R, Allen CM, Winden Mvd, Karhausen K (2002) Integral modeling of texture evolution in multiple pass hot rolling in aluminium alloys. Mat Sci Forum 396–402:379–386

23. Phaniraj MD, Behera BB, Lahiri AK (2006) Thermo-mechanical modeling of two phase rolling and microstructural evolution in the hot strip mill Part II. - Microstructure evolution. Mat Proc Tech 178(1–3):388–394

24. Zhao D, Cheng C, Anbajagane R, Dong H, Suarez FS (1997) Three-dimensional computer simulation of alloy 718: Ingot breakdown by cogging. In: Loria EA (ed). Proceedings of the 4th International Symposium on Superalloys 718, 625, 706 and Various Derivatives, Pittsburgh, Pennsylvania, June, 15–18 1997. The Minerals, Metals & Materials Society (TMS), Warrendale. pp 163–172

25. Hirch J, Karhausen K, Kopp R (1994) Microstructural simulation during hot rolling of Al-Mg Alloys. In: Proceedings of the 4th International Conference on Aluminium Alloys, Atlanta, Georgia, Georgia Institute of Technology, School of Materials Science & Engineering, Atlanta. pp 476–483

26. Serajzadeh S (2003) Prediction of microstructural changes during hot rod rolling. Int J Mach Tools Manf 43(14):1487–1495

27. Serajzadeh S (2005) Thermomechanical modeling of hot slab rolling. Mat Sci Tech 21(1):93–102

28. Zhou LX, Baker TN (1995) Effects of dynamic and metadynamic recrystallization on microstructures of wrought IN-718 due to hot deformation. Mat Sci Engng A 196(1–2):89–95

29. Serajzadeh S (2004) Prediction of dynamic recrystallization kinetics during hot rolling. Model Sim Mat Sci Engng 12(6):1185–1200

30. Semiatin SL, Weaver DS, Fagin PN, Glavicic MG, Goetz RL, Frey ND, Kramb RC, Antony MM (2004) Deformation and recrystallization behavior during hot working of a coarse-grain, nickel-base superalloy ingot material. Metall Mater Trans A 35(2):679–693

31. Karhausen K, Kopp R, de Souza MM (1991) Numerical simulation method for designing thermomechanical treatments, illustrated by bar rolling. Scandinavian J Metall 20(6):351–363

32. Pauskar P, Shivpuri R (1999) Microstructure and mechanics interaction in the modeling of hot rolling of rods. CIRP Annals Manuf Tech 48(1):101–104

33. Yeom JT, Lee CS, Kima JH, Park NK (2007) Finite-element analysis of microstructure evolution in the cogging of an alloy 718 ingot. Mat Sci Eng A 449–451:722–726

34. Subramanian K (2009) Microstructure evolution during multi-stand rolling of nickel-base superalloy. Dissertation, University of North Carolina at Charlotte

35. Guest RP, Tin S (2005) The dynamic and metadynamic recrystallisation of the in 718. In: Loria EA (ed). Proceedings of the 6th international symposium on superalloys 718, 625, 706 and various derivatives, Pittsburgh PA, October 2–5 2005. The Minerals, Metals & Materials Society (TMS), Warrendale

36. Zhang JM, Gao ZY, Zhuang JY, Zhong ZY (1999) Mathematical modeling of the hot deformation behaviour of superalloy in718. Metall Mater Trans A 30(10):2701–2712

6

Reconstruction of three-dimensional anisotropic microstructures from two-dimensional micrographs imaged on orthogonal planes

Veera Sundararaghavan

Correspondence:
veeras@umich.edu
Aerospace Engineering, University
of Michigan, Ann Arbor, MI 48109,
USA

Abstract

A pervasive method for reconstructing microstructures from two-dimensional microstructures imaged on orthogonal planes is presented. The algorithm reconstructs 3D images through matching of 3D slices at different voxels to the representative 2D micrographs and an optimization procedure that ensures patches from the 2D micrographs meshed together seamlessly in the 3D image. We show that the method effectively models the three-dimensional features in the microstructure using three cases (i) disperse spheres, (ii) anisotropic lamellar microstructure, and (iii) a polycrystalline microstructure. The method is validated by comparing the point probability functions of the reconstructed images to the original 2D image, as well as by comparing the elastic properties of reconstructed image to the experimental data.

Keywords: Microstructure; Markov random field; Ising model; Sampling; Reconstruction; Statistical descriptors

Background

Three-dimensional microstructural information is essential for understanding the relationships between the material structure and its properties. Three-dimensional microstructures experimentally characterized by serial sectioning or X-ray computed tomography are expensive for routine applications due to the time and effort involved. The direct problem of measuring 2D surface images using optical or micro-diffraction methods is relatively easier. Using these 2D images, inverse models could be developed that would allow the generation of full 3D microstructural maps and speeding-up the development of microstructure databases for the purposes of microstructure selection and design.

An inverse problem of specific interest in this paper is the reconstruction of 3D microstructures from three orthogonal 2D sectional images taken along the x-, y-, and z-planes. The information contained in these three 2D micrographs is in the form of pixels containing colors corresponding to different constituent phases. The outcome of the inverse problem is a 3D microstructure containing voxels colored consistently such that any arbitrary x-, y-, or z-slice 'looks' similar to the corresponding input micrographs. This reconstruction problem leads to anisotropic microstructures, which is in contrast to other such works in literature that use a single reference (2D) image and make assumptions of microstructural isotropy, i.e., slices in every direction look similar to a single input image [1]. The most popular among these methods involves matching statistical features like two-point correlation functions of a single planar image to a random 3D image using optimization procedures like simulated annealing [2,3]. Extension of these methods to achieve anisotropic microstructures has been proposed in the past using directionally dependent statistical features [4]. However, these methods are restricted to simple two-phase microstructures and are not applicable to more complex microstructures such as metallic polycrystals.

The approach proposed here involves maximizing the similarity between the solid microstructure and the 2D sectional microstructures by minimizing a neighborhood cost function. This cost function ensures that the local neighborhood on 2D slices taken along the x-, y-, or z-directions through the 3D microstructure is similar to some neighborhood in the 2D micrograph imaged along that plane. The approach is similar to those proposed in the computer graphics community [5] based on Markov random field assumption. This assumption simply states that microstructures have a stationary probability distribution or, in other words, different windows taken from a large microstructure 'look alike'. To synthesize a voxel in the 3D image, the window in the 2D micrograph that best matches the unknown voxel's neighborhood is chosen. The color of the voxel is decided based on the color indicated by the matching window in the 2D input image. The result is a simple method for generating 3D microstructures from 2D micrographs that generates visually striking 3D reconstructions of anisotropic microstructures, is computationally efficient, and is applicable to diverse microstructures.

Methods

Mathematical modeling of microstructures as Markov random fields

Some of the early attempts at microstructure modeling were based on Ising models [6]. In the Ising model, a $N \times N$ lattice (L) is constructed with values X_i assigned for each particle i on the lattice, $i \in [1, .., N^2]$. In an Ising model, X_i is a binary variable equal to either $+1$ or -1 (e.g., magnetic moment [6]). In this work, the values X_i may contain any one of G color levels in the range $\{0, 1, .., G-1\}$ (following the integer range extension of the Ising model by Besag [7]). A *coloring* of L denoted by X maps each particle in the lattice L to a particular value in the set $\{0, 1, .., G-1\}$. Ising models fall under the umbrella of *undirected graph models* in probability theory. In order to rewrite the Ising model as a graph, we assign neighbors to particles and link pairs of neighbors using a bond as shown in Figure 1a. The rule to assign neighbors is based on a *pairwise Markov property*. A particle j is said to be a neighbor of particle i only if the conditional probability of the value X_i given all other particles (except (i,j), i.e., $p(X_i|X_1, X_2, .., X_{i-1}, X_{i+1}, .., X_{j-1}, X_{j+1}, .., X_{N^2})$) depends on the value X_j.

Note that the above definition does not warrant the neighbor particles to be close in distance, although this is widely employed for physical reasons. For example, in the classical Ising model, each particle is bonded to the next nearest neighbor as shown in Figure 1a. In this work, we assume that a microstructure is a higher-order Ising model (Figure 1b). The particles of the microstructure correspond to pixels of the 2D image (or voxels in 3D). The neighborhood of a pixel is modeled using a square window around that pixel and bonding the center pixel to every other pixel within the window. The window size is a parameter that is chosen based on the scale of the biggest regular feature

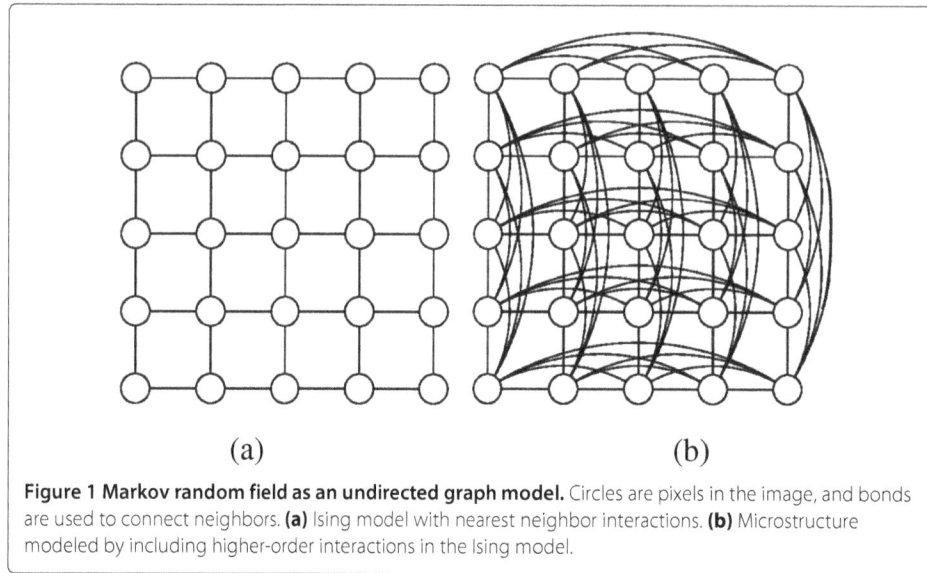

Figure 1 Markov random field as an undirected graph model. Circles are pixels in the image, and bonds are used to connect neighbors. **(a)** Ising model with nearest neighbor interactions. **(b)** Microstructure modeled by including higher-order interactions in the Ising model.

(e.g., grain size). Using this graph structure, a *Markov random field* can be defined as the joint probability density $P(X)$ on the set of all possible colorings X, subject to a *local Markov property*. The local Markov property states that the probability of value X_i, given its neighbors, is conditionally independent of the values of all other particles. In other words, $P(X_i|\text{all particles except } i) = p(X_i|\text{neighbors of particle } i)$. The microstructures are obtained by sampling the Markov random field $P(X)$. In this paper, we present an algorithm to sample 3D microstructures by selecting the color of pixel X_i by sampling the conditional probability density $p(X_i|\text{neighbors of voxel } i)$ from available 2D experimental data.

Algorithm

In the following discussion, let S^x, S^y, and S^z denote the set of orthogonal (x, y, and z, respectively) slices of the microstructure. Let V denote the solid (3D) microstructure. The color of voxel v in the 3D microstructure is denoted by V_v. In addition to the color (e.g., RGB triplet), the vector V_v may also contain other values including grain orientation and phase index. In this work, the color is represented using G color levels in the range $\{0, 1, .., G-1\}$, each of which maps to an RGB triplet. The number of color levels is chosen based on the microstructure to be reconstructed, e.g., for binary images $G = 2$.

Recall our Markovian assumption which states that the probability distribution of the color for a pixel given the colors of its spatial neighborhood is independent of the rest of the image. The vectors denoting the spatial neighborhood of voxel v in the slices orthogonal to the x, y, and z axis, respectively, are denoted as V_v^x, V_v^y, and V_v^z (see Figure 2). The neighborhood is taken over a small user-assigned window around the voxel v. Let $S^{x,w}$, $S^{y,w}$, and $S^{z,w}$ denote a window of the same size in the input 2D micrographs. In order to find the coloring of voxel v based on the neighbor voxels in the x-plane, one needs to compute the conditional probability density $p(V_v|\text{color of } x\text{-plane neighbors of } v)$. Explicit construction of such a probability density is often computationally intractable. Instead, the most likely value of v is identified by first finding a window $S^{x,w}$ that is most similar to V_v^x in the input 2D micrograph. This window is denoted by S_v^x (see Figure 2). Similarly,

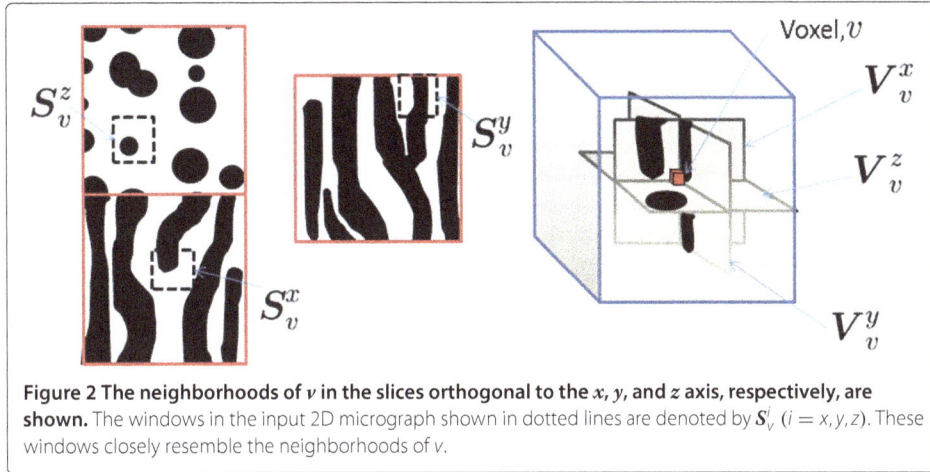

Figure 2 The neighborhoods of v in the slices orthogonal to the x, y, and z axis, respectively, are shown. The windows in the input 2D micrograph shown in dotted lines are denoted by S_v^i ($i = x, y, z$). These windows closely resemble the neighborhoods of v.

matching windows to the y- and z- plane neighborhoods of voxel v in the corresponding 2D sectional image (denoted as S_v^y, S_v^z) are found. Each of these matching windows S_v^x, S_v^y, and S_v^z may have different coloring of the center pixel. Thus, we need an optimization methodology to effectively merge these disparate values and identify a unique coloring for voxel v. The optimization approach is described next.

Let the value $V_{v,u}^x$ denote the color of voxel u in the neighborhood V_v^x. Similarly, the values $S_{v,u}^x$ and $S_u^{x,w}$, respectively, denote the color of pixel u in the window S_v^x and $S^{x,w}$. The 3D microstructure is synthesized by posing the problem as a L^2 minimization of the energy [8]:

$$E(V) = \sum_{i \in \{x,y,z\}} \sum_v \sum_u \omega_{v,u}^i \| V_{v,u}^i - S_{v,u}^i \|^2. \tag{1}$$

Here, $\omega_{v,u}^i$ denotes a per pixel weight. In order to preserve the short-range correlations of the microstructure as much as possible, the weight for the nearby pixel is taken to be greater than those of the pixels farther away (Gaussian weighting is used).

The optimization is carried out in two steps. In the first step, the energy is minimized with respect to S_v^i. In this step, we assume that the most likely sample from the conditional probability distribution of the center pixel in the 3D image (e.g., $p(V_v|\text{colors of } x\text{-plane neighbors of } v)$) is the center pixel of a best matching window in an experimentally obtained 2D slice on the corresponding plane. The best matching neighborhood of voxel v along the x-plane is selected by solving the following problem:

$$S_v^x = \arg\min_{S^{x,w}} \sum_u \omega_{v,u}^x \| V_{v,u}^x - S_u^{x,w} \|^2. \tag{2}$$

This is an exhaustive search that compares all the windows in the input 2D micrograph to the corresponding x-slice neighborhood of voxel v and identifies a window that leads to a minimum weighted squared distance. In this process, for 2D images of size 64×64 with a 16×16 neighborhood window, a matrix of size $16^2 \times (64-16)^2$ is built containing all possible neighborhoods of pixels that have a complete 16^2 window around it. The column in this matrix that has a minimum distance to the 3D slice V_v^x is then found through a k-nearest neighbor algorithm [9]. Note that we are only given a limited (in this work, a single) 2D experimental sample along each cross-section, which means that the best match may not be an exact match for V_v^x.

Thus, for each voxel v, a set of three best matching neighborhoods are obtained, possibly with different colors corresponding to the center pixel. A unique value of v thus needs to be found by weighting colors pertaining to location v not only in the matching windows of voxel v but also in its neighbors. This is exactly done in the second step of the optimization procedure, where the optimal color of voxel v is computed by setting the derivative of the energy function with respect to V_v to zero. This leads to a simple weighted average expression for the color of voxel v:

$$V_v = \left(\sum_{i \in \{x,y,z\}} \sum_u \omega_{u,v}^i S_{u,v}^i \right) \Big/ \left(\sum_{i \in \{x,y,z\}} \sum_u \omega_{u,v}^i \right). \tag{3}$$

Note that the subscripts u and v are switched in the above expression as compared to Equation 1. This implies that the optimal color of the voxel v is the weighted average of the colors at locations corresponding to voxel v in the best matching windows (S_u^i) of voxels (u) in the solid microstructure. Since V_v changes after this step, the set of closest input neighborhoods S_v^i will also change. Hence, these two steps were repeated until convergence, i.e., until the set S_v^i stops changing. As a starting condition, a random color from the input 2D images is assigned to each voxel v. The process is carried out in a multiresolution (or multigrid) fashion [10]: starting with a coarse voxel mesh and interpolating the results to a finer mesh once the coarser 3D image has converged to a local minimum. Three resolution levels (16^3, 32^3, and 64^3) were used. Synthesizing a 64^3 solid microstructure took between 10 and 15 min on a 3-GHz desktop computer, with about two-thirds of the time spend in step 1 (search) algorithm.

Results and discussion

The approach has been demonstrated for three test cases with 2D images corresponding to

1. Case 1. An isotropic distribution of solid circles;
2. Case 2. An anisotropic case with solid circles in the z-slice (similar to case 1) but an interconnected lamellar structure in the x- and y-slices;
3. Case 3. A polycrystalline microstructure.

In case 1, all three slices (x, y, and z) were assigned to the same 2D image depicted in Figure 3a. The resulting 3D microstructure is expected to be a random distribution of spheres. The 3D microstructure obtained by our approach is shown in Figure 3b. The internal structure of the solid microstructure is shown via slices in the x-plane at different distances from the origin. Various slices 'look' similar to the input image as expected from the Markov random field assumption. Case 2 builds upon this case by introducing anisotropy in the x- and y-planes. Three 2D images corresponding to x, y, and z-slices (as shown in Figure 4a) were used in the reconstruction. An interconnected lamellar structure was used in the x- and z-planes, while the z-plane image allowed merging of the solid circles to allow for a more complex microstructure. In the algorithm, we match the 2D images with all three orthogonal slices through every voxel. The resulting anisotropic 3D microstructure shown in Figure 4b is quite complex. The y-axis slices as shown in Figure 4c show the depth profile of various solid circles seen at the top surface, with intricate internal structure revealed.

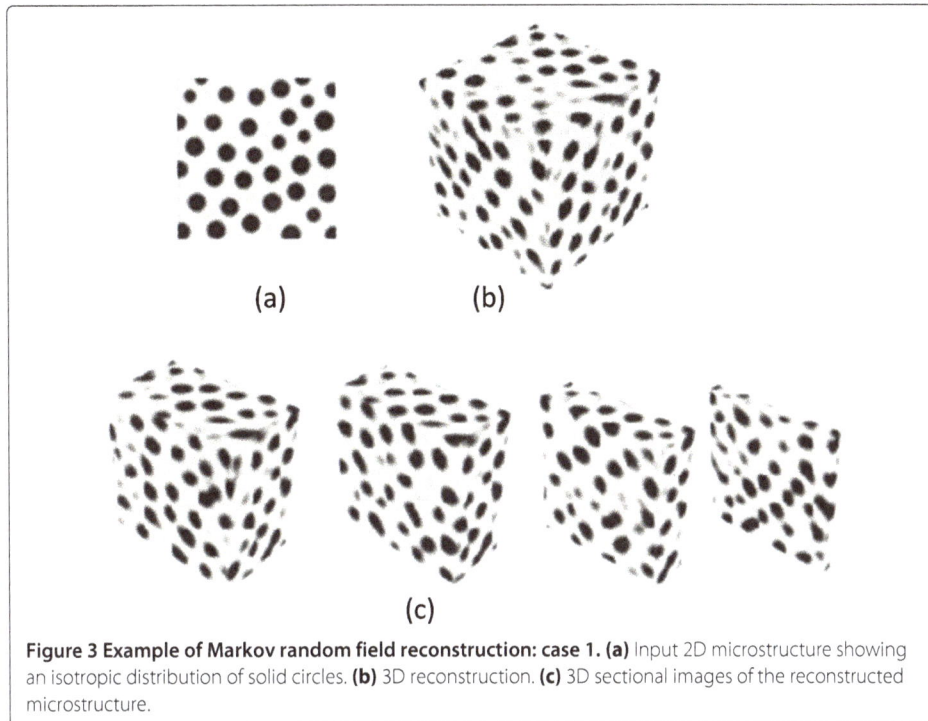

Figure 3 Example of Markov random field reconstruction: case 1. (a) Input 2D microstructure showing an isotropic distribution of solid circles. **(b)** 3D reconstruction. **(c)** 3D sectional images of the reconstructed microstructure.

In the last example, a polycrystalline microstructure was employed to show the applicability of the algorithm to cases beyond two-phase media. The microstructure is equiaxed, and all three slices were assigned to the same 2D image shown in Figure 5a. The resulting 3D microstructure is shown in Figure 5b, and its internal structure revealed through the x-axis slices in Figure 5c. The results show that the grains built by the algorithm are also equiaxed with a variety of 3D shapes identified by the algorithm.

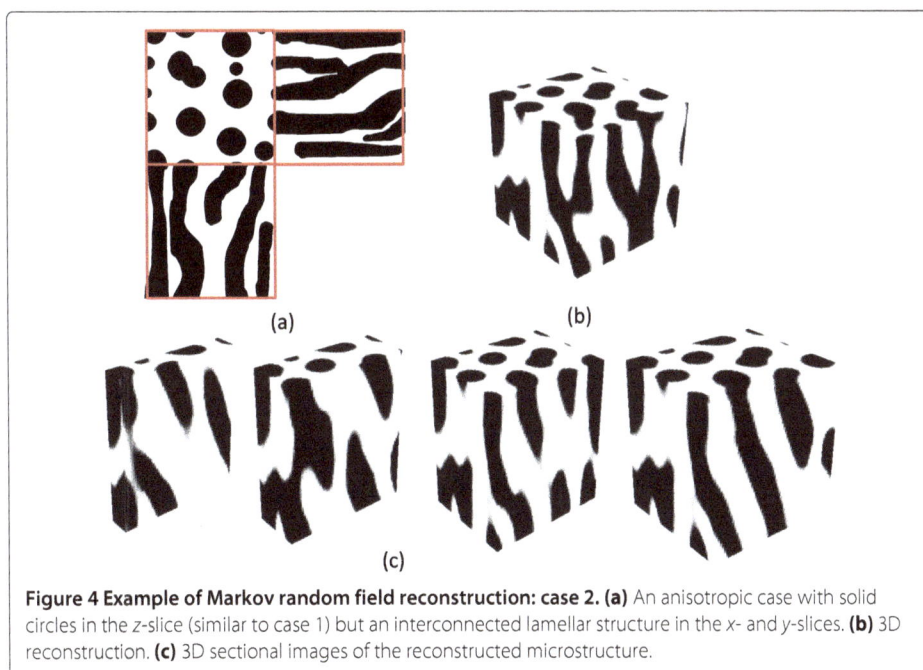

Figure 4 Example of Markov random field reconstruction: case 2. (a) An anisotropic case with solid circles in the z-slice (similar to case 1) but an interconnected lamellar structure in the x- and y-slices. **(b)** 3D reconstruction. **(c)** 3D sectional images of the reconstructed microstructure.

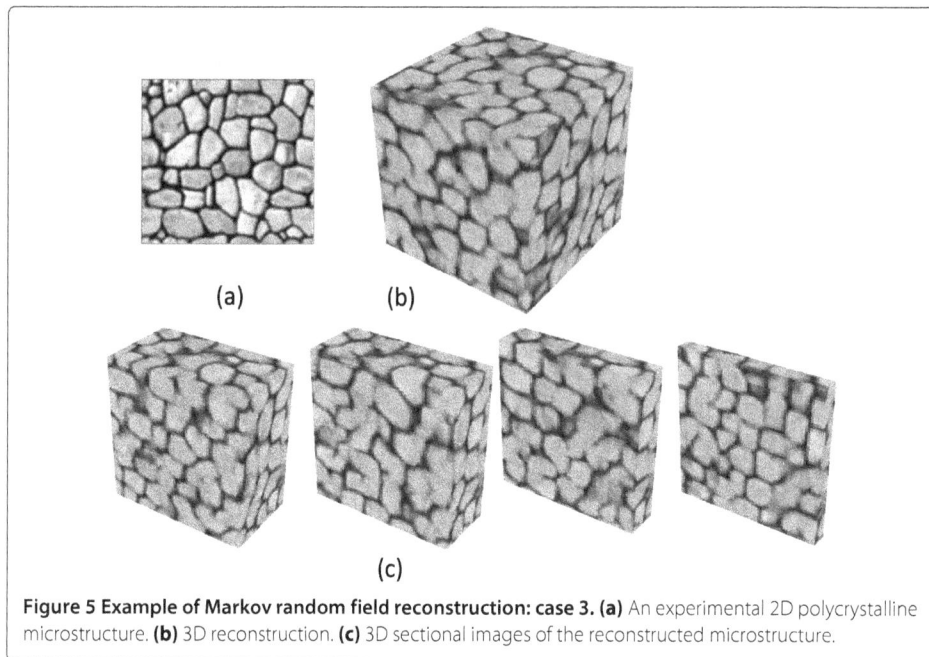

Figure 5 Example of Markov random field reconstruction: case 3. (a) An experimental 2D polycrystalline microstructure. **(b)** 3D reconstruction. **(c)** 3D sectional images of the reconstructed microstructure.

However, some of grain boundaries do not show up well in the slices which is primarily attributed to the lower resolution of the 3D image (64^3) compared to the original input image.

Validation tests

For testing the validity of the 3D reconstructions, quantitative comparisons were made between the original 2D image and the reconstructed image, through comparison of the statistical correlation functions as described in [11]. Rotationally invariant probability functions are employed as the microstructural features. Rotationally invariant N-point correlation measure ($S_{(N)}^i$) can be interpreted as the probability of finding the N vertices of a polyhedron separated by relative distances $x_1, x_2, ..., x_N$ in phase i when tossed, without regard to orientation, in the microstructure. The simplest of these probability functions is the one-point function, $S_{(1)}$, which is just the volume fraction (V) of phase i. The two-point correlation measure, $S_{(2)}^i(r)$, can be obtained by randomly placing line segments of length r within the microstructure and counting the fraction of times the end points fall in phase i. These statistical descriptors occur in rigorous expressions for the effective electromagnetic, mechanical, and transport properties like effective conductivity, magnetic permeability, effective elastic modulus, Poisson's ratio, and fluid permeability of such microstructures ([4,12]). All the required correlation measures needed for comparison and property bound calculation are obtained using a Monte Carlo sampling procedure [1]. The procedure involves initially selecting a large number of initial points in the microstructure. For every initial point, several end points at various distances are randomly sampled, and the number of successes (of all points falling in the ith phase) are counted to obtain the required correlation measures. Statistical measures were extracted from the microstructures by sampling 15,000 initial points.

Two- and three-point correlations of the isotropic distribution of solid circles

The statistical features of the 2D distribution of solid circles from Figure 3 and its 3D reconstruction were compared. The original 2D image was a square of side 64 μm and had a phase 1 (white phase) volume fraction of 70%. The comparison of two-point probability $(S_{(2)}^1)$ and the three-point probability function $S_{(3)}^1$ is shown in Figure 6a,b, respectively. The three-point probability measure $S_{(3)}^1(r, s, t)$ is depicted in a feature vector format with the distances (r, s, t)μm indicated for key points in Figure 6b. The first points in both graphs (Figure 6) show the volume fraction of white phase for 2D image as well as the reconstructed image. The decay in the two-point correlation function is identical for the reconstructed image up until 3μm, showing excellent reproduction of the short-range correlation. The same aspect can also be seen from comparing the short-range correlation in the three-point probability function (Figure 6b). Although the longer range correlations match qualitatively, there is a drift seen as the distance between pixels increases. Both the excellent match in short-range correlation and the small drift in the long-range correlation can be explained based on the reconstruction algorithm, which models a stronger interaction of a center pixel to pixels in its immediate local neighborhood than pixels farther away. In effect, the algorithm gives a stronger weighting towards matching the short-range correlations in the microstructure.

Elastic properties of two-phase composite

The experimental data in [13] provides a high-resolution planar microstructure image (Figure 7a) of a silver-tungsten composite with porous tungsten matrix and molten silver (volume fraction of silver phase $p = 20\%$). The microstructure has been employed for several reconstruction studies [14,15]. A 657×657 pixel region of the microstructure corresponding to 204-μm square area was converted to a black-and-white image for distinguishing the two phases. This was done by selecting a threshold color below in which phases were set to white (the silver phase), and the rest of the image was set to black (the tungsten phase). The final black-and-white image is shown in the inset of Figure 7a. A 64-μm square cell within this image was chosen to reconstruct the 3D image.

An instance of the reconstructed microstructure is shown in Figure 7b,c with the distribution of each phase shown separately. The auto-correlation function for the silver phase

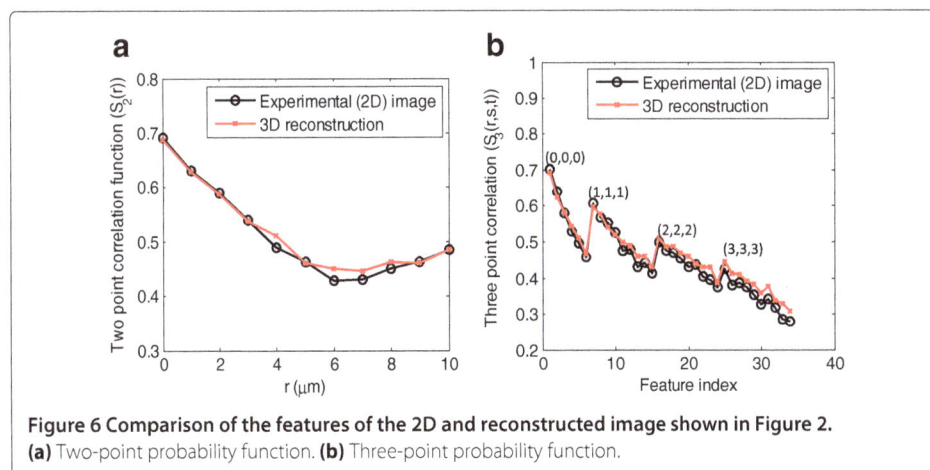

Figure 6 Comparison of the features of the 2D and reconstructed image shown in Figure 2.
(a) Two-point probability function. **(b)** Three-point probability function.

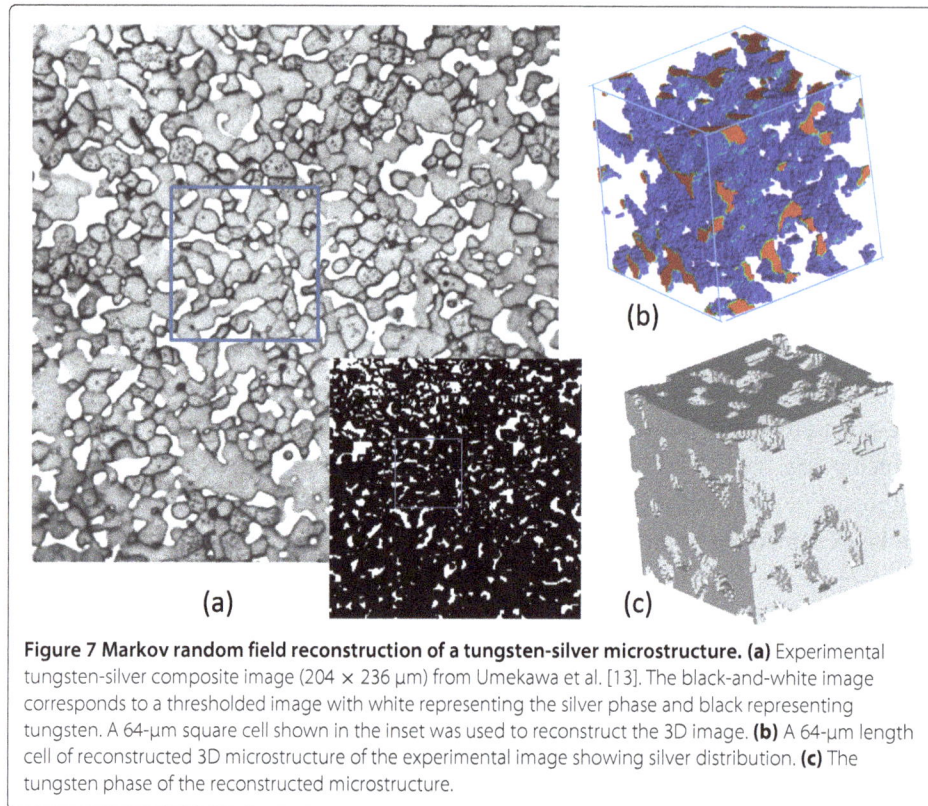

Figure 7 Markov random field reconstruction of a tungsten-silver microstructure. (a) Experimental tungsten-silver composite image (204 × 236 μm) from Umekawa et al. [13]. The black-and-white image corresponds to a thresholded image with white representing the silver phase and black representing tungsten. A 64-μm square cell shown in the inset was used to reconstruct the 3D image. **(b)** A 64-μm length cell of reconstructed 3D microstructure of the experimental image showing silver distribution. **(c)** The tungsten phase of the reconstructed microstructure.

$\gamma(r) = \frac{S^1_{(2)}(r) - p^2}{p - p^2}$ of the reconstructed 3D microstructure and the experimental image are compared in Figure 8a showing excellent match of short-range correlations with a small difference seen in longer range correlations. Short-range correlations carry the greatest weightage in determining mechanical properties such as elastic modulus (e.g., [4]), although long-range correlations have been found to be important for phenomena such as surface roughening during plastic deformation [16]. To test if the elastic properties are well captured in the reconstructed 3D microstructure, we compared against the experimental data from [13] of the elastic modulus as a function of temperature. The elastic properties of the individual components at different temperatures are available from [14]

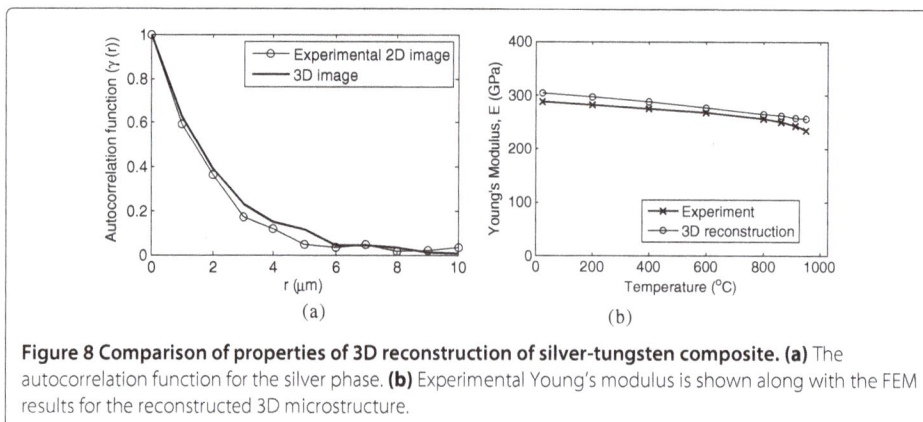

Figure 8 Comparison of properties of 3D reconstruction of silver-tungsten composite. (a) The autocorrelation function for the silver phase. **(b)** Experimental Young's modulus is shown along with the FEM results for the reconstructed 3D microstructure.

Table 1 Elastic properties of silver and tungsten phases as a function of temperature (from [14])

T (°C)	E_{silver} (GPa)	ν_{silver}	$E_{tungsten}$ (GPa)	$\nu_{tungsten}$
25	71	0.36	400	0.28
200	69	0.36	392	0.28
400	63	0.36	383	0.28
600	54	0.36	373	0.28
800	45	0.37	363	0.28
860	42	0.37	361	0.28
910	39	0.37	359	0.28
950	37	0.37	357	0.28

and are listed in Table 1. The data was used within a finite-element simulation to compute the elastic modulus of the reconstructed microstructure using the method described in [17]. The computed properties of the reconstructed 3D microstructure closely follow the experimentally measured Young's modulus from [13] as shown in Figure 8b with an average error from the experimental data of about 5%.

Conclusions

A method for reconstructing diverse microstructure from two-dimensional microstructures imaged on orthogonal planes is presented. The algorithm reconstructs 3D images through matching of 3D slices at different voxels to the representative 2D micrographs. This is posed as an iterative optimization problem where the first step involves searching of patches in the 2D micrographs that look alike to the 3D voxel neighborhood, followed by a second step involving the optimization of an energy function that ensures various patches from the 2D micrographs meshed together seamlessly in the 3D image. The method is particularly promising for anisotropic cases where the x-, y-, and z-slices look different. The results demonstrate that the method can effectively model three-dimensional features in the microstructure including complex interconnectivity of the features and complex shapes that are not intuitive at first sight. The approach can be useful to rapidly build a library of 3D microstructures for modeling purposes from 2D micrographs. Although, this preliminary study shows significant promise as to the feasibility of the approach, future work will focus on increasing the resolution of the reconstruction and code optimization. In addition, future work will focus on more rigorous testing of the stereological features (e.g., grain-size histograms) and other engineering properties (yield strength) of reconstructed anisotropic microstructures.

Availability of supporting data

The executables and data files for the methodology described here are available upon request.

Competing interests
The authors declare that they have no competing interests.

Author's information
VS is an Associate Professor at the University of Michigan and is a lifetime member of The Minerals, Metals and Materials Society (TMS). VS developed the methods and examples shown in this work.

Acknowledgements
The author would like to acknowledge the Air Force Office of Scientific Research, MURI contract FA9550-12-1-0458, for the financial support.

References
1. Sundararaghavan V, Zabaras N (2005) Classification and reconstruction of three-dimensional microstructures using support vector machines. Compu Mater Sci 32:223–239
2. Yeong CLY, Torquato S (1998) Reconstructing random media II. Three-dimensional media from two-dimensional cuts. Phys Rev E 58(1):224–233
3. Manwart C, Torquato S, Hilfer R (2000) Stochastic reconstruction of sandstones. Phys Rev E 62:893–899
4. Torquato S (2002) Random heterogeneous materials: microstructure and macroscopic properties. Springer, New York
5. Efros A, Leung T (1999) Texture synthesis by non-parametric sampling. Int Conf Comput Vis 2:1033–1038
6. Ising E (1925) Beitrag zur Theorie des Ferromagnetismus. Zeitschrift Physik 31:253–258
7. Besag J (1974) Spatial interaction and the statistical analysis of lattice systems. J R Stat Soc. Series B (Methodological) 36(2):192–236
8. Kwatra V, Essa I, Bobick A, Kwatra N (2005) Texture optimization for example-based synthesis. ACM Trans Graph (Proc. SIGGRAPH) 24(3):795–802
9. Altman NS (1992) An introduction to kernel and nearest-neighbor nonparametric regression. The American Statistician 46(3):175–185
10. Kopf J, Fu C-W, Cohen-Or D, Deussen O, Lischinski D, Wong T-T (2007) Solid texture synthesis from 2D exemplars. Proc SIGGRAPH 2:1–9
11. Xu H, Dikin DA, Burkhart C, Chen W (2014) Descriptor-based methodology for statistical characterization and 3D reconstruction of microstructural materials. Comput Mater Sci 85:206–216
12. Quintanilla J (1999) Microstructure and properties of random heterogeneous materials: a review of theoretical results. Polymer Engg Sci 39:559–585
13. Umekawa S, Kotfila R, Sherby OD (1965) Elastic properties of a tungsten-silver composite above and below the melting point of silver. J Mech Phys Solids 13(4):229–230
14. Roberts AP, Garboczi EJ (1999) Elastic properties of a tungsten-silver composite by reconstruction and computation. J Mech Phys Solids 47:2029–2055
15. Roberts AP, Torquato S (1999) Chord-distribution functions of three-dimensional random media: approximate first-passage times of Gaussian processes. Phys Rev E 59(5):4953–4963
16. Lee PS, Piehler HR, Rollett AD, Adams BL (2002) Texture clustering and long-range disorientation representation methods: application to 6022 aluminum sheet. Metallurgical Mater Trans A 33(12):3709–3718
17. Garboczi EJ (1998) NIST Internal Report 6269. Chapter 2. http://ciks.cbt.nist.gov/garboczi/. Accessed 24 Nov 2013

7

A software framework for data dimensionality reduction: application to chemical crystallography

Sai Kiranmayee Samudrala[1][*], Prasanna Venkataraman Balachandran[2], Jaroslaw Zola[3], Krishna Rajan[4][*] and Baskar Ganapathysubramanian[5][*]

*Correspondence:
ssamudrala@me.gatech.edu;
krajan@iastate.edu;
baskarg@iastate.edu
[1] School of Mechanical Engineering, Georgia Tech, Atlanta, GA 30332-0405, USA
[4] Department of Materials Science and Engineering, Iowa State University, Ames, IA 50011, USA
[5] Department of Mechanical Engineering, Iowa State University, Ames, IA 50011, USA
Full list of author information is available at the end of the article

Abstract

Materials science research has witnessed an increasing use of data mining techniques in establishing process-structure-property relationships. Significant advances in high-throughput experiments and computational capability have resulted in the generation of huge amounts of data. Various statistical methods are currently employed to reduce the noise, redundancy, and the dimensionality of the data to make analysis more tractable. Popular methods for reduction (like principal component analysis) assume a linear relationship between the input and output variables. Recent developments in non-linear reduction (neural networks, self-organizing maps), though successful, have computational issues associated with convergence and scalability. Another significant barrier to use dimensionality reduction techniques in materials science is the lack of ease of use owing to their complex mathematical formulations. This paper reviews various spectral-based techniques that efficiently unravel linear and non-linear structures in the data which can subsequently be used to tractably investigate process-structure-property relationships. In addition, we describe techniques (based on graph-theoretic analysis) to estimate the optimal dimensionality of the low-dimensional parametric representation. We show how these techniques can be packaged into a modular, computationally scalable software framework with a graphical user interface - Scalable Extensible Toolkit for Dimensionality Reduction (SETDiR). This interface helps to separate out the mathematics and computational aspects from the materials science applications, thus significantly enhancing utility to the materials science community. The applicability of this framework in constructing reduced order models of complicated materials dataset is illustrated with an example dataset of apatites described in structural descriptor space. Cluster analysis of the low-dimensional plots yielded interesting insights into the correlation between several structural descriptors like ionic radius and covalence with characteristic properties like apatite stability. This information is crucial as it can promote the use of apatite materials as a potential host system for immobilizing toxic elements.

Keywords: Non-linear dimensionality reduction; Process-structure-property; Apatites; Materials science; High-throughput analysis

Background

Using data mining techniques to probe and establish process-structure-property relationships has witnessed a growing interest owing to its ability to accelerate the process of tailoring materials by design. Before the advent of data mining techniques, scientists used a variety of empirical and diagrammatic techniques [1], like pettifor maps [2], to establish relationships between structure and mechanical properties. Pettifor maps, one of the earliest graphical representation techniques, is exceedingly efficient except that it requires a thorough understanding and intuition about the materials. Recent progress in computational capabilities has seen the advent of more complicated paradigms - so-called virtual interrogation techniques - which span from first-principles calculations to multi-scale models [3-7]. These complex multi-physics and/or statistical techniques and simulations [8,9] result in an integrated set of tools which can predict the relationships between chemical, microstructural, and mechanical properties producing an exponentially large collection of data. Simultaneously, experimental methods - combinatorial materials synthesis [10,11], high-throughput experimentation, atom probe tomography - allow synthesis and screening of a large number of materials while generating huge amounts of multivariate data.

A key challenge is then to efficiently probe this large data to extract correlations between structure and property. This data explosion has motivated the use of data mining techniques in materials science to explore, design, and tailor materials and structures. A key stage in this process is to *reduce the size of the data, while minimizing the loss of information during this data reduction*. This process is called data dimensionality reduction. By definition, dimensionality reduction (DR) is the process of reducing the dimensionality of the given set of (usually unordered) data points and extracting the low-dimensional (or parameter space) embedding with a desired property (for example, distance, topology, etc.) being preserved throughout the process. Examples for DR methods are principal component analysis (PCA) [12], Isomap [13], Hessian locally linear embedding (hLLE) [14], etc. Applying DR methods enables visualization of the high-dimensional data and also estimates the optimal number of dimensions required to represent the data without considerable loss of information. Additionally, burgeoning cyberinfrastructure-based tools and collaborations sustained by the government's recent Materials Genome Initiative (MGI) provides a great platform to leverage the data dimensionality reduction tools. This will enable integration of information obtained from the individual high-throughput simulations and experimentation efforts in various domains (e.g., mechanical, electrical, electro-magnetic, etc.) and at multiple length-scales (macro-meso-micro-nano) in a fashion as never seen before [15].

Data dimensionality reduction is not a novel concept. Page [16] describes different techniques of data reduction and their applicability for establishing process-structure-property relationships. Statistical methods like PCA [17] and factor analysis (FA) [18] have been used on materials data generated by first-principles calculations or by experimental methods. However, dimensionality reduction techniques like PCA or factor analysis to establish process-structure-property relationships traditionally assume a linear relationship among the variables. This is often not strictly valid; the data usually lies on a non-linear manifold (or surface) [13,19]. Non-linear dimensionality reduction (NLDR) techniques can be applied to unravel the non-linear structure from unordered data. An example of such application for constructing a low-dimensional stochastic representation of property variations in random heterogenous media is [19]. Another exciting application of data dimensionality reduction is in combination with quantum mechanics-based calculations to predict the structure [20-22]. For a more mathematical list of linear and non-linear DR techniques, the interested reader can consult [23,24].

In this paper, the theory and mathematics behind various linear and non-linear dimensionality reduction methods is explained. The mathematical aspects of dimensionality reduction are packaged into an easy-to-use software framework called Scalable Extensible Toolkit for Dimensionality Reduction (SETDiR) which (a) provides a user-friendly interface that successfully abstracts user from the mathematical intricacies, (b) allows for easy post-processing of the data, and (c) represents the data in a visual format and allows the user to store the output in standard digital image format (eg: JPEG), thus making data more tractable and providing an intuitive understanding of the data. We conclude by applying the techniques discussed on a dataset of apatites [25-29] described using several structural descriptors. This paper is seen as an extension of our recent work [30]. Apatites $(A_4^I A_6^{II}(BO_4)_6 X_2)$ have the ability to accommodate numerous chemical substitutions and hence represent a unique family of crystal chemistries with properties catering many technological applications, such as toxic element immobilization, luminescence, and electrolytes for intermediate temperature solid oxide fuel cells, to name a few [25-29].

The outline of the paper is as follows: The section 'Methods: dimensionality reduction' briefly describes the concepts of DR, algorithms, and the dimensionality estimators that can be used to estimate the dimensionality. The software framework, SETDiR, developed to apply DR techniques is described in the section 'Software: SETDiR'. The section 'Results and discussion' discusses the interpretation of low-dimensional results obtained by applying SETDiR to the apatite dataset.

Methods: dimensionality reduction

The problem of dimensionality reduction can be formulated as follows. Consider a set of data, X. This set consists of n data points, x_i. Each of the data points x_i is vectorized to form a 'column' vector of size D. Usually, D is large. Thus, $X = \{x_0, x_1, \ldots, x_{n-1}\}$ of n points, where $x_i \in \mathbb{R}^D$ and $D \gg 1$. Visualizing and analyzing correlations, patterns, and connections within high-dimensional dataset is difficult. Hence, we are interested in finding a set of *equivalent low-dimensional points*, $Y = \{y_0, y_1, \ldots, y_{n-1}\}$, that exhibit the same correlations, patterns, and connections as the high-dimensional data. This is mathematically posed as

Find $Y = \{y_0, y_1, \ldots, y_{n-1}\}$, such that $y_i \in \mathbb{R}^d$, $d \ll D$ and $\forall_{i,j} |x_i - x_j|_h = |y_i - y_j|_h$. Here, $|a - b|_h$ denotes a specific norm that captures properties, connections, or correlations we want to preserve during dimensionality reduction [23].

For instance, by defining h as Euclidean norm, we preserve Euclidean distance, thus obtaining a reduction equivalent to the standard technique of PCA [12]. Similarly, defining h to be the angular distance (or conformal distance [31]) results in locally linear embedding (LLE) [32] that preserves local angles between points. In a typical application [33,34], x_i represents a state of the analyzed system, e.g., temperature field, concentration distribution, or characteristic properties of a system. Such state description can be derived from experimental sensor data or can be the result of a numerical simulation. However, irrespective of the source, it is characterized by high dimensionality, that is D is typically of the order of 10^2 to 10^6 [35,36]. While x_i represents just a single state of the system, contemporary data acquisition setups deliver large collections of such observations, which correspond to the temporal or parametric evolution of the system [33]. Thus, the cardinality n of the resulting set X is usually large ($n \sim 10^2$ to 10^5). Intuitively, information obfuscation increases with the data dimensionality. Therefore, in the process of DR, we seek as small a dimension d as possible, given the constraints induced by the norm $|a - b|_h$ [23]. Routinely, $d < 4$ as it permits, for instance, visualization of the set Y.

The key mathematical idea underpinning DR can be explained as follows: We encode the desired information about X, i.e., topology or distance, in its entirety by considering all pairs of points in X. This encoding is represented as a matrix $A_{n \times n}$. Next, we subject matrix A to unitary transformation V, i.e., transformation that preserves the norm of A (thus, preserving connectivities and correlations in the data), to obtain its sparsest form Λ, where $A = V\Lambda V^T$. Here, $\Lambda_{n \times n}$ is a diagonal matrix with rapidly diminishing entries. As a result, it is sufficient to consider only a small, d, number of entries of Λ to capture all the information encoded in A. These d entries constitute the set Y. The above procedure hinges on the fact that unitary transformations preserve original properties of A [37]. Note also, that it requires a method to construct matrix A in the first place. Indeed, what differentiates various spectral data dimensionality methods is the way information is encoded in A.

We focus on four different DR methods: (a) PCA, a linear DR method; (b) Isomap, a non-linear isometry-preserving DR method; (c) LLE, a non-linear conformal-preserving DR method; and (d) Hessian LLE, a topology-preserving DR method.

Principal component analysis

PCA is a powerful and a popular DR strategy due to its simplicity and ease in implementation. It is based on the premise that the high-dimensional data is a linear combination of a set of hidden low-dimensional axes. PCA then extracts the latent parameters or low-dimensional axes by reorienting the axes of the high-dimensional space in such a way that the variance of the variables is maximized [23].

PCA algorithm

1. Compute the pair-wise Euclidean distance for all points in the input data X. Store it as a matrix $[E]$.
2. Construct a matrix $[W^*]$ such that the elements of $[W^*]$ are -0.5 times the square of the elements of the euclidean distance matrix $[E]$.
3. Find the dissimilarity matrix $[A]$ by double centering $[W^*]$:

$$[A] = \left[H^T\right][W^*][H] \tag{1}$$

$$H_{ij} = \begin{cases} (1 - 1/n) \ \forall \ \mathbf{i} = \mathbf{j}, \\ (-1/n) \ \forall \ \mathbf{i} \neq \mathbf{j}. \end{cases} \tag{2}$$

4. Solve for the largest d eigenpairs of $[A]$:

$$[A] = [U][\Lambda]\left[U^T\right]. \tag{3}$$

5. Construct the low-dimensional representation in $\mathbb{R}^d$ from the eigenpairs:

$$[Y] = [I][\Lambda]^{1/2}\left[U^T\right]. \tag{4}$$

The functionality of the identity matrix is to extract the most important d-dimensions from the eigenpairs of $[A]$.

The limitation of PCA is that it assumes the data lies on a linear space and hence performs poorly on the data that are inherently non-linear. In these cases, PCA also tends to over-estimate the dimensionality of the data.

Isomap

Isomap relaxes the assumption of PCA that the data lies on a linear space. A classic example of a non-linear manifold is the Swiss roll. Figure 1 shows how PCA tries to fit the best linear plane while Isomap unravels the low-dimensional surface. Isomap essentially smooths out the non-linear manifold into a corresponding linear space and subsequently applies PCA. This smoothing out can intuitively be understood in the context of the spiral, where the ends of the spiral are pulled out to straighten the spiral into a straight line. Isomap accomplishes this objective mathematically by ensuring that the geodesic distance between data points are preserved under transformations. The geodesic distance is the distance measured along the curved surface on which the points rest [23]. Since it preserves (geodesic) distances, Isomap is an isometry (distance-preserving) transformation. The underlying mathematics of the Isomap algorithm assumes that the data lies on a manifold which is convex (but not necessarily linear). Note that both PCA and Isomap are isometric mappings; PCA preserves pair-wise Euclidean distances of the points while Isomap preserves the geodesic distance.

Isomap algorithm

1. Compute the pair-wise Euclidean distance matrix $[E]$ from the input data X.
2. Compute the k-nearest neighbors of each point from the distance matrix $[E]$.
3. Compute the pair-wise geodesic distance matrix $[G]$ from $[E]$. This is done using Floyd's algorithm [38].
4. Construct a matrix $[W^*]$ such that the elements of $[W^*]$ are -0.5 times the square of the elements of the geodesic distance matrix $[G]$.
5. Find the dissimilarity matrix $[A]$ by double centering $[W^*]$:

$$[A] = \left[H^T \right] \left[W^* \right] [H] \tag{5}$$

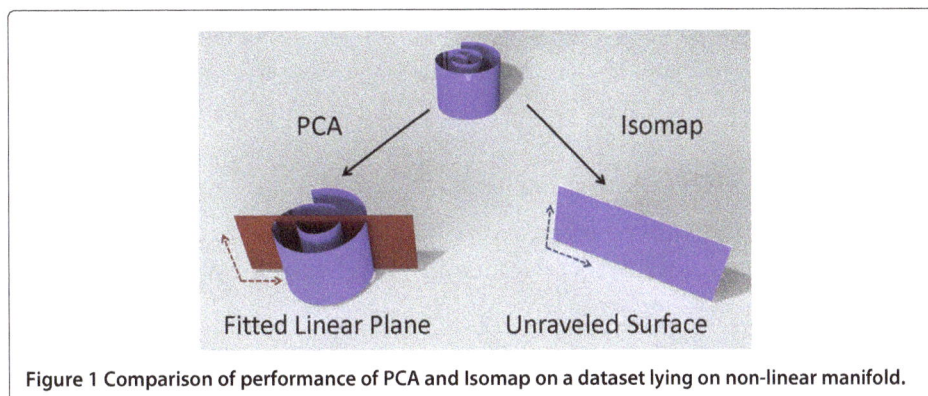

Figure 1 Comparison of performance of PCA and Isomap on a dataset lying on non-linear manifold.

$$H_{ij} = \begin{cases} (1 - 1/n) \; \forall \; \mathbf{i} = \mathbf{j}, \\ (-1/n) \; \forall \; \mathbf{i} \neq \mathbf{j}. \end{cases} \tag{6}$$

6. Solve for the largest d eigenpairs of A:

$$[A] = [U][\Lambda]\left[U^T\right]. \tag{7}$$

7. Construct the low-dimensional representation in $\mathbb{R}^d$ from the eigenpairs:

$$[Y] = [I][\Lambda]^{1/2}\left[U^T\right]. \tag{8}$$

The non-linearity in the data is accounted for by using geodesic distance metric. The graph distance is used to approximate the geodesic distance [39]. Graph distance between a pair of points in a graph (V, E) is the shortest path connecting the two given points. The graph distances are calculated using the well-known Floyd's algorithm [38].

Locally linear embedding

In contrast to PCA and Isomap methods which preserve distances, LLE preserves the local topology (or local orientation, or angles between data points). LLE uses the notion that locally a non-linear manifold (or curve) is well-approximated by a linear curve. In other words, the manifold is locally linear and hence can be represented as a patchwork of linear curves. The algorithm first divides the manifold into patches and reconstructs each point in the patch based on the information (or weights) obtained from its neighbors (i.e., infer how a specific point is located with respect to its neighbors). This process extracts the local topology of the data. Finally, the algorithm reconstructs the global structure by combining individual patches and finding an optimized, low-dimensional representation. Numerically, local topology information is constructed by finding the k-nearest neighbors of each data point and reconstructing each point from the information about the weights of the neighbors. The global reconstruction from the local patches is accomplished by assimilating the individual weight matrices to form a global weight matrix $[W]$ and evaluating the smallest eigenvalues of normalized global weight matrix $[A]$.

LLE algorithm

1. For each of the n input vectors from $X = \{x_0, x_1, \ldots, x_{n-1}\}$:

    (a) Find the k-nearest neighbors of the data point x_i.

    (b) Construct the local covariance or Gram matrix $\mathbf{G_i}$

    $$g_{r,s}(i) = (x_i - x_r)^T (x_i - x_s) \tag{9}$$

    where x_r and x_s are neighbors of x_i.

    (c) Weight vector, w_i is computed by solving the linear system:

    $$\mathbf{G_i} w_i = \mathbf{1} \tag{10}$$

    where $\mathbf{1}$ is a $k \times 1$ vector of ones.

2. Using the vectors w_i, build the sparse matrix W. The (i, j) of W is zero if x_i and x_j are not neighbors. If x_i and x_j are neighbors, then $W(i, j)$ takes the values of the corresponding with vector, $w_i(j)$.

3. From W, build A:

    $$[A] = (I - W)^T (I - W). \tag{11}$$

4. Compute the eigenpairs (corresponding to the smallest eigenvalues) for A:

$$[A] = [U][\Lambda]\left[U^T\right]. \tag{12}$$

5. Compute the low-dimensional points in $\mathbb{R}^d$ from the smallest eigenpairs.

Hessian LLE

Hessian LLE [14] (hLLE) is a modification of LLE and Laplacian Eigenmaps [40]. Mathematically, hLLE replaces the Laplacian (first derivative) operator with a Hessian (second derivative) operator over the graph. hLLE constructs patches, performs a local PCA on each patch, constructs a global Hessian from the eigenvectors thus obtained, and finally finds the low-dimensional representation from the eigenpairs of the Hessian. hLLE is a topology preservation method and assumes that the manifold is locally linear.

hLLE algorithm

1. At each given point x_i, construct a $k \times n$ neighborhood matrix M_i such that each row, j, of the matrix represents a point

$$x_j = x_j - \bar{x}_i, \tag{13}$$

   where $\bar{x}_i$ is the mean of k neighboring points.
2. Perform singular value decomposition (SVD) of the M_i to obtain the SVD matrices, U, V, D.
3. Construct the $(N * d(d+1)/2)$ local Hessian matrix $[H]^i$ such that the first column is a vector of all ones and the next d columns are the columns of U followed by the products of all the d columns of $[U]$.
4. Compute Gram-Schmidt orthogonalization [37] on the local Hessians $[H]^i$ and assimilate the last $d(d+1)/2$ orthonormal vectors of each to construct the global Hessian matrix $[A]$ [14].
5. Compute the eigenpairs (corresponding to the smallest eigenvalues) of the Hessian matrix:

$$[A] = [W][\Lambda][W]^T. \tag{14}$$

6. Compute the low-dimensional points $[Y]$ in $\mathbb{R}^d$ from the eigenpairs:

$$[Y] = [W]\left([W]^T[W]\right)^{-1/2}. \tag{15}$$

An important point to note here is that, as discussed in the section 'Methods: dimensionality reduction', matrix $[A]$ encodes the required information for each of the DR techniques, and the construction of this matrix is what differentiates a spectral DR method from the rest. Matrix $[A]$ is a normalized Euclidean matrix in the case of PCA, a normalized geodesic matrix in the case of Isomap, a normalized Hessian matrix for hLLE, and so on.

Dimensionality estimators

A key step in constructing the low-dimensional points from the data is the choice of the low dimensionality or optimal dimensionality d. Methods like PCA and Isomap have an implicit technique to estimate the low dimensionality (approximately) using scree plots. We introduce a graph-based technique that rigorously estimates the latent dimensionality of the data, which can be used in conjunction with the scree plot.

Dimensionality from the scree plot

Scree plot is a plot of the eigenvalues with the eigenvalues arranged in decreasing order of their magnitude. Scree plots obtained from PCA and Isomap (distance-preserving methods) give an estimate of the dimensionality. A heuristic method of identifying the dimensionality is by identifying the elbow in the scree plot. A more quantitative estimate of dimensionality is estimated by choosing a value for $p_{var}(d)$ that ensures a threshold of the minimum percentage variability. If $\lambda_1 >= \lambda_2 >= \ldots \lambda_n$ are the individual eigenvalues arranged in descending order, the percentage variability ($p_{var}(d)$) covered by considering first d eigenvalues is given by

$$p_{var}(d) = 100 \times \frac{\sum\limits_{i=1}^{d} \lambda_i}{\sum\limits_{i=1}^{n} \lambda_i} \tag{16}$$

A usual approach is to choose a d that takes 95% of the variability into account.

Geodesic minimal spanning tree estimator

We have recently utilized a dimensionality estimator based on the BHH theorem (Breadwood-Halton-Hammersley Theorem) [41]. This theorem states that the rate of convergence of the length of minimal spanning tree[a] gives a measure of the latent dimensionality. This theorem allows one to express the dimensionality (d) of an unordered dataset as a function of the length of geodesic minimal spanning tree (GMST) of the graph of the dataset. Specifically, the slope of a $\log(n)$ vs. $\log(L_n)$ plot constructed by calculating the GMST length (L_n) with respect to increasing size of randomly chosen data points (n) provides an estimate of the dimensionality: $d = \frac{1}{(1-m)}$, where m is the slope of the log-log plot [19].

Correlation dimension

Correlation dimension is a space-filling dimension which is derived from a more generic fractal dimension by assigning a value of $q = 2$ in

$$C(\mu, \epsilon) = \int \left[\mu \bar{B}_e(z) \right]^{q-1} d\mu(z) \tag{17}$$

where μ is a Borel probability measure on a metric space $\mathbb{Z}$. $\bar{B}_e(z)$ is a closed ball of radius ϵ centered on z.

Numerical definition of correlation dimension is given by

$$d_{cor}(\epsilon_1, \epsilon_2) = \frac{\log(\hat{C}_2(\epsilon_2)) - \log(\hat{C}_2(\epsilon_1))}{\log(\epsilon_2) - \log(\epsilon_1)} \tag{18}$$

where $\hat{C}_2(\epsilon_2)$ is a measure of proportion of distances less than ϵ [23,42]. Intuitively, these ϵ values are like window ranges through which one zooms through the data. Too small ϵ will render the data as individual points, while too huge ϵ will make the entire dataset look like a single fuzzy spot. Hence, correlation dimension is sensitive to the epsilon values. One important point to note, however, is that the correlation dimension provides the user with a lower bound of the optimal dimensionality.

Software: SETDiR

These DR techniques are packaged into a modular, scalable framework for ease of use by the materials science community. We call this package, *SETDiR*. This framework contains two major components:

1. Core functionality: developed using C++
2. User interface: developed based on Java (Swing)

Figure 2 describes the scope of the functionality of both modules in SETDiR.

Core functionality

Functionality is developed using object-oriented C++ programming language. It implements the following methods: PCA, Isomap, LLE, and dimensionality estimators like GMST and correlation dimension estimators [23].

User interface

A graphical user interface (shown in Figure 3) is developed using Java™ Swings Components with the following features which make it user-friendly:

1. Abstracts the user from the mathematical and programming details.
2. Displays the results graphically and enhances the visualization of low-dimensional points.
3. Easy post-processing of results: in-built cluster analysis, ability to save plots as image files.
4. Organized settings tabs: Based on the niche of the user, the solver settings are organized as Basic User and Advanced User tabs which abstract a new or a naive user from, otherwise overwhelming, details.

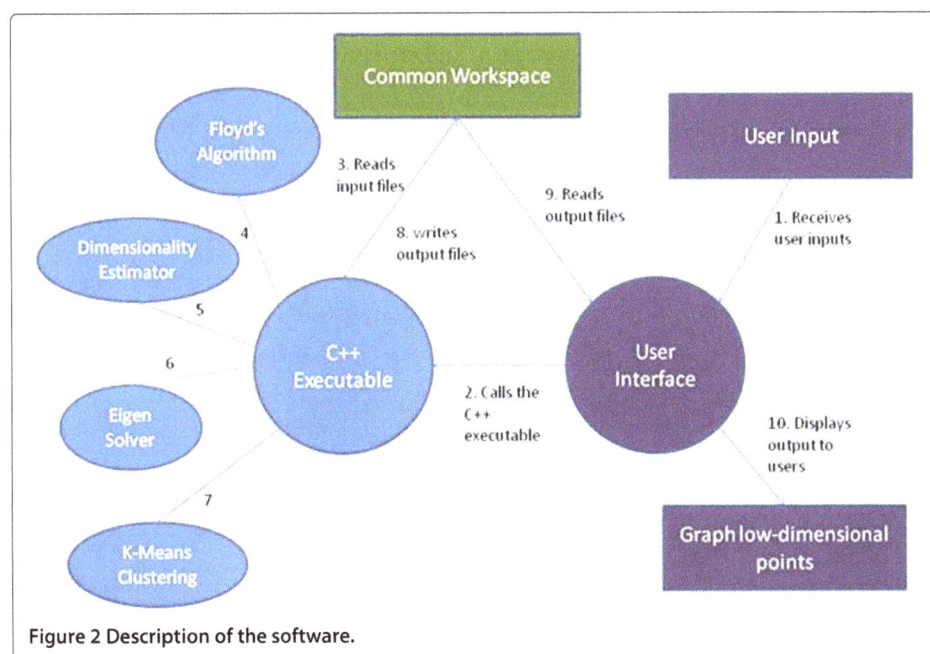

Figure 2 Description of the software.

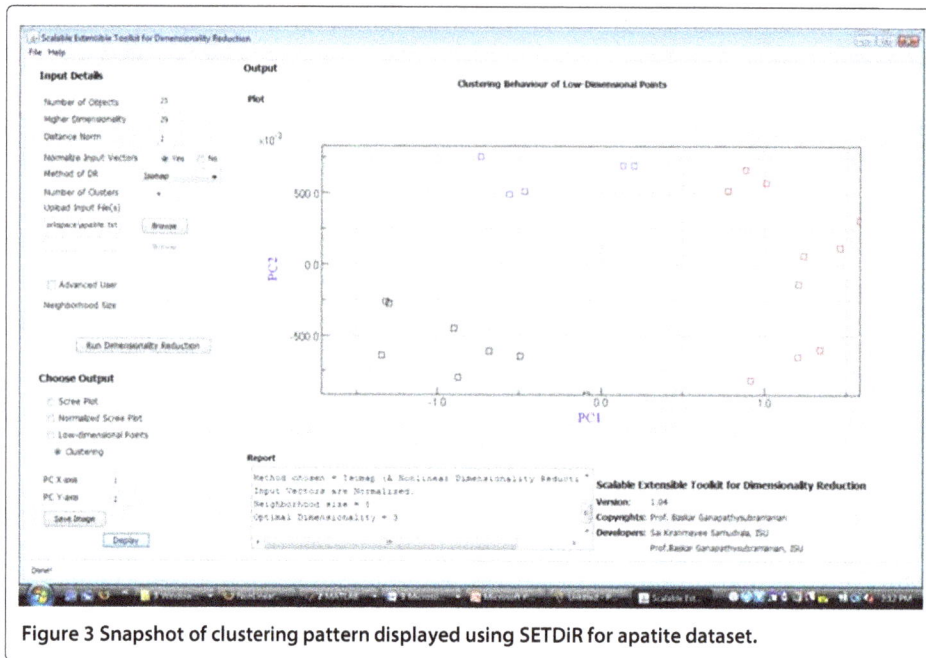

Figure 3 Snapshot of clustering pattern displayed using SETDiR for apatite dataset.

This framework can be downloaded from SETDiR (http://setdir.engineering.iastate.edu/doku.php?id=download). A more detailed discussion of the parallel features of the code is deferred to another publication. We next showcase the framework and the mathematical strategies on the apatite dataset.

Results and discussion

In this section of the paper, we compare and contrast the algorithms on an interesting dataset of apatites with immense technological and scientific significance. Apatites have the ability to accommodate numerous chemical substitutions and exhibit a broad range of multifunctional properties. The rich chemical and structural diversity provides a fertile ground for the synthesis of technologically relevant compounds [25-29]. Chemically, apatites are conveniently described by the general formula $A_4^I A_6^{II}(BO_4)_6 X_2$, where A^I and A^{II} are distinct crystallographic sites that usually accommodate larger monovalent (Na$^+$, Li$^+$, etc.), divalent (Ca^{2+}, Sr^{2+}, Ba^{2+}, Pb^{2+}, etc.), and trivalent (Y^{3+}, Ce^{3+}, La^{3+}, etc.), B-site is occupied by smaller tetrahedrally coordinated cations (Si^{4+}, P^{5+}, V^{5+}, Cr^{5+}, etc.), and the X-site is occupied by halides (F$^-$, Cl$^-$, Br$^-$), oxides, and hydroxides. Establishing the relationship between the microscopic properties of apatite complexes with those of the macroscopic properties can help us in gaining an understanding and promote the use of apatites in various technological applications. For example, information about the relative stability of the apatite complexes can promote the utilization of apatites as a suitable host material for immobilizing toxic elements such as lead, cadmium, and mercury (i.e., by identifying an apatite chemical composition that contain at least one of the aforementioned toxic elements and yet remaining thermodynamically stable). DR techniques offer unique insights into the originally intractable high-dimensional datasets by enabling visual clustering and pattern association, thereby establishing process-structure-property relationship for chemically complex solids such as apatites.

Apatite data description

The crystal structure of the aristotype $P6_3/m$ $Ca_4^I Ca_6^{II}(PO_4)_6F_2$ apatite with hexagonal unit cell is shown in the Figure 4 with the atoms projected along the (001) axis. The polyhedral representation of $A^I O_6$ and BO_4 structural units are clearly shown with the Ca^{II}-site (pink atoms) and F-site (green atoms) occupying the tunnel. Thin black line represents the unit-cell of the hexagonal lattice.

The sample apatite dataset considered consists of 25 different compositions described using 29 structural descriptors. These structural descriptors, when modified, affect the crystal structure [44]. Therefore, by establishing the relationship between the crystal structure and these structural descriptors and analyzing the clustering of different compositions, conclusions can be drawn about how the changes in these structural descriptors (defining the atomic features) could affect the macroscopic properties (such as elastic modulus, band gap, and conductivity). The bond length, bond angle, lattice constants, and total energy data are taken from the work of Mercier et al. [26]; the ionic radii data are taken from the work of Shannon [45] and the electronegativity data is based on the Pauling's scale [46]. The ionic radii of A^I-site (r_{A^I}) has a coordination number nine and A^{II}-site ($r_{A^{II}}$) has a coordination number seven (when the X-site is F^-) or eight (when the X-site is Cl^- or Br^-). Our database describes Ca, Ba, Sr, Pb, Hg, Zn, and Cd in the A-site; P, As, Cr, V, and Mn in the B-site; and F, Cl, and Br in the X-site. The 25 compounds considered in this study belong to the aristotype $P6_3/m$ hexagonal space group. We utilize SETDiR on the apatite data and present some of the results below. More information regarding the source of the apatite data can be found in [44]. A preliminary analysis (focusing only on PCA) can be found in [30].

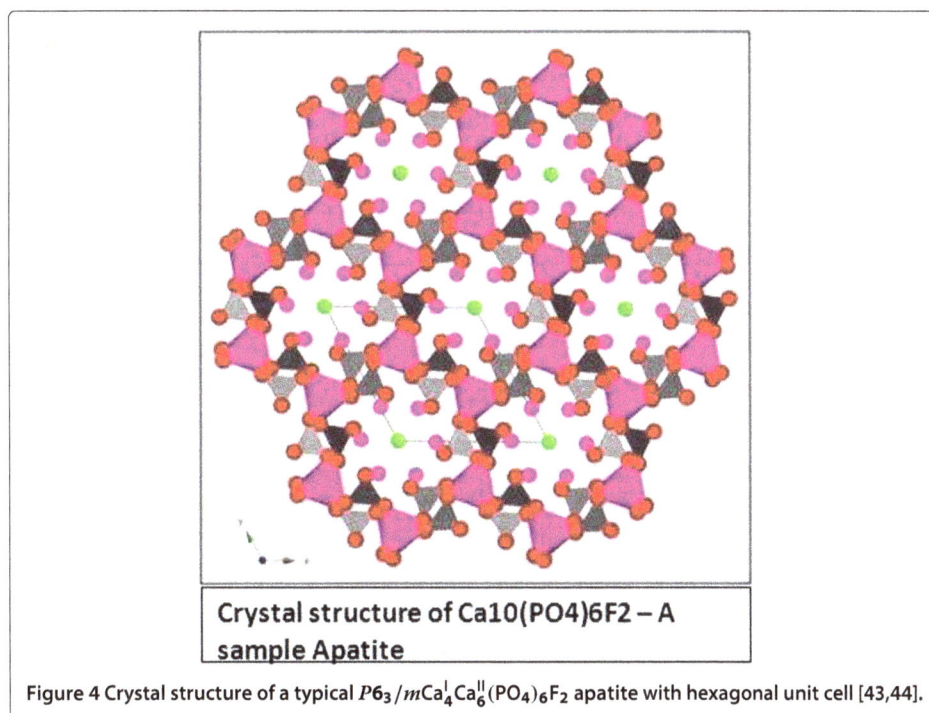

Crystal structure of Ca10(PO4)6F2 – A sample Apatite

Figure 4 Crystal structure of a typical $P6_3/m Ca_4^I Ca_6^{II}(PO_4)_6F_2$ apatite with hexagonal unit cell [43,44].

Dimensionality estimation

SETDiR first estimates the dimensionality using the scree plot. A scree plot is a plot of eigenvalue indices vs. eigenvalues. The occurrence of an elbow (or a sharp drop in eigenvalues) in a scree plot gives the estimate of the dimensionality of the data. Figure 5 displays the scree plots when the input vectors $\{x_0, x_1, \ldots, x_{n-1}\}$ were normalized with respect to that when they were not normalized. We plot for comparison the eigenvalues that are obtained from both PCA and Isomap. This plot shows how the second eigenvalue collapses to zero when the input vectors are not normalized and hence emphasizes the importance of normalization of input vectors[b]. It is also interesting to compare the eigenvalues of PCA and Isomap for normalized input: PCA being a linear method overestimates the dimensionality as 5, while Isomap estimates it to be 3. SETDiR subsequently uses the geodesic minimal spanning tree method to estimate the dimensionality of the apatite data. This method gives a rigorous estimate of 3 (Figure 6), which matches the outcome of the more heuristic scree plot estimate.

Low-dimensional plots

In this section, we discuss the visual interpretation of the low-dimensional plots obtained by applying the dimensionality reduction techniques - PCA, Isomap, LLE, and hLLE - to a set of apatites described using structural descriptors. Figure 7 (left) shows the 2D plot between principal components 2 and 3. The reason for showing this plot is that PC2-PC3 map captures pattern that is similar to Isomap components 1 and 2. While we find associations among compounds that are similar to those as shown in Figure 7 (right), the nature of information is manifested differently. This is mainly attributed to the differences in the underlying mathematics of the two techniques, where PCA is essentially a linear technique and, on the other hand, Isomap is a non-linear technique. To further interpret the hidden information captured by Isomap classification map (Figure 7), we have focused on the three regions separately.

Figure 7 (right) shows a two-dimensional classification map with isomap components 1 and 2 in the orthogonal axes. The two-dimensional classification map groups various apatite compounds into three distinct regions that capture various interactions between

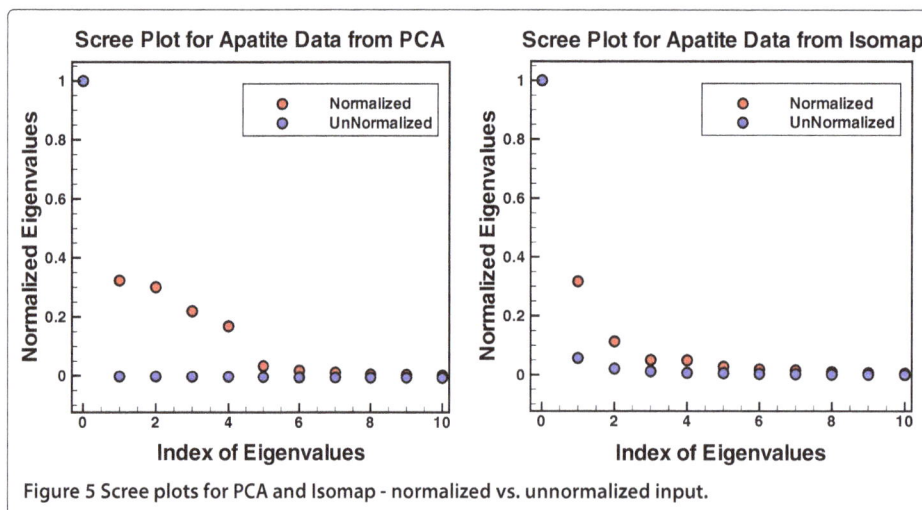

Figure 5 Scree plots for PCA and Isomap - normalized vs. unnormalized input.

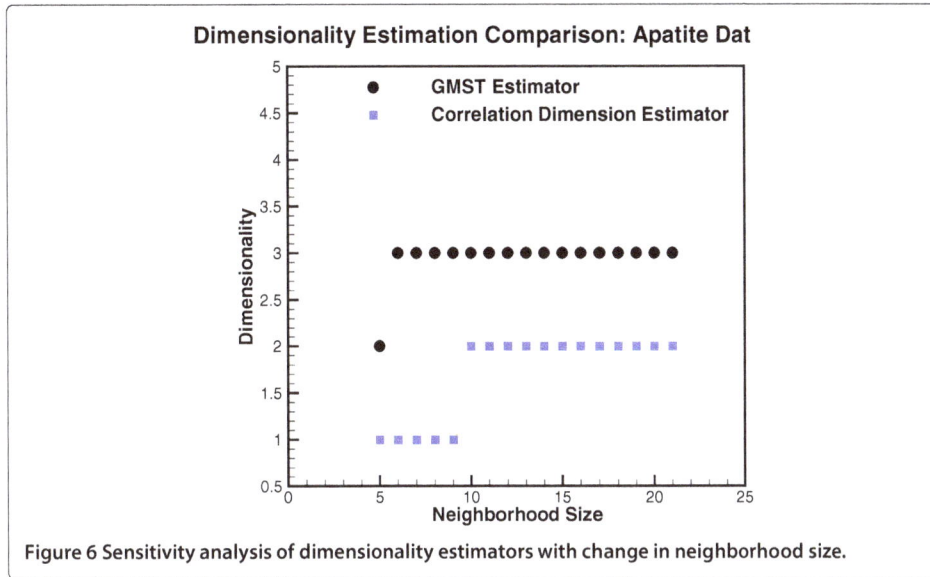

Figure 6 Sensitivity analysis of dimensionality estimators with change in neighborhood size.

A-, B-, and X-site ions in complex apatite crystal structure. Region 1 corresponds to apatite compounds with fluoride (F) ion in the X-site. All apatite compounds in this region contain only F in the X-site but has different A-site (Ca, Sr, Pb, Ba, Cd, Zn) and B-site elements (P, Mn, V). Therefore, this unique region classifies F-apatites from Cl and Br-apatites. Region 2 belongs to apatite compounds with phosphorus (P) ion in the B-site and contains Cl and Br ions in the X-site. The uniqueness of this region is manifested mainly due to the presence of only smaller P ions in the B-site. Similarly, region 3 belongs to apatite compounds with Cl ions in the X-site and contains larger B-site Cr, V, and As cations.

Figure 8 (right) presents the results from hLLE. It can be observed that the compounds that have highly covalent A-site cation (e.g., Hg^{2+} and Pb^{2+}) and highly covalent B-site cation (P^{5+}) clearly separate out from the rest. An exception to this rule is $Pb_{10}(CrO_4)_6Cl_2$. Our PCA-derived structure map also revealed similar pattern - i.e.,

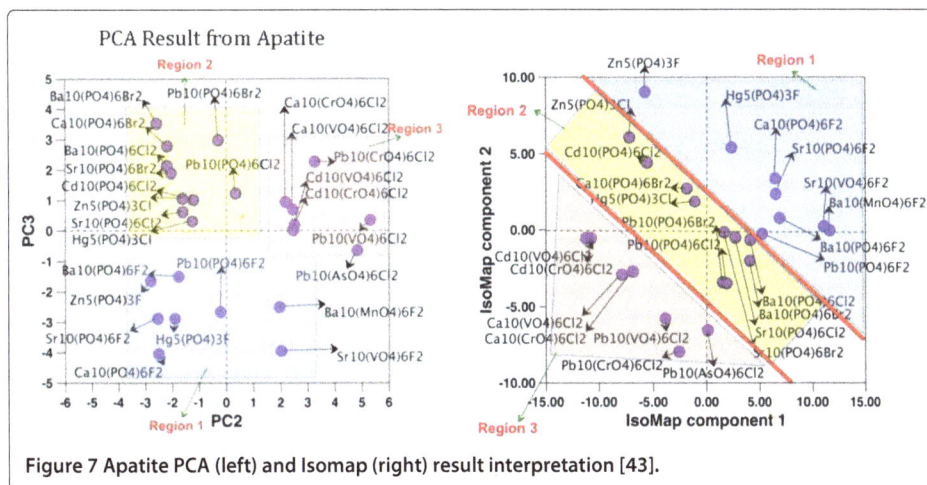

Figure 7 Apatite PCA (left) and Isomap (right) result interpretation [43].

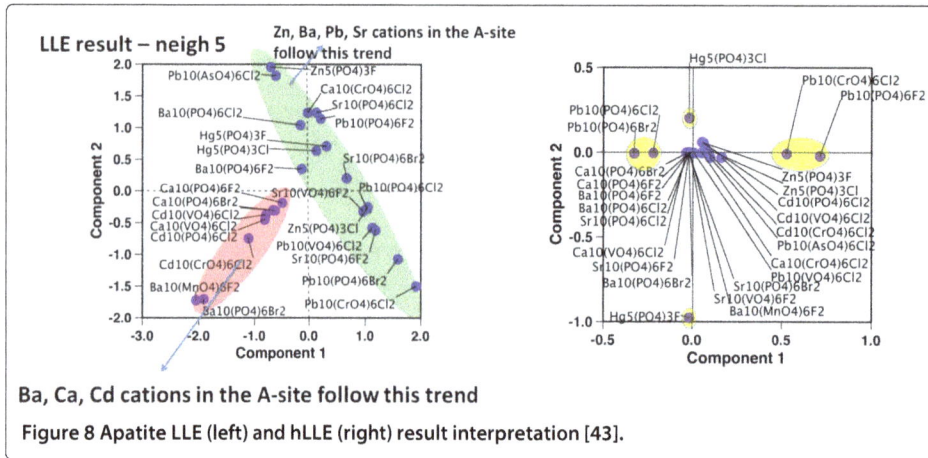

Figure 8 Apatite LLE (left) and hLLE (right) result interpretation [43].

$Pb_{10}(CrO_4)_6Cl_2$ was found to not obey the general trend [44]. Note that the presence of Cr cation in the B-site has been known to cause structural distortions in apatites.

For example, $Sr_{10}(PO_4)_6Cl_2$ has a $P6_3/m$ symmetry, whereas $Sr_{10}(CrO_4)_6Cl_2$ has a distorted $P6_3$ symmetry [28]. Based on the previous PCA work [44], we attribute the cause for this exception to two bond distortion angles: (i) rotation angle of A^{II}-A^{II}-A^{II} triangular units and the angle that bond A^I-O_1 makes with the c-axis. Compared to Hessian LLE, we cannot find any clear pattern with respect to chemical bonding in the LLE result Figure 8 (left).

Figure 9 shows a zoomed-in plot of the Hessian LLE result.[c] Around the origin, we can find two clusters of compounds: (i) one on the left with negative component 1 value corresponding to compounds that have ionic alkaline earth metal cations in the A-site and (ii) one on the right with positive component 1 value corresponding to compounds that have covalent A-site cations. An exception here is $Ca_{10}(CrO_4)_6Cl_2$, which is found among the covalent A-site cluster indicating that $Ca_{10}(CrO_4)_6Cl_2$ may have a distorted symmetry. It is important to recognize that neither PCA nor Isomap identifies $Ca_{10}(CrO_4)_6Cl_2$ as an exception. Compared to hLLE, we do not find any intriguing insights from the LLE analysis and therefore, we do not discuss LLE results.

One needs to explore different manifold methods to fully understand high dimensional correlations and mappings. Hence, in the following section, we shall explore the impact of the Isomap analysis.

In Figure 10 region 1, the ionic radii of A-site elements increases along the direction shown, with Zn^{2+} cation being the smallest and Ba^{2+} being the largest. Note that this A-site ionic radii trend is not clearly seen in the PC2-PC3 classification map (Figure 7). One of the key outcomes from Figure 10 is the identification that $Pb_{10}(PO_4)_6F_2$ compound is an outlier. In terms of Shannon's ionic radii scale, Pb^{2+} is larger than Ca^{2+} but smaller than Sr^{2+} cation. Ideally (assuming apatites as ionic crystals), the relative position of $Pb_{10}(PO_4)_6F_2$ should have been between $Ca_{10}(PO_4)_6F_2$ and $Sr_{10}(PO_4)_6F_2$ compounds in the map. However, this was not the case. The physical reason behind this observation could be attributed to the electronic structure of Pb^{2+} ions [47]. The theoretical electronic structure calculations indicate that in the atom-projected density of states curves, the Pb^{2+} ions have active $6s^2$ lone-pair electrons that hybridize with oxygen $2p$ electrons resulting in a strong covalent bond formation. Indeed, recent density functional theory

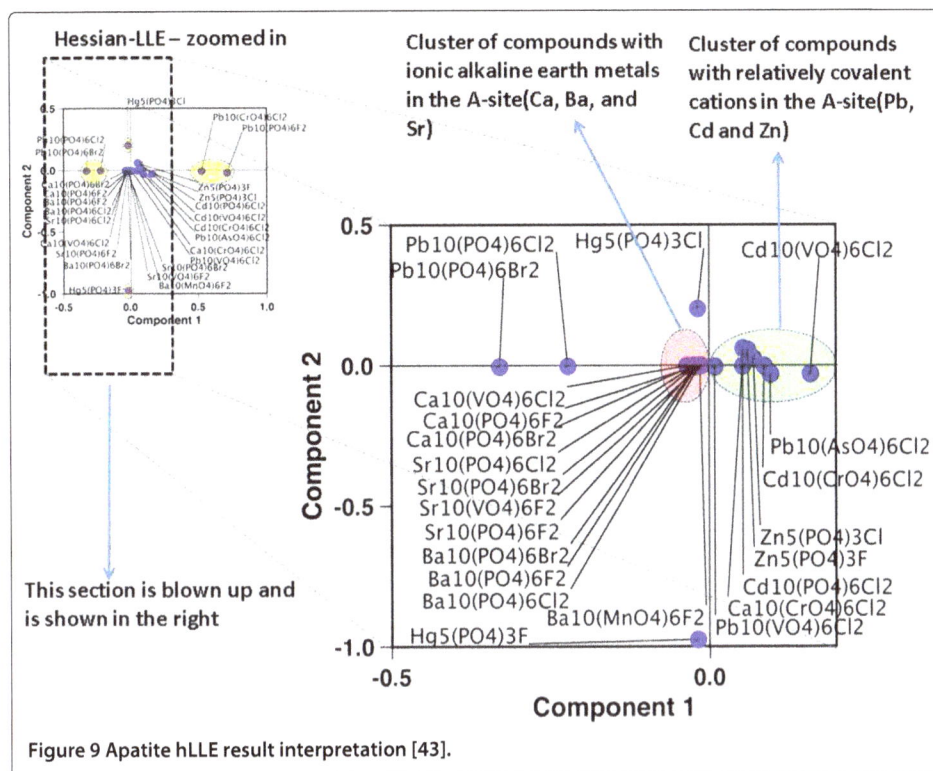

Figure 9 Apatite hLLE result interpretation [43].

(DFT) calculations [48] show that the electronic band gap (at the generalized gradient approximation (GGA) level) for $Pb_{10}(PO_4)_6F_2$ is 3.7 eV, which is approximately 2 eV smaller compared to $Ca_{10}(PO_4)_6F_2$ (5.67 eV) and $Sr_{10}(PO_4)_6F_2$ compounds (5.35 eV).

In our dataset, the electronic structure information of A-site elements was quantified using Pauling's electronegativity data. While PCA captures this behavior, the dominating effect of the electronic structure of Pb^{2+} ions is more transparent within the mathematical framework of non-linear Isomap analysis.

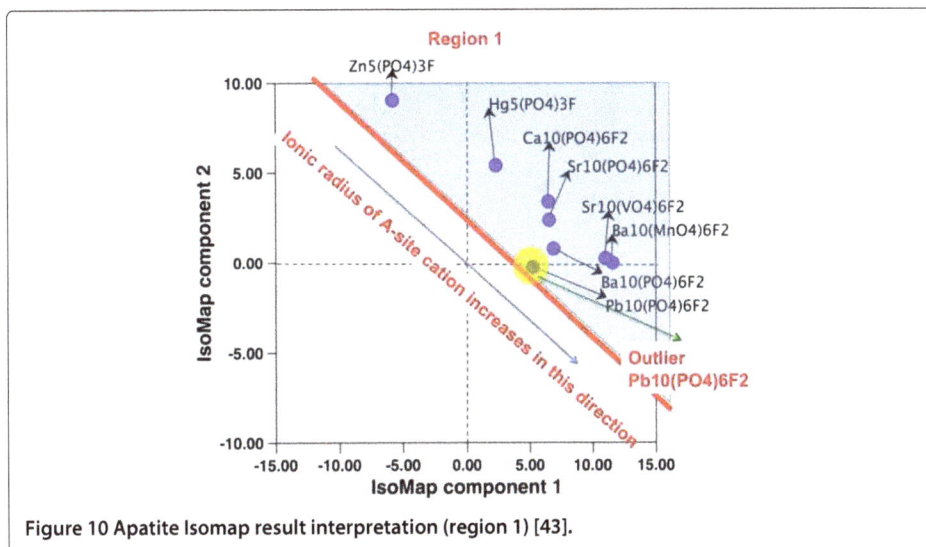

Figure 10 Apatite Isomap result interpretation (region 1) [43].

Besides, from Figure 10, it can also be inferred that the bond distortions of Zn apatite is different from other compounds. This trend correlates well with the non-existence of $Zn_{10}(PO_4)_6F_2$ compounds due to the difficulty in experimental synthesis [49]. On the other hand, the relative correlation position of $Hg_{10}(PO_4)_6F_2$ compound offer intriguing insights. In fact, the uniqueness of $Hg_{10}(PO_4)_6F_2$ chemistry was previously detected in a PCA-derived structure map [44], which clearly identified the composition as an outlier among other isostructural compounds. Guided by this original insight from PCA, recently, Balachandran $et\ al.$ [48] showed using DFT calculations that the ground state structure of $Hg_{10}(PO_4)_6F_2$ is triclinic (space group $P\bar{1}$). Although the ionic size of Hg^{2+} is very close to that of Ca^{2+}, the aristotype $P6_3/m$ symmetry distorts to $P\bar{1}$ symmetry in $Hg_{10}(PO_4)_6F_2$ due to the mixing of fully occupied Hg-$5d^{10}$ orbitals with the empty Hg-$6s^0$ orbitals. This mixing is unavailable to the $Ca_{10}(PO_4)_6F_2$ compound, because it does not have orbitals of appropriate symmetry.

In Figure 11, region 2 is highlighted where we find a clear trend of apatite compounds with respect to the ionic radii of A-site elements. Similar to region 1, Pb apatites manifest themselves as outliers in region 2. The unique electronic structure of Pb^{2+} cations in forming a covalent bond with oxygen $2p$-states is identified as the reason for the deviation of Pb apatites from the expected trend. The covalent bonding among Pb compounds appear to be independent of X-site anion, when the B-site is occupied by phosphorus cations. In Figure 11, $Hg_{10}(PO_4)_6Cl_2$ compound is found to be closely associated with $Ca_{10}(PO_4)_6Br_2$ indicating some similarity in the bond distortions of the two compounds. In comparing the relative correlation position of all Cl-containing apatites (except Pb-based compounds) in region 2, we predict $Hg_{10}(PO_4)_6Cl_2$ to have a stable apatite structure type (in sharp contrast to $Hg_{10}(PO_4)_6F_2$).

Figure 12 describes region 3 where we find clusters of apatite compounds with Cl ions in the X-site and contain larger V, Cr, and As cations in the B-site. The ionic radius of A-site element increases in the direction as shown in the figure, and in this case, the Pb apatites are not outliers. The presence of large V, Cr, and As cations (compared to smaller P cations in regions 1 and 2) in the B-site were identified as the reason for this behavior.

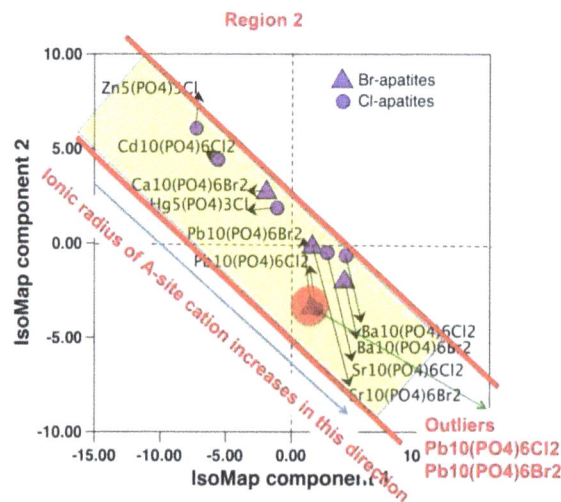

Figure 11 Apatite Isomap result interpretation (region 2) [43].

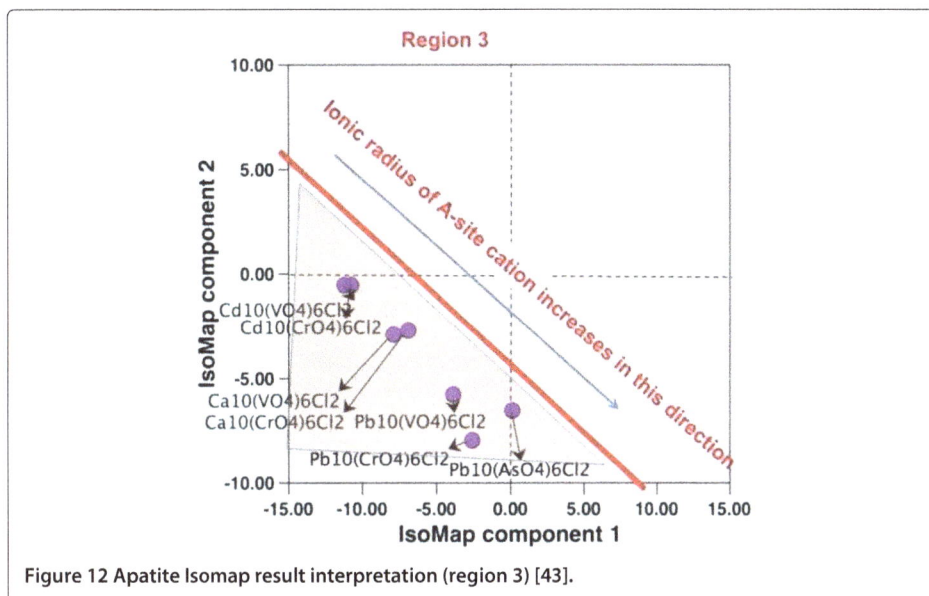

Figure 12 Apatite Isomap result interpretation (region 3) [43].

Region 3 also identifies the existence of complex relationship between A-site and B-site chemistries in Cl apatites.

Several topological observations can be made on the data. Firstly, since low-dimensional points obtained are different for both Isomap and PCA, it could be interpreted that the apatite data lie on a non-linear manifold in the embedding space. However, a counter argument can be made based on the fact that PC2-PC3 plot shows similar trends and clustering as that in Isomap1-Isomap2. One possible reason for this happening could be due to the existence of outliers dominating and deviating the first PCA component (PC1) while Isomap being unaffected by this outlier; in which case, the data could actually be lying on a linear manifold. Secondly, the different clustering phenomena observed along different dimensionality reduction techniques might imply that the pattern/features seen in PCA and Isomap clusters are a function of the distance preserved, while those in hLLE and LLE is a function of the topology preserved. Hence, these chosen features represented by these clusters happen to be preserved all along the dimensionality reduction process from the embedded space to the lower-dimensional space.

Conclusions

In this paper, we have detailed a mathematical framework of various data dimensionality reduction techniques for constructing reduced order models of complicated datasets and discussed the key questions involved in data selection. We introduced the basic principles behind data dimensionality reduction[d]. The techniques are packaged into a modular, computational scalable software framework with a graphical user interface - SETDiR. This interface helps to separate out the mathematics and computational aspects from the scientific applications, thus significantly enhancing utility of DR techniques to the scientific community. The applicability of this framework in constructing reduced order models of complicated materials dataset is illustrated with an example dataset of apatites. SETDiR was applied to a dataset of 25 apatites being described by

29 of its structural descriptors. The corresponding low-dimensional plots revealed previously unappreciated insights into the correlation between structural descriptors like ionic radius, bond covalence, etc., with properties such as apatite compound formability and crystal symmetry. The plots also uncovered that the shape of the surface on which the data lies could be non-linear. This information is crucial as it can promote the use of apatite materials as a potential host lattice for immobilizing toxic elements.

Availability of supporting data
Information regarding the source of the apatite data can be found in [44].

Endnotes
[a] A tree is a graph where each pair of vertices is connected exactly with one path. A spanning tree of a graph $G(V, E)$ is a sub-graph that traces all the vertices in the graph. A minimal spanning tree (MST) of a weighted graph $G(V, E, W)$ is a spanning tree with a minimal sum of the edge weights (length of the MST) along the tree. A geodesic minimal spanning tree (GMST) is an MST with edge weight representing geodesic distance. Computationally, GMST is computed using Prim's (greedy) algorithm [50].

[b] Normalization of a variable is forcing a limit of $[-1, 1]$ or $[0, 1]$ to an existing limit of $[a, b]$ of a variable by dividing the sequence of numbers with the maximum absolute value of the sequence.

[c] Hessian LLE is highly sensitive to neighborhood size and is much more sensitive to the input estimated dimensionality. Incorrect input of estimated dimensionality implies construction of tangent planes of incorrect dimensions which, in turn, implies sub-optimal low-dimensional representation.

[d] A comprehensive catalogue of non-linear dimensionality reduction techniques along with the mathematical prerequisites for understanding dimensionality reduction could be found in [23].

Competing interests
The authors declare that they have no competing interests.

Authors' contributions
SKS, BG, and JZ formulated the mathematical framework. SKS and JZ implemented the mathematical framework. SKS and PVB performed the model reduction on the apatite data to extract the low-dimensional representation. PVB and KR interpreted the results. SKS, PVB, BG, JZ, and KR discussed the results and wrote the paper. All authors read and approved the final manuscript.

Acknowledgements
We gratefully acknowledge the support from the National Science Foundation (NSF) grant CDI- NSF-CDI -PHY 09-41576. KR acknowledges the support from NSF: DMR- 13-07811 and DMS-11-25909, Department of Homeland Security/NSF-ARI Program: CMMI 09-389018; Army Research Office grant W911NF-10-0397, Air Force Office of Scientific Research SFA9550-12-1-0456, and the Wilkinson Professorship of Interdisciplinary Engineering. BG also acknowledges the support from NSF CAREER CMMI-11-49365.

Author details
[1] School of Mechanical Engineering, Georgia Tech, Atlanta, GA 30332-0405, USA. [2] Department of Materials Science, Drexel University, Philadelphia, PA 19104, USA. [3] Rutgers Discovery Informatics Institute, Rutgers University, Piscataway, NJ 08854, USA. [4] Department of Materials Science and Engineering, Iowa State University, Ames, IA 50011, USA. [5] Department of Mechanical Engineering, Iowa State University, Ames, IA 50011, USA.

References
1. Rabe KM, Phillips JC, Villars P, Brown ID (1992) Global multinary structural chemistry of stable quasicrystals, high-t_c ferroelectrics, and high-t_c superconductors. Phys Rev B 45:7650–7676
2. Morgan D, Rodgers J, Ceder G (2003) Automatic construction, implementation and assessment of pettifor maps. J Phys: Condens Matter 15(25):4361

3. Chawla N, Ganesh VV, Wunsch B (2004) Three-dimensional (3d) microstructure visualization and finite element modeling of the mechanical behavior of SiC particle reinforced aluminum composites. Scripta Materialia 51(2):161–165

4. Langer SA, Jr. Fuller ER, Carter WC (2001) OOF: an image-based finite-element analysis of material microstructures. Comput Sci Eng 3(3):15–23

5. Liu ZK, Chen LQ, Raghavan P, Du Q, Sofo JO, Langer SA, Wolverton C (2004) An integrated framework for multi-scale materials simulation and design. J Comput Aided Mater Des 11:183–199

6. van Rietbergen B, Weinans H, Huiskes R, Odgaard A (1995) A new method to determine trabecular bone elastic properties and loading using micromechanical finite-element models. J Biomech 28(1):69–81

7. Yue ZQ, Chen S, Tham LG (2003) Finite element modeling of geomaterials using digital image processing. Comput Geotechnics 30(5):375–397

8. McVeigh C, Liu WK (2008) Linking microstructure and properties through a predictive multiresolution continuum. Comput Methods Appl Mech Eng 197(4142):3268–3290

9. Zabaras N, Sundararaghavan V, Sankaran S (2006) An information-theoretic approach for obtaining property PDFs from macro specifications of microstructural variability. TMS Lett 3:1–2

10. Meredith JC, Smith AP, Karim A, Amis EJ (2000) Combinatorial materials science for polymer thin-film dewetting. Macromolecules 33(26):9747–9756

11. Takeuchi I, Lauterbach J, Fasolka MJ (2005) Combinatorial materials synthesis. Mater Today 8(10):18–26

12. Lumley JL (1967) The structure of inhomogeneous turbulent flows. Atmospheric turbulence and radio wave propagation 166–178

13. Tenenbaum JB, de Silva V, Langford JC (2000) A global geometric framework for nonlinear dimensionality reduction. Science 290(5500):2319–2323

14. Donoho DL, Grimes C (2003) Hessian eigenmaps: new locally linear embedding techniques for high-dimensional data. Proc Natl Acad Sci 100:5591–5596

15. Jain A, Ong SP, Hautier G, Chen W, Richards WD, Dacek S, Cholia S, Gunter D, Skinner D, Ceder G, Persson K (2013) The materials project: a materials genome approach to accelerating materials innovation. APL Mater 1(1):011002

16. Page YL (2006) Data mining in and around crystal structure databases. MRS Bulletin 31:991–994

17. Rajan K, Suh C, Mendez PF (2009) Principal component analysis and dimensional analysis as materials informatics tools to reduce dimensionality in materials science and engineering. Stat Anal Data Mining 1(6):361–371

18. Brasca R, Vergara LI, Passeggi MCG, Ferrona J (2007) Chemical changes of titanium and titanium dioxide under electron bombardment. Mat Res 10:283–288

19. Ganapathysubramanian B, Zabaras N (2008) A non-linear dimension reduction methodology for generating data-driven stochastic input models. J Comput Phys 227(13):6612–6637

20. Curtarolo S, Morgan D, Persson K, Rodgers J, Ceder G (2003) Predicting crystal structures with data mining of quantum calculations. Phys Rev Lett 91:135503

21. Fischer CC, Tibbetts KJ, Morgan D, Ceder G (2006) Predicting crystal structure by merging data mining with quantum mechanics. Nat Mater 5(8):641–646

22. Morgan D, Ceder G, Curtarolo S (2005) High-throughput and data mining with ab initio methods. Meas Sci Technol 16(1):296

23. Lee JA, Verleysen M (2007) Nonlinear dimensionality reduction. Springer

24. Van der Maaten LJP, Postma EO, Van Den Herik HJ (2009) Dimensionality reduction: a comparative review

25. Elliott JC (1994) Structure and chemistry of the apatites and other calcium orthophosphates, volume 4. Elsevier, Amsterdam

26. Mercier PHJ, Le Page Y, Whitfield PS, Mitchell LD, Davidson IJ, White TJ (2005) Geometrical parameterization of the crystal chemistry of P63/m apatites: comparison with experimental data and ab initio results. Acta Crystallogr Sect B: Structural Sci 61(6):635–655

27. Pramana SS, Klooster WT, White TJ (2008) A taxonomy of apatite frameworks for the crystal chemical design of fuel cell electrolytes. J Solid State Chem 181(8):1717–1722

28. White T, Ferraris C, Kim J, Madhavi S (2005) Apatite–an adaptive framework structure. Rev Mineralogy Geochem 57(1):307–401

29. White TJ, Dong ZL (2003) Structural derivation and crystal chemistry of apatites. Acta Crystallogr Sect B: Structural Sci 59(1):1–16

30. Samudrala S, Rajan K, Ganapathysubramanian B (2013) Data dimensionality reduction in materials science In: Informatics for materials science and engineering: data-driven discovery for accelerated experimentation and application. Elsevier Science

31. Bergman S (1950) The kernel function and conformal mapping. Am Math Soc

32. Roweis ST, Saul LK (2000) Nonlinear dimensionality reduction by locally linear embedding. Science 290(5500):2323–2326

33. Fontanini A, Olsen M, Ganapathysubramanian B (2011) Thermal comparison between ceiling diffusers and fabric ductwork diffusers for green buildings, Energy and Buildings 43(11):2973–2987. ISSN 0378–7788. http://dx.doi.org/10.1016/j.enbuild.2011.07.005

34. Amini H, Sollier E, Masaeli M, Xie Y, Ganapathysubramanian B, Stone HA, Di Carlo D (2013) Engineering fluid flow using sequenced microstructures. Nature Communications 4:2013

35. Guo Q (2013) Incorporating stochastic analysis in wind turbine design: data-driven random temporal-spatial parameterization and uncertainty quantication. Graduate Theses and Dissertations. Paper 13206. http://lib.dr.iastate.edu/etd/13206

36. Wodo O, Tirthapura S, Chaudhary S, Ganapathysubramanian B (2012) A novel graph based formulation for characterizing morphology with application to organic solar cells. Org Electron:1105–1113

37. Golub GH, Van Loan CF (1996) Matrix computations. The John Hopkins University Press

38. Floyd RW (1962) Algorithm 97: shortest path. Commun ACM 5(6):345

39. Bernstein M, De Silva V, Langford JC, Tenenbaum JB (2000) Graph approximations to geodesics on embedded manifolds. Technical report, Department of Psychology, Stanford University

40. Belkin M, Niyogi P (2003) Laplacian eigenmaps for dimensionality reduction and data representation. Neural Comput 15(6):1373–1396

41. Beardwood J, Halton JH, Hammersley JM (1959) The shortest path through many points. Math Proc Camb Philos Soc 55:299–327

42. Grassberger P, Procaccia I (1983) Measuring the strangeness of strange attractors. Phys D: Nonlinear Phenomena 9(12):189–208

43. Balachandran PV (2011) Statistical learning for chemical crystallography. PhD thesis, Iowa State University

44. Balachandran PV, Rajan K (2012) Structure maps for $A^I_4 A^{II}_6 (BO_4)_6 X_2$ apatite compounds via data mining. Acta Crystallogr Sect B 68(1):24–33

45. Shannon RD (1976) Revised effective ionic radii and systematic studies of interatomic distances in halides and chalcogenides. Acta Crystallographic Sect A: Crystal Phys Diffraction Theor Gen Crystallography 32(5):751–767

46. Pauling L (1960) The nature of the chemical bond and the structure of molecules and crystals: an introduction to modern structural chemistry, vol 18. Cornell University Press

47. Matsunaga K, Inamori H, Murata H (2008) Theoretical trend of ion exchange ability with divalent cations in hydroxyapatite. Phys Rev B 78:094101

48. Balachandran PV, Rajan K, Rondinelli JM (2014) Electronically driven structural transitions in $A_{10}(PO_4)_6 F_2$ apatites (A = Ca, Sr, Pb, Cd and Hg). Acta Crystallogr Sect B 70: 612–615

49. Flora NJ, Hamilton KW, Schaeffer RW, Yoder CH (2004) A comparative study of the synthesis of calcium, strontium, barium, cadmium, and lead apatites in aqueous solution. Synthesis Reactivity Inorganic Metal-organic Chem 34(3):503–521

50. Prim RC (1957) Shortest connection networks and some generalizations. Bell Syst Tech J 36(6):1389–1401

8

The development of phase-based property data using the CALPHAD method and infrastructure needs

Carelyn E Campbell[1*], Ursula R Kattner[1] and Zi-Kui Liu[2]

* Correspondence:
Carelyn.campbell@nist.gov
[1]Materials Science and Engineering
Division, National Institute of
Standards and Technology,
Gaithersburg, MD 20899, USA
Full list of author information is
available at the end of the article

Abstract

Initially, the CALPHAD (Calculation of Phase Diagrams) method was established as a
tool for treating thermodynamics and phase equilibria of multicomponent systems.
Since then the method has been successfully applied to diffusion mobilities in
multicomponent systems, creating the foundation for simulation of diffusion processes
in these systems. Recently, the CALPHAD method has been expanded to other
phase-based properties, including molar volumes and elastic constants, and has the
potential to treat electrical and thermal conductivity and even two-phase properties,
such as interfacial energies. Advances in the CALPHAD method or new information
on specific systems frequently require that already assessed systems be re-assessed.
Therefore, the next generation of CALPHAD necessitates data repositories so that
when new models are developed or new experimental and computational information
becomes available the relevant low-order (unary, binary, and ternary) systems can be
re-assessed efficiently to develop the new multicomponent descriptions. The present
work outlines data and infrastructure needs for efficient CALPHAD assessments and
updates, highlighting the requirement for data repositories with flexible data formats
that can be accessed by a variety of tools and that can evolve as data needs change.
Within these repositories, the data must be stored with the appropriate metadata to
enable the evaluation of the confidence of the stored data.

Keywords: CALPHAD; Thermodynamics; Diffusion; Property data; Data and file
repositories; Materials data infrastructure

Review

The first efforts using computational methods to describe Gibbs energy functions to
represent the phases and describe phase equilibria were made more than 60 years ago
as reviewed in [1]. However, only after computers became available did these efforts
become systematic. In 1970, Kaufman and Bernstein [2] presented a collection of analytical thermodynamic descriptions of the Gibbs energy of the phases of binary and
ternary systems as functions of temperature and concentration. These descriptions
could be used for the calculation of phase equilibria and thermochemical properties in
a large number of systems. This collection established the CALPHAD method as a
valuable tool for the treatment of multicomponent[a] phase equilibria and spawned the
development of several software packages and databases with collections of thermodynamic descriptions of multicomponent systems [1]. However, it soon became

obvious that more sophisticated models were needed to account for the structure (long and short range order) of these phases when describing the concentration dependence. With the development of more physics-based model descriptions [3] the reproducibility of the experimental information improved, but also required the revisions of systems that had been already described by Kaufman and Bernstein [2].

Kaufman and Bernstein also presented the concept of "lattice stabilities" to describe the thermochemical properties of the stoichiometric and often hypothetical end-member phases in solid solutions. Although it is straightforward to obtain functions for the stable forms of an element from their thermochemical properties, obtaining these quantities for end-members that are not stable proved to be a challenge. In the early days of CALPHAD these quantities were frequently estimated from the extrapolation of the properties of the solution phases to the non-stable[b] end-member phase. This practice resulted in different quantities being used by different authors for the same non-stable end-member form of an element or phase. This was a severe impediment to the development of descriptions for multicomponent systems since only one description of an element (phase) in a specific structure can be used, *e.g.*, the description of end-member A must be the same in all systems A-B, A-C, *etc.* This problem was greatly alleviated by the publication of the lattice stabilities of the elements by Dinsdale [4]. These lattice stabilities allowed the development of descriptions that were consistent with respect to the description of the elements, but also made the re-assessment of older descriptions necessary.

The approach used by the CALPHAD method lends itself to the modeling of other phase properties. It has been successfully employed for the description of diffusion mobilities [5,6], molar volumes [7], and elastic constants [8]. Since the descriptions of these properties are also based on the properties of the end-members and the models being used, these properties are also strongly affected by any change to the model.

The strengths of the CALPHAD method are that the data obtained from the calculation with a CALPHAD description are self-consistent and that these descriptions can be used for the extrapolation [9] of multicomponent systems. This makes the CALPHAD method attractive for the prediction of materials properties [10] and the coupling with materials simulation codes, such as solidification simulations [11] or phase field simulations [12]. With this ability to describe multicomponent materials properties as functions of composition and temperature, CALPHAD-based approaches have served as the foundation for several successful materials design programs [13,14].

For the construction of functional databases to describe multicomponent systems it is imperative that the same models are used for the description of the temperature, pressure and the concentration dependence of the Gibbs energy of each phase in all of the constitutive binary and ternary systems. Therefore, any modification of a constitutive subsystem has a compounding effect on the description of a multicomponent system because it affects every description of systems that includes this subsystem; *e.g.*, a change in the binary A-B system affects the description of the ternary systems A-B-C, A-B-D, *etc.* This compounding effect also makes re-assessments of higher component subsystems necessary. Figure 1 demonstrates this compounding effect illustrating the number of binary and ternary re-assessments needed if one unary description is changed in a six-component system. Since the inception of the CALPHAD method, models, functions for end-member phases (including the quantities of the pure elements) and the descriptions of the systems have been constantly evolving and re-assessments are continuously needed [15].

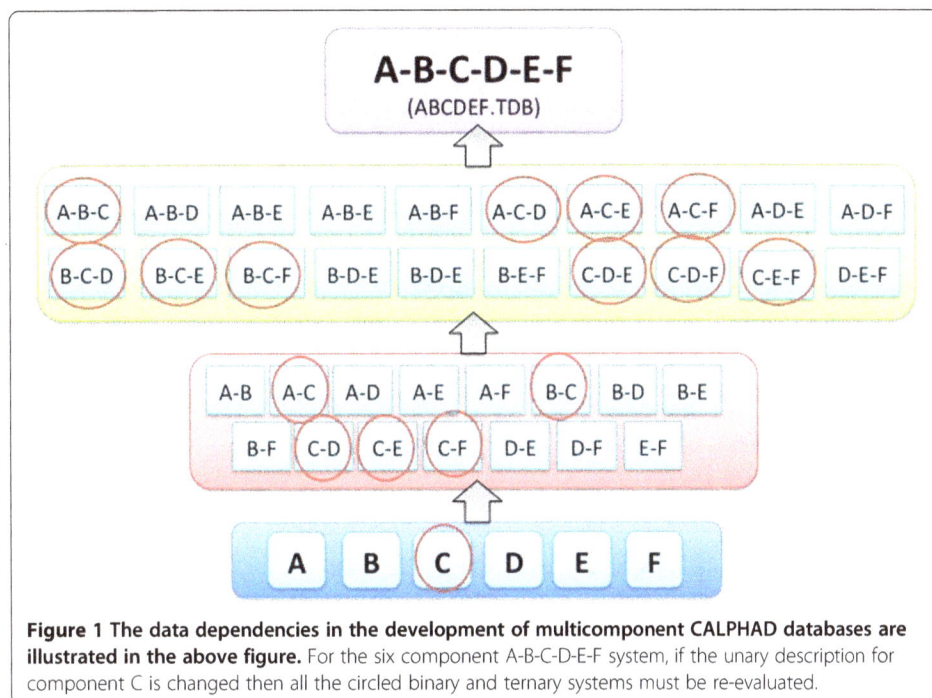

Figure 1 The data dependencies in the development of multicomponent CALPHAD databases are illustrated in the above figure. For the six component A-B-C-D-E-F system, if the unary description for component C is changed then all the circled binary and ternary systems must be re-evaluated.

These re-assessments are needed not only as the result of the availability of improved model descriptions or lattice stabilities for end-member phases, but also due to the availability of new experimental and theoretical data that improve the knowledge of a system. Although each re-assessment is progress, it is also a setback in the development of the description of multicomponent systems because of the previously mentioned compounding effect that delays the growth of these databases and the extension to the modeling of other phase properties using the CALPHAD method. As a result of this the CALPHAD community has become increasingly reluctant to adopt changes in models and end-member phase properties, even when it is known that updates need to be executed. This clearly demonstrates the urgent need for tools and data repositories to streamline the re-assessment process and make it more efficient.

The purpose of this article is to give a perspective of future development of the CALPHAD method within the concept of the Materials Genome Initiative [16] and to identify data repositories, data compilation and software needs.

Thermodynamics

Thermodynamics defines the state of a system when interacting with the surroundings, based on the first and second laws of thermodynamics. The first law of thermodynamics describes those interactions, and the second law of thermodynamics governs the evolution inside the system. Consequently, how the system is controlled from the surroundings dictates the state function used to describe the system. Under typical experimental conditions, temperature, pressure, and mass are the variables being controlled, thus the Gibbs energy is the characteristic state function widely used in materials science and engineering. The Gibbs energy of a system with more than one phase is the sum of the Gibbs energies of individual phases. The Gibbs energy of a stoichiometric phase, *i.e.*, a completely ordered phase, can be simply described by its Gibbs energy of formation from its components.

For a solution phase, the chemical activity of a component, i, is defined as $a_i = f_i x_i$ with f_i being the activity coefficient and x_i the mole fraction, and its value depends on the reference state where $a_i = 1$. A commonly used reference state is the stable element reference (SER) state, *i.e.*, its stable structure at 298 K and 10^5 Pa, with its chemical potential denoted by μ_i^{SER}, though it is sometimes more convenient to use a pure component in the same structure of the solution as the reference state of chemical activity, with its chemical potential denoted by $\mu_i^{o\alpha}$, where α represents a specific structure/phase. The Gibbs energy can thus be written using both reference states of chemical activities as follows.

$$G_m^a = \sum_i x_i^a \mu_i^a = \sum_i \left(x_i^a \mu_i^{SER} + RT_{x_i}^a \ln x_i^a \right) + RT \sum_i x_i^a \ln f_i^{aSER}$$
$$= \sum_i \left(x_i^a \mu_i^{oa} + RT_{x_i}^a \ln x_i^a \right) + RT \sum_i x_i^a \ln f_i^a \tag{1}$$

With.

$$^{ex}G_m^a = RT \sum_i x_i^a \ln f_i^a = RT \sum_i x_i^a \ln f_i^{aSER} - \sum_i x_i^a \left(\mu_i^{oa} - \mu_i^{SER} \right)$$
$$= {}^{ex}G_m^{SER} - \sum_i x_i^a \left(\mu_i^{oa} - \mu_i^{SER} \right) \tag{2}$$

where $^{ex}G_m^\alpha$ and $^{ex}G_m^{SER}$ are the excess Gibbs energy of mixing in J/mol, and $\mu_i^{oa} - \mu_i^{SER}$ is the lattice stability of component i in structure α with respect to the SER state, T is the temperature in K and R is the gas constant in J/(mol K). It is self-evident that the lattice stability is extremely important as it dictates the value of Gibbs energy of mixing. Any change in lattice stability will result in a different excess Gibbs energy of mixing to maintain the same Gibbs energy of the phase. The excess Gibbs energy of mixing is commonly modeled using the Muggianu extension of the Redlich-Kister formalism due to its symmetrical characteristics when applied to multicomponent systems [17,18], *i.e.*,

$$^{ex}G_m^a = \sum_i \sum_j x_i^a x_j^a \sum_{m=o}^n {}^m L_{ij} \left(x_i^a - x_j^a \right)^m + \sum_i \sum_j \sum_k x_i^a x_j^a x_k^a \left(x_i^a L_i + x_j^a L_j + x_k^a L_k \right) \tag{3}$$

where the summation of i, j, k is over all components, and $^m L_{ij}$, L_i, L_j, and L_k are the excess binary and ternary mixing terms and are temperature-dependent using a form similar to that shown in [4], *i.e.*,

$$^m L_{ij} = {}^m a_{ij} + {}^m b_{ij} T + {}^m c_{ij} T \ln T + {}^m d_{ij} T^2 + {}^m e_{ij} T^3 + {}^m f_{ij}/T \tag{4}$$

where the coefficients ($a,b,c,d,e,$ and f) are model parameters evaluated from thermochemical and phase equilibrium data, typically with $m \le 2$ and only $^m a_{ij}$ and $^m b_{ij}$ being used. Model parameters often need to be re-evaluated when new data become available, particularly data for metastable phases and metastable states. New data at low temperatures now accessible by first-principles calculations based on density functional theory (DFT) [19] also necessitate the revision of model parameters to extend and validate the Gibbs energy functions to 0 K. The Einstein and Debye models have been proposed to describe the heat capacity [20], and the implementation of these models into CALPHAD descriptions is being investigated by several groups [21-23]. Vřešťál *et al.* [24] developed an extension of the Dinsdale [4] Gibbs energy functions of the pure elements to 0 K employing an extended Einstein model. New model descriptions for the temperature and pressure dependence of the pure elements have been discussed at a recent workshop [25-29].

For solution phases with short-range and long-range ordering additional order-parameters are needed to fully define the state of the phase. Long-range ordering is commonly treated by sublattice models [30] with the Gibbs energy of mixing defined in the individual sublattices (This is discussed in detail in the Section "Crystallography"). The cluster site approximation [31] treats long-range and short-range ordering; however, sublattice models were shown capable of including both long-range and short-range ordering [32,33]. Magnetic ordering has been treated by adding a magnetic contribution to the total Gibbs energy. This magnetic contribution is based on the local magnetic moment theory of Curie and Weiss [34,35]. However, it has been recognized that current descriptions of the magnetic contributions to the Gibbs energy still cannot accurately describe all the observed properties [21,27,36,37], and a concept of mixture of individual magnetic microstates has been proposed [38-41].

Modeling of a liquid remains a challenge, particularly the liquid properties below the melting temperature. A two-state model [42] has been suggested to describe the temperature dependence of the Gibbs energy but this model has been demonstrated for only a few cases [36,43] and is partially supported by the recent DFT-based molecular dynamic simulations [44]. Short-range ordering in the liquid phase can be described using a quasi-chemical model [45], the associate model [46,47] or the liquid ionic sublattice model [48].

Currently, CALPHAD modeling is based on the Gibbs energy, which describes a thermodynamic system at constant temperature and pressure, as most experimental data are collected under these conditions. However, describing a thermodynamic system at constant temperature and volume using the Helmholtz energy may be useful for direct implementation of data from first-principles calculations and coupling with diffusion simulations of systems with large molar volume differences. Although, Lu and Chen [49] have investigated the implementation of the Helmholtz energy within the CALPHAD approach for fcc-Cu, this approach has yet to be tested for multicomponent, multiphase systems.

Crystallography

The basis for any CALPHAD description is the model selection. The models for temperature and pressure dependence are usually selected for the entire system while the model for the concentration dependence is specific for each phase. The modeling of ordered phases with homogeneity ranges or interstitial solutions are usually described with the sublattice model. In the early days of the CALPHAD method, the specifics of the sublattice model were chosen in part out of convenience to describe the appearance of a phase in the phase diagram without much consideration of the physical properties of the phase. However, for accurate description of the phase properties and reliable extrapolation to multicomponent systems it is imperative that the crystal structure of the phase is considered when the model is chosen. The choice of model has not only implications on the thermodynamic modeling, but also on the modeling of diffusion, molar volume and other phase-based properties.

The basic premise of the sublattice model is that a sublattice is assigned for each distinct site in the crystal structure. Each of the sublattices can be occupied by one or more species, which are usually elements, ions or a vacancy forming regular type solutions. End-members of a phase represent the phase with only one species on each sublattice. In a full thermodynamic description of a multicomponent phase the number of parameters to describe the end-members and regular type interactions could become excessively large. For example, if all the elements in the system were allowed to occupy each sublattice in a phase, the number of end-member phases would be n^k where n is the number of elements and k is the number of sublattices. One option is to generate these parameters from first-principles calculations, and the other option is to reduce the number of parameters by introducing constraints or decreasing the number of sublattices. Constraints are usually based on nearest neighbor and, maybe, second nearest neighbor configurations while similarities, such as co-ordination number, point symmetry *etc.*, and the changes in site fraction upon deviation from the ideal stoichiometry are used as criterion for combining different sublattices. Therefore, the use of crystallographic information is crucial for selecting the proper model descriptions.

The selection of the model to describe the concentration dependence of the Gibbs energy function of a phase is typically made during the assessment of a binary or ternary system. Consideration of the phase properties only in the system being assessed may result in a model description that is not suitable to describe the phase correctly in a multicomponent system. For example, the substitutional behavior on two or more distinct but similar sites could be very similar in the A-B system while in the A-C or A-B-C system the substitutional behavior may be quite different. If in the assessment of the A-B system the model is simplified by combining the similar sites into one sublattice, this model would not be capable of describing the substitutional behavior in the A-C or A-B-C systems. Therefore, it is essential to examine all the mechanisms of deviation from stoichiometry of a phase in all systems where this phase occurs before any combinations of sublattices are used for its model description. Therefore, site occupation data are also important data for the refinement of the model parameters.

The crystallographic information needed to correctly determine a phase can be found in classic compilations, such as Pearson's Handbook [50], or in electronic databases, such as ICSD [51], Pauling File [52] or Pearson's Crystal Data [53]. The advantage of electronic databases is that they are easily searchable and allow with relatively little effort the extraction of phase information for all elemental combinations for which data for this phase are available. The electronic databases also offer the advantage that they can be combined with other software that could be developed to assist in model selection. Since the model of the crystal structure is critical to the development of functional databases for multicomponent systems, unique identifiers for the crystal structure that are linked to the description of the individual phases are needed. Such an identifier could be the space group and Wyckoff sequence as recommended by Brown *et al.* [54] and adopted by the International Union of Pure and Applied Chemistry (IUPAC).

Phase diagram data

Important phase diagram data for the development and verification of a CALPHAD type description are phase boundary data. These data are either determined from isothermal measurements or scans of phase changes with heating and cooling experiments. In most of these experimental measurements the pressure is kept constant. Isothermal measurements are usually carried out on samples that have been heat treated and quenched; only in rare cases is an isothermal measurement made *in situ*. These measurements can be either quantitative phase analysis, such as electron probe microanalysis (EPMA) or energy dispersive x-ray analysis (EDX), or semi-quantitative observations of phase changes with slightly different compositions in metallographic or x-ray diffraction samples. Scans with temperature can be simple thermal analysis (TA), differential thermal analysis (DTA) or differential scanning calorimetry (DSC). With these latter methods the signal caused by the enthalpy change in the material resulting from the phase change is recorded. Zhao [55] provides an overview of the experimental methods available for the determination of phase diagrams.

For an assessment of the thermodynamic description using an optimization procedure the experimental error associated with these data needs to be input. In addition to the experimental error itself other factors that affect accuracy must also be considered, such as material purity and heat treatment history of the sample. Therefore, it is imperative that a repository of experimental phase diagram data contains all this information. Unfortunately not all experimental work reported in the literature contains these data in sufficient detail and some of these accuracies need to be estimated during an assessment. When auxiliary estimates on data accuracy have been made in an assessment this also needs to be documented.

Thermochemical data

The availability of thermochemical data is extremely valuable for the development of a thermodynamic description of a system. However, in general the only quantities that can be directly measured are enthalpies and chemical potentials while all other quantities, such as heat capacity or entropies, are derived from these measurements. The majority of enthalpies are determined with a variety of calorimetric methods [56,57]. The chemical potentials are usually determined from electromotive force (emf) [58] or vapor pressure measurements [59,60]. Depending on the specific method the values are obtained either through direct or indirect measurements. In cases where the values stem from indirect measurement methods, the values may need to be used in their converted form; however, in all other cases the directly measured value should be used. Unfortunately, the originally measured data are not always reported and the information given may be inadequate to reconstruct the originally measured data. The reported values may carry errors that have accumulated from measurement to measurement, and also may be affected by the choice of reference, which is used to correct for temperature differences at different stages of the experiment. Even though it is desirable to use only data from direct measurements these data may not be available. For particular systems, the melting points of the components or other thermochemical properties of the system may limit the use of some direct measurement techniques. Thus, when using indirectly obtained data, the way these data are derived must be documented and taken into account when the accuracy of these data is evaluated.

A data repository of experimental phase diagram data and thermochemical data and all the relevant metadata, including documentation on the error estimation, is needed to improve the accuracy, efficiency, and reproducibility of the CALPHAD assessments.

Diffusion

CALPHAD-based diffusion mobility databases have been developed based on the phenomenological theory of diffusion for multicomponent systems developed by Onsager [5,6]. Assuming irreversible thermodynamics and an exchange mechanism in a crystalline material, the diffusional flux, J_k, of a component is assumed to be proportional to a thermodynamic force, for an isothermal isobaric system,

$$J_k = -\sum_{i}^{n-1} D_{ki} \frac{\partial c_i}{\partial z} \tag{5}$$

where D_{ki} is the diffusivity (m^2/s), c_i is the concentration of component i (Note: where x_i is the mole fraction of component i, and V_m is the molar volume of the phase), and z is the diffusion distance. Rewriting Eq 5 in terms of the more convenient quantities, the chemical potentials μ_i, and using the chain rule, the flux and diffusivity can be expressed as

$$J_k = \sum_{i}^{n-1} L'_{ki} \frac{\partial \mu_i}{\partial c_j} \frac{\partial c_j}{\partial z} \quad \text{and} \quad D_{kj} = -\sum_{i}^{n-1} L'_{ki} \frac{\partial \mu_j}{\partial c_j} \tag{6}$$

where L'_{ik} is a proportionality factor which is dependent of the mobility of individual components, and $\partial \mu_i / \partial c_j$ are the thermodynamic factors. Thus, the diffusivities are defined by a pure thermodynamic factor and pure kinetic factor (L'_{ik}). The kinetic factor, L'_{ik}, is defined by the atomic mobilities as follows:

$$L'_{kj} = \sum_{i=1}^{n} (\delta_{ik} - c_k V_i) c_i y_{Va} M_i \tag{7}$$

Where δ_{ik} represents the Kronecker delta and $=1$ when $i = k$ and $= 0$ when $i \neq k$, and V_i is the partial volume of element i, and y_{Va} is the fraction of vacant lattice sites on the sublattice where i is dissolved. The atomic mobility, M_i, is defined by an Arrhenius function:

$$M_i = M_i^o \exp\left(\frac{-Q_i}{RT}\right) \frac{1}{RT} \tag{8}$$

where Q_i and M_i^o, the activation energy and frequency factor, respectively, are expanded using Redlich-Kister polynomials to express the composition and temperature dependence (similar to the excess Gibbs energy in Eq 3.) of the mobility and are the functions stored in the diffusion mobility databases [5,6]. It should be noted by assuming the mobility matrix is diagonal in a lattice fixed frame of reference, as followed in the Darken approximation, the coupling between fluxes, otherwise known as the vacancy wind (Manning-effect) is neglected. For additional information on the inclusion of these factors see the following work [61,62].

When radioactive tracers are used in a homogeneous material, the concentration gradient approaches zero and tracer diffusivity is defined as a function of the kinetic factors, which are proportional to the atomic mobility. Thus, the tracer diffusivity is independent of a concentration gradient, $D_i^* = RTM_i$. While the temperature dependence of the diffusivity of many species can be described by an Arrhenius relation, there are some materials that have anomalous behavior at low temperatures, such as β-Ti and β-Zr. These diffusivities could be described using a two-exponential fit [63].

Tracer diffusivity data from single crystal material are the preferred experimental data on which to build a diffusion mobility database, as these data are not dependent on a thermodynamic factor or influenced by the Gibbs energy. However, for a given system, it is often difficult to find enough tracer diffusivity data to complete a diffusion mobility description (some data sources of diffusivity data, mostly tracer and impurity data, found in [64-66]). Thus, chemical and intrinsic diffusivity data are used to complete the diffusion assessments. Using these data requires a thermodynamic database to calculate the concentration gradients and to determine the diffusion mobilities [67-69]. In the volume-fixed frame of reference for substitutional elements the chemical diffusivity, $^VD_{kj}$, is defined as

$$^VD_{kj} = \sum_{i=1}^{n}(\delta_{ik}-x_k)x_iM_i\frac{\partial\mu_i}{\partial x_j}V_m \tag{9}$$

The derivatives of the chemical potential can be calculated using a CALPHAD-based description of the thermodynamics. The intrinsic diffusivity, $^LD_{kj}$, is defined in the lattice-fixed frame of reference, where the sum of the diffusion fluxes equals the vacancy flux, as:

$$^LD_{kj} = \sum_{i=1}^{n}\delta_{xi}x_iM_k\frac{\partial\mu_k}{\partial\mu_j} \tag{10}$$

Similar to the thermodynamic and phase diagram data, it is essential when reporting diffusivity data to include not only the measurement error, but also the type of material used, the initial composition/s and relative errors, and the methods used to analyze the data.

Just as the CALPHAD thermodynamic databases require a consistent set of reference lattice stabilities, the diffusion mobility databases also require a consistent set of reference tracer-diffusion mobilities for the pure elements. However, unlike the lattice stabilities, which have been established, the reference tracer-diffusion mobilities have yet to be established. A reference tracer-diffusion mobility database for the pure elements is in the process of being developed in conjunction with NIST and a group of international researchers [70]. The reference mobilities are determined by reviewing all of the available experimental data and assessments for a given element. When multiple assessments exist for the same element, weight mean statistics are used to determine which assessment best represents the critically reviewed experimental data [71]. The reference mobility database also requires values for the non-stable end-members. These quantities may be estimated using diffusion correlations [72,73] or using data from first-principle calculations [19,74-76] or atomistic simulations [77], when possible.

The choice of sublattice model is also important in the development of robust diffusion mobility databases. Even if only tracer diffusivity data are used to developed the diffusion mobility description, eventually the diffusion and thermodynamic descriptions will be used together to predict diffusional controlled phase transformations and thus the same phase model must be used by both the diffusion and thermodynamic database. The choice of phase model must accurately represent the crystal structure and the active diffusion mechanism. For example, in modeling the austenite (fcc) phase for a steel, carbon should not be placed on the substitutional sublattice, but rather on the interstitial sublattice.

Physical phase-based properties related to the thermodynamics

Molar volume and thermal expansion

Recently, the CALPHAD approach has been applied to more phase-based properties, such as molar volume and elastic moduli, and databases of assessments of functions for these properties are in the process of being developed. Again essential to the development of these CALPHAD-based descriptions are reference experimental data and a reference database for the pure end-member quantities in various phases. Molar volumes for the stable crystalline phases and compounds can be determined using high accuracy lattice parameter measurements, such as x-ray or neutron diffraction. For liquids, molar volumes may be measured directly, determined from measured volume changes occurring during melting or solidification, or derived from changes in the melting temperature as a function temperature. Databases can be developed assuming pressure independent volumes below 1 GPa [78]. Above pressures of 1 GPa, the pressure dependent volumes should be included by using the appropriate equation of state (EOS) [79] (This will be discussed further in the next section). Hallstedt et al. [78] recommended that metastable end points can be extrapolated from binary lattice parameter data using Vegard's law, assuming that a linear composition dependence holds. In using this process as many binary systems as possible should be considered to determine an average extrapolated metastable end-member quantity. Low solubility in a given phase may limit the application of this extrapolation method. When this occurs, using first-principles calculations to estimate the lattice parameter is the preferred method.

The temperature dependence of the molar volume (thermal expansion) and heat capacity are related in both the Mie-Grüneisen and Einstein models. The molar volume, is defined as function of linear expansion (CLE), α, as the following:

$$V_{\mathrm{m}} = V_{\mathrm{o}} \exp\left(\int_{T_{\mathrm{o}}}^{T} 3\alpha dT\right) + \Delta V_{\mathrm{m}}^{\mathrm{magn}}(T) \tag{11}$$

Where the first term represents the nonmagnetic molar volume and the second term, ΔV_m^{magn}, is the magnetic contribution to the molar volume. The variable is the molar volume at the reference temperature, T_0, usually 298.15 K. Lu et al. [7,80] applied the CALPHAD method to expand the nonmagnetic volumetric expansivity by defining:

$$3\alpha = a + bT + cT^2 + dT^{-2} \tag{12}$$

where the constants a, b, c, and d are evaluated using available data from lattice parameter expansion data determined using diffraction methods or measured length changes determined using dilatometric and interferometric methods. (It should be noted that high temperature diffraction data are preferred as vacancy contributions can be minimized.) The magnetic contributions to the molar volume are treated by Fernández Guillermet [81]. Based on this formalism Lu *et al.* developed descriptions for a number of fcc, bcc and hcp elements [7]. Lu *et al.* [7] acknowledged that the current modeling efforts do not describe the Invar effect well in Fe-Ni alloys and additional modeling efforts are needed [41]. Lu *et al.* [7] demonstrated that the Debye-Grüneisen model is useful to approximate thermal expansion coefficient for metastable and unstable phases and systems where experimental data are lacking. Kim *et al.* [82] demonstrate the use of first-principles to develop multicomponent composition thermal expansion coefficients descriptions for Ni-Al based systems.

Alternatively, Hallstedt *et al.* [78] noted that the thermal expansion contribution to the volume is generally small and thus, a simple polynomial may be sufficient at this time.

$$V_\mathrm{m} = V_\mathrm{o} \left[1 + \frac{B}{V_\mathrm{o}} \left(T - T_\mathrm{o} \right) + \frac{C}{V_\mathrm{o}} \left(T^2 - T_\mathrm{o}^2 \right) \right] \tag{13}$$

Assuming.

$$\alpha = \frac{1}{V} \frac{\mathrm{d}V}{\mathrm{d}T} \approx \frac{1}{V} \left(B + 2CT \right) \tag{14}$$

For metastable phases, Hallstedt *et al.* [78] suggested using the same thermal expansion descriptions that correspond to the stable phases. If thermal expansion data for a liquid phase are not available, it has been recommended to assume that the thermal expansion is proportional to the volume of the liquid using the same proportionality constant that is defined for the solid phase. The additivity rule can be used as approximation for intermediate phases with missing thermal expansion data [78]. Hallstedt demonstrated the feasibility of this approach for Al, Mg, and Si [83].

Recently, Zhang *et al.* [84] demonstrated yet another approach for modeling the thermal expansion. This approach was used to evaluate 42 pure metallic elements and was based on Debye-Grüneisen model. The approached showed potential to for use to extrapolate thermal expansion coefficients when sufficient temperature data is unavailable.

Elastic properties

To apply a CALPHAD approach to modeling the elastic properties requires using an EOS, which can be expressed in the form $V = V(T,P)$ and can be integrated, and data on the bulk moduli. The data needed to describe the EOS include the volume, $V_\mathrm{o}(T)$, compressibility, $K_\mathrm{o}(T)$, and the pressure dependence of the compressibility at a reference temperature [78]. Most of the bulk moduli for the elements and many compounds are known. When experimental data are not available first-principles calculations, using volume functions fitted to an appropriate EOS, can be employed [85,86]. The temperature dependence can be obtained through the thermal expansion assuming that the temperature dependence of elastic stiffness coefficients mainly results from the volume change as a function of temperature. This approach has been demonstrated by Liu *et al.* [8] for the Mg-Al system.

As noted previously to describe the volume at pressures above 1 GPa, the pressure dependence must be included in the EOS [78]. Lu *et al.* [79] developed such an EOS based on the work of Jacobs and Oonk [87], which extrapolates well to at least 6000 K and 200 GPa.

$$V_m(T,P) = V_m(T,P_o) + c\ln\left(\frac{k_o(T,P)}{k(T,P_o)}\right) \tag{15}$$

Where $c = c(T)$ is defined by the pressure data, κ is the isothermal compressibility, and κ_o is the isothermal compressibility at P_o. The Gibbs energy is then expressed as.

$$G(T,P) = G(T) + \frac{c(T)}{k(T,P_o)}\left(\exp\left(\frac{V_m(T,P_o)-V_m(T,P)}{c(T)}\right)-1\right). \tag{16}$$

The temperature dependence of $V_m(T,P_o)$ is given by Eq. 13. Other EOS have also been considered [88-90]; however, all of these EOS have had some compatibility issues with the CALPHAD method that results in the incorrect prediction of high pressure thermophysical properties. Brosh *et al.* [91] proposed another EOS which attempts to address some of these capability issues; however, an EOS which satisfactorily accounts for the compatibility issues with CALPHAD has yet to be identified. Hammerschmidt *et al.* [28] give a summary of the calculation of the elastic constants and point out that the magnetic properties of a phase also need to be taken into account for their treatment.

Physical base properties independent of thermodynamics

The phase-based properties that have been addressed thus far are either thermodynamic quantities or use the thermodynamics to fully describe the property. However, the CALPHAD method has been demonstrated to apply to other properties.

Electrical resistivity and thermal conductivity

The electrical resistivity of a phase is composed of contributions from scattering of conducting electrons and static lattice defects. The electrical conductivity, the inverse of the electrical resistivity, ρ, is related to the thermal conductivity via Wiedemann-Franz law, which states:

$$\lambda = {}^{LT}/\rho \tag{17}$$

where λ is the thermal conductivity, L is the Lorentz number and T is the absolute temperature (K). However it should be noted when the heat is conducted by phonons, as in an insulating phase, the assumptions made in the Wiedemann-Franz law no longer apply. The temperature of these quantities can be described by linking to the heat capacity as demonstrated, by Grimvall [92], or by using a simple polynomial as shown in the work by Terada *et al.* [93].

Using a CALPHAD approach the composition dependence of these quantities can be expressed. For example, the electrical resistivity in a metallic binary solid solution can be expressed as a sum of the pure component contributions and a mixing term, as proposed by Nordheim for a metallic binary solid regular solution [94]:

$$\rho_\alpha = (1-x)\rho_A + x\rho_B + kx(1-x) \tag{18}$$

where ρ_α is the electrical resistivity of phase α, ρ_A and ρ_B are the electrical resistivities of components A and B in the α phase, and k is an excess mixing term.

Terada *et al.* [95,93] have reviewed the thermal conductivity in a wide range of inter-metallic compounds and have found that the Wiedemann-Franz law applies for not only stoichiometric intermetallic compounds, but also for off-stoichiometric compositions, such as NiAl and CoAl. Using the work of Terada *et al.* [93] and others [96] should provide a starting basis for thermal conductivity and electrical resistivity CALPHAD-based databases.

For insulating materials, such as oxides and semiconductors, Gheribi and Chartrand [97] developed a model based on the Debye model to predict thermal conductivity. This CALPHAD-based model, which simultaneously optimizes the thermal expansion, heat capacity and bulk modulus, also correctly predicts the thermal conductivity as a function of temperature.

In addition to the phase-based properties described in this work, the CALPHAD approach could be used to describe the composition and temperature of optical, thermo-electric, acoustic properties of materials.

Interphase properties

The ability to simulate microstructure evolution requires not only input of many phase-based properties (thermodynamic, diffusion, volume), but also many interphase dependent properties. The physical properties of molten alloys are dependent on the surface tension (*i.e.*, for solder and solidification applications). Using Bulter's equation, which defines an equilibrium between the bulk and monolayer surface, Picha *et al.* [98] and Tanaka *et al.* [99] have developed CALPHAD-based descriptions for various low melting binary metal systems and molten salt systems. The interfacial energy between two solid phases is an essential input for simulating precipitation and coarsening processes; however, these values are often difficult to measure and calculate. Establishing a database of available measured interfacial energies and methods to extrapolate these quantities is critical for the use of many precipitation codes. Currently, composition independent interfacial energies between phases are used within most precipitation and coarsening simulation tools [100-102]; however, if sufficient data were collected and stored composition-dependent interfacial energies CALPHAD databases might be developed. Efforts to estimate the composition dependent interfacial energies for a given set of solid phases were reviewed by Costae Silva *et al.* [103], who noted that while thermodynamic approaches can provide reasonable estimates, fundamental first-principles approaches [104] should be used as the reference values. More recently, Shi and Luo [105] developed a method to estimate the interfacial energy associated with a grain boundary between two phases. Combining these and other methods with data available from coarsening experiments [106,107] and single-sensor DTA nucleation cooling experiments [101] a CALPHAD based databases could be established, where the interfacial energies are defined by two phases.

Grain boundary diffusion models, thermal migration models [108], and nucleation models are also used to predict composition dependent microstructure evolution and require additional material dependent input, such as the grain boundary diffusion coefficients, dislocation densities, and nucleation kinetics. Eventually composition and temperature dependent databases for these quantities will also be required. Thus, experimental databases documenting these inputs should also be established; however, the specific format for these data is yet to be determined.

Data repositories and tools

Repositories

Based on the evolution of CALPHAD modeling and multicomponent database development, it is now clear that data repositories are essential for improving the efficiency of future modeling and assessment work. Without these repositories, the effort needed to improve current descriptions will continue to be burdensome and continue to slow the implementation of new models. As CALPHAD modeling expands to other phase-based properties there will be even greater need for these data repositories. For these repositories to be effective, these data need to be well documented. This documentation needs to be sufficient such that the user does not need to obtain the original work to determine the error associated with the data for the application of interest.

Although there are currently many various data compilations (such as [109,110]), it is rare that these data can be easily manipulated into the desired format. Some of these data compilations, especially those in printed form, are rarely updated. The advantages of electronic compilations include the possibility of easier data manipulation and the ability to easily update with new information.

NIST has been charged with developing materials data infrastructure and data repositories to support the Materials Genome Initiative (MGI). NIST has initiated several different efforts in the areas of data capture, data dissemination, and data processing to develop this infrastructure, as illustrated in Figure 2. As a first step, files used in the assessment of the description of a given system should be collected and stored electronically in a public repository, such as that initiated by NIST and Kent State [111] (http://nist.matdl.org/). Figure 3 is screen shot of this repository, which

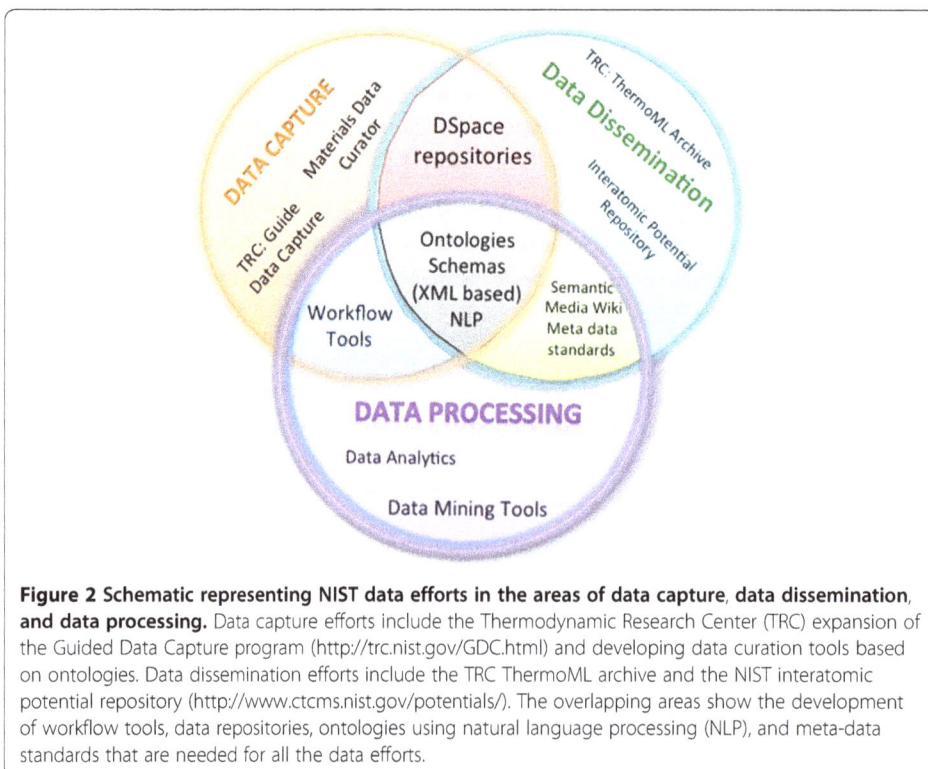

Figure 2 Schematic representing NIST data efforts in the areas of data capture, data dissemination, and data processing. Data capture efforts include the Thermodynamic Research Center (TRC) expansion of the Guided Data Capture program (http://trc.nist.gov/GDC.html) and developing data curation tools based on ontologies. Data dissemination efforts include the TRC ThermoML archive and the NIST interatomic potential repository (http://www.ctcms.nist.gov/potentials/). The overlapping areas show the development of workflow tools, data repositories, ontologies using natural language processing (NLP), and meta-data standards that are needed for all the data efforts.

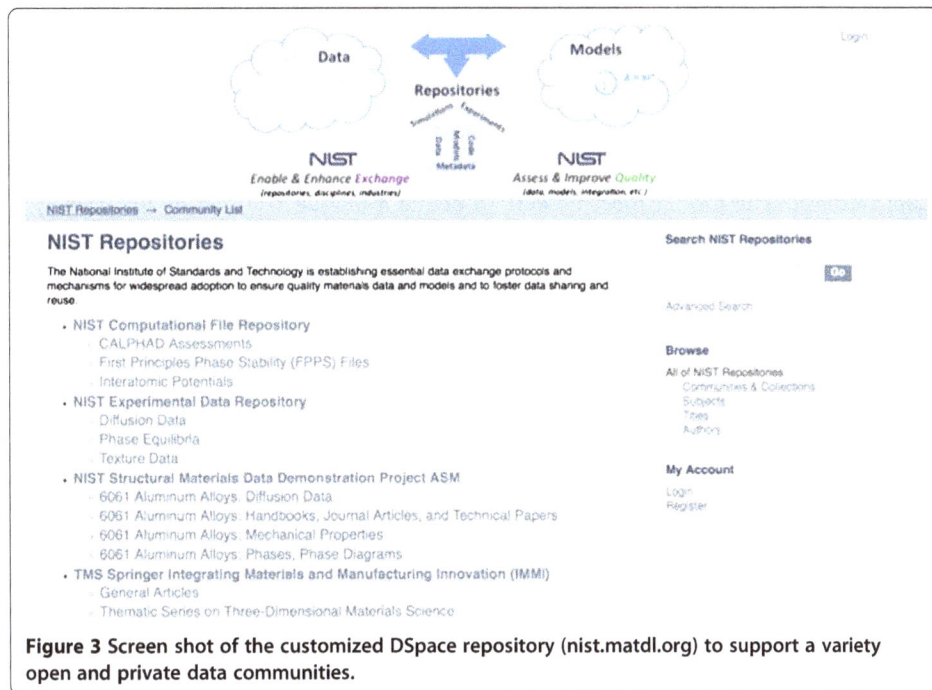

Figure 3 Screen shot of the customized DSpace repository (nist.matdl.org) to support a variety open and private data communities.

is a customized version of the open source software DSpace (http://www.dspace.org/), showing the development of different materials-based data communities. Authors are asked to contribute the files from their work. These files can then be linked to the originally published work. It is advantageous that the files used to assess the description can be directly associated to the files containing the description. Figure 4 illustrates the relationship the evaluated data files and the files containing the functional descriptions. Although these files are software specific, translation tools can be used to make the data available to other software codes.

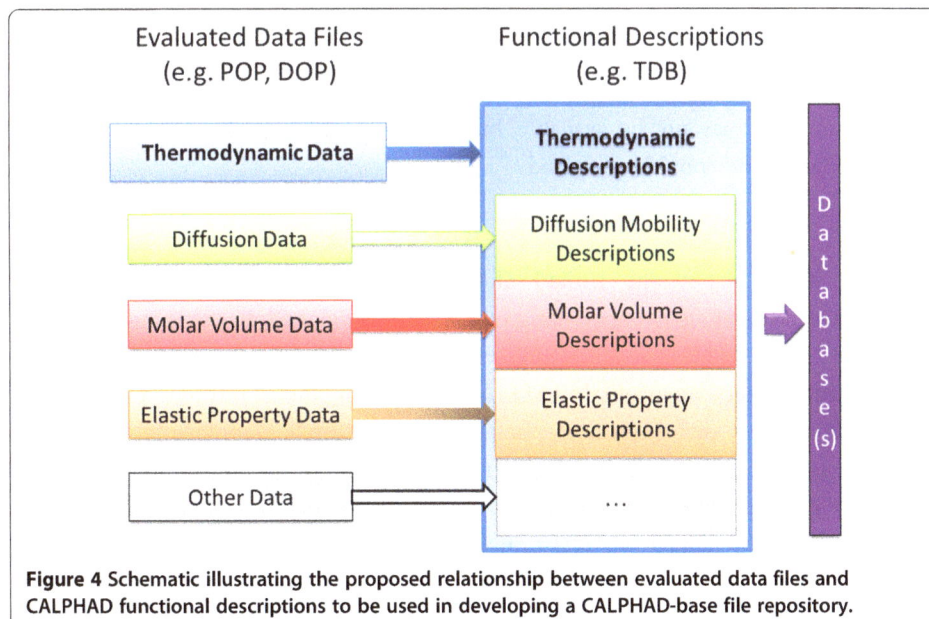

Figure 4 Schematic illustrating the proposed relationship between evaluated data files and CALPHAD functional descriptions to be used in developing a CALPHAD-base file repository.

Data may be collected in repositories either as single value points or as a data series. Most data can be stored as single value points; however, some data must be stored as a data series. For example, a diffusion profile is a data series where the individual data points must be stored together to be meaningful. The format for the repository, while not specific, must be flexible enough to accommodate extensions while maintaining easy data retrieval. The format must be able to define a unique material in an unambiguous way such that the same material identifier can be used by other databases and tools searching for information on that specific material. In addition, phase names must also be unique and unambiguous. Examples of currently used identifiers for chemicals include the Chemical Abstracts Registry (CAS) number [112] and International Chemical Identifier (InChI) [113]. An example for defining phase names is given by Schmid-Fetzer *et al.* [114]. One possibility for a flexible data format is the use of various extensible markup languages. Some of the languages to be consider for use include MatML specifically designed for the interchange of materials data [115] and ThermoML for thermochemical and thermophysical property data [116-118]. Other choices to consider are the HDF5 (Hierarchical Data Format) format used by the Materials Data Environment [119] or data-interchange formats such as, JSON (Java Script Object Notation) [120] or BSON (Binary Script Object Notation) [121]. Regardless of the choice of data format, the data stored should ideally include descriptions of the appropriate metadata. These metadata would include information describing how the individual datum was obtained. These descriptions could include information on material purity, alloy preparation, *etc.*, for experimental data and information on the method and input used, *etc.*, for calculated data. For phase data it is essential that the data are linked to the crystal structure of the phase. Following the recommendations of the CCN (Commission on Crystallographic Nomenclature), the crystal structure should be inputted using the format prescribed by Brown *et al.* [54]. A preliminary summary of the information needed for data typically used in CALPHAD assessments is shown in Table 1. If all entries in Table 1 are provided, this would represent an ideal data description case. In reality, it is accepted that information to all entries in Table 1 may not be known. For example, a diffusion couple composition profile may be entered for a given system: ideally, the profile should be entered with a corresponding micrograph of the microstructure and the compositions should be identified as single phase or multiphase. However, this information may not be available, but the composition profile still provides valuable information and should be entered.

ESPEI

In a project supported by the United States Automotive Materials Partnership (USAMP) of Department of Energy, Shang *et al.* [122] developed an relational infrastructure, ESPEI (Extensible Self-optimizing Phase Equilibrium Infrastructure), based on SQL (structured query language). In ESPEI, all data are organized through the phase identifiers with one for each individual phase-based on crystallographic information. There are currently about 300 phase identifiers in ESPEI covering known symmetries designated by space groups. The input data are tabulated for each phase in three groups: thermochemical data from first-principles calculations, thermochemical data from experiments, and phase equilibrium data. Each data point is divided into multiple entries with one for each element in the phase, which enables the extension to

Table 1 General summary of information needed to define data used in CALPHAD assessments

Data type entries	Descriptors for entries	Additional Sub-descriptors
Elements present		
Type of value (e.g. enthalpy, heat of formation, phase boundary, diffusivity, lattice parameter, bulk moduli)	Experimental or computational method	
	Type of measurement (direct or indirect)	
Number of phases present		
Datum value and error	Type (single value or series)	
	Units	
	Actual value (s) and error (s)	
Phase Information for each phase	Phase name	
	Composition and fraction and errors	
	Crystal structure (this input will follow the format prescribed by the CCN) or amorphous	
	Lattice parameter	
Temperature and error	Units	
Pressure and error	Units	
Metadata		
Type of material	Bulk composition	
	Material purity	
Sample preparation	Sample preparation	
	Microstructure information	Single Crystal
		Polycrystalline (grain size, dislocation density)
		Non-crystalline
Data manipulation details (if any, e.g. reference state corrections, analysis method to determine interdiffusion coefficient)		
Reporting format (raw data, digitized data, other)		
Reference (DOI or text ; one must be present)		
Additional information		

multi-component systems. In addition to the value and uncertainty, each entry details the independent variables and their uncertainties such as temperature, pressure, composition, magnetic data, and other miscellaneous information. The ESPEI SQL data are searchable by elements. An automation procedure was developed to model Gibbs energy functions of individual phases in binary and ternary systems using the three groups of data mentioned previously.

The data flow in ESPEI is illustrated in Figure 5. ESPEI heavily depends on thermochemical data from DFT-based first-principles calculations for both stoichiometric and solution phases at finite temperatures [19] due to the lack of experimental measurements of thermochemical properties of individual phases. Even though the

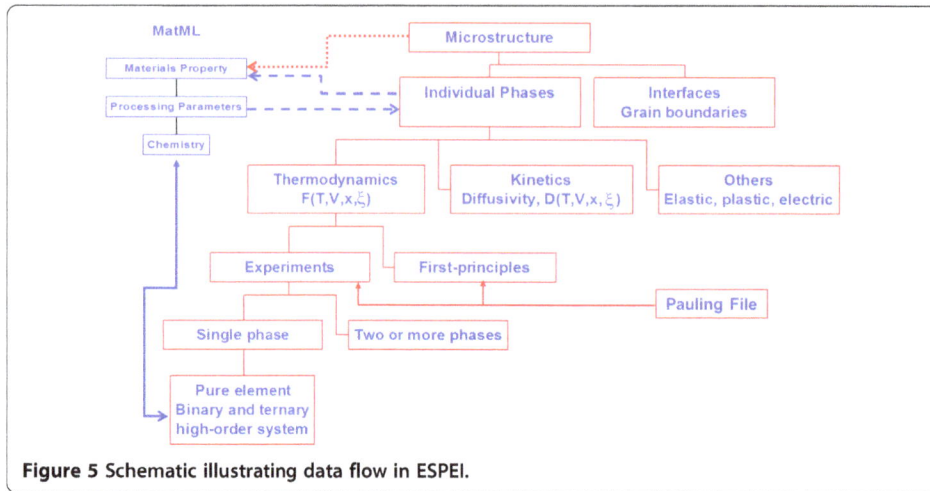

Figure 5 Schematic illustrating data flow in ESPEI.

discrepancies on lattice stability between the classic CALPHAD modeling and DFT-based first-principles calculations still exist [123] progresses have been made to narrow the differences such as bcc Ti [124] and fcc W [125], even for liquid solution phases [44]. The efficient special quasirandom structures (SQS) approach [126-128] is particularly useful in predicting the enthalpy and entropy of mixing in solid solution phases using phonon or Debye models [129,130]. Consequently, the thermodynamic model parameters of all individual phases can be evaluated solely from DFT-based first-principles calculations. On the other hand, the uncertainty in DFT-based first principles calculations prohibits accurate predictions of phase equilibrium temperature and compositions between two or more phases, and model parameters are further refined using phase equilibrium data from experiments.

Conclusions

The CALPHAD method is a powerful tool for assessing the properties of multicomponent, multiphase materials that can readily be coupled with materials simulation tools. However, the method is currently lacking strategies for straightforward implementation of new models and new data to update databases. The basis for efficient update methods is the availability of the data that are needed for determination of the functions used by the CALPHAD method.

The strengths of the CALPHAD method are that the data obtained from the calculation with a CALPHAD description are self-consistent and that descriptions of constitutive subsystems can be combined for the extrapolation [9] of multicomponent systems. However, from the assessment of the descriptions to the prediction of the properties in a multicomponent system, the confidentiality interval of the calculated data is not available from the calculations. Although several attempts [131,132] have been made to quantify the reliability of these calculated data none of these methods has been tested for its general applicability for results from calculations of multicomponent systems. For full implementation into an Integrated Computational Materials Engineering (ICME) approach, information on the confidence of the multicomponent predictions is needed to design materials within specified property tolerances. While the CALPHAD approach only requires unary, binary, and ternary data, multicomponent data are essential for providing verification and confidence assessments of the multicomponent material property predictions. Thus, it is important that multicomponent property data are also published and stored in data repositories [133].

As outlined in the 2008 National Research Council report on ICME [134], a data repository infrastructure is needed. The report suggests following the example of the human genome project (HGP) [135]. This would require that data from government sponsored research be published in publically available databases as recently mandated by the recent US government executive order [136]. Likewise, materials-oriented research journals should require that the reported data be deposited in a publically available database before the research is accepted for publication. The development of a cyber-infrastructure is required as a first step, within the framework of the Materials Genome Initiative [16]. This infrastructure is needed to host repositories, to provide tools to enable efficient population of reference data repositories, and to integrate the variety of different repositories and tools currently being developed. NIST has initiated several projects and collaborations to develop this infrastructure and host some of these repositories (nist.matdl.org and materialsdata.nist.gov/dspace/xmlui).

Endnotes

[a]In this article the term "multicomponent" is used to describe systems with more than 3 components.

[b]The term "non-stable" is used if a distinction between metastable and unstable is not available or not needed.

Competing interests
The authors declare that they have no competing interests.

Authors' contribution
CEC reviewed the data needed to develop the diffusion mobility, molar volume and thermal expansion descriptions. CEC also contributed the discussion on expanding the CALPHAD approach to other phase-based properties that are independent of the thermodynamics, including the thermal conductivity and interphase properties. URK reviewed the development of the CALPHAD approach and the crystallographic, phase equilibria, thermochemical data needed to develop the thermodynamic descriptions. CEC reviewed the data needed to develop the diffusion mobility, molar volume and thermal expansion descriptions. Z-KL contributed the discussion on elastic properties and the development of the ESPEI. URK and CEC both contributed to the discussion on data repositories and needed data infrastructure. All the authors collaborated in presenting the motivation of this work and in the discussion of all the work presented. All authors have read and approved the final manuscript.

Acknowledgments
The work by ZK Liu is partially supported by the National Science Foundation (NSF) through grant DMR-1006557.

Author details
[1]Materials Science and Engineering Division, National Institute of Standards and Technology, Gaithersburg, MD 20899, USA. [2]Department of Materials Science and Engineering, The Pennsylvania State University, University Park, PA 16802, USA.

References
1. Saunders N, Miodownik AP (1998) CALPHAD Calculation of Phase Diagrams: A Comprehensive Guide. Pergamon Materials Series. Elsevier Science Inc, New York
2. Kaufman L, Bernstein H (1970) Computer Calculation of Phase Diagrams. Academic Press, London
3. Lukas HL, Fries SG, Sundman B (2007) Computational Thermodynamics: The CALPHAD Method. Cambridge University Press, Cambridge
4. Dinsdale AT (1991) SGTE Data for Pure Elements. CALPHAD 15(4):317–425
5. Ågren J (1982) Diffusion in Phases with Several Components and Sublattices. J Phys Chem Solids 43(5):421–430
6. Andersson J-O, Ågren J (1992) Models for numerical treatment of multicomponent diffusion in simple phases. J Appl Phys 72(4):1350–1355

7.   Lu X-G, Selleby M, Sundman B (2005) Assessments of molar volume and thermal expansion for selected bcc, fcc, and hcp metallic elements. CALPHAD 29:68–89

8.   Liu ZK, Zhang H, Ganeshan S, Wang Y, Mathaudhu SN (2010) Computational modeling of effects of alloying elements on elastic coefficients. Scr Mater 63(7):686–691. doi:10.1016/j.scriptamat.2010.03.049

9.   Hillert M (2007) Phase Equilibria, Phase Diagrams and Phase Transformations, 2nd edn. Cambridge University Press, Cambridge

10. Saunders N, Kucherenko S, Li X, Miodownik AP, Schille J (2001) A new computer program for predicting materials properties. J Phase Equil 22(4):463–469. doi:10.1361/105497101770333036

11. Guo J, Samonds M (2007) Alloy Thermal Physical Property Prediction Coupled Computational Thermodynamics with Back Diffusion Consideration. J Phase Equil Diffus 28(1):58–63. doi:10.1007/s11669-006-9005-6

12. Steinbach I, Böttger B, Eiken J, Warnken N, Fries SG (2007) CALPHAD and phase-field modeling: A successful liaison. J Phase Equil Diffus 28(1):101–106

13. Olson GB (2013) Genomic materials design: The ferrous frontier. Acta Mater 61(3):771–781. doi:10.1016/j.actamat.2012.10.045

14. Kang Y-B, Aliravci C, Spencer P, Eriksson G, Fuerst C, Chartrand P, Pelton A (2009) Thermodynamic and volumetric databases and software for magnesium alloys. JOM 61(5):75–82. doi:10.1007/s11837-009-0076-9

15. Kaufman L, Ågren J (2014) CALPHAD, first and second generation – Birth of the materials genome. Scr Mater 70(1):3–6.

16. National Science and Technology Council (2011) Materials Genome Initiative for Global Competitiveness. Washington DC, http://www.whitehouse.gov/mgi

17. Redlich O, Kister AT (1948) Algebraic Representations of Thermodynamic Properties and the Classification of Solutions. Ind Eng Chem 40(2):345–348

18. Muggianu YM, Gambino M, Bros JP (1975) Enthalpies Of Formation Of Liquid Alloys Bismuth-Gallium-Tin At 723 k - Choice Of An Analytical Representation Of Integral And Partial Thermodynamic Functions Of Mixing For This Ternary-System. J Chim Phys-Chim Biol 72(1):83–88

19. Liu ZK (2009) First-Principles Calculations and CALPHAD Modeling of Thermodynamics. J Phase Equil Diffus 30(5):517–534. doi:10.1007/s11669-009-9570-6

20. Chase MW, Ansara I, Dinsdale A, Eriksson G, Grimvall G, Höglund L, Yokokawa H (1995) Group 1: Heat capacity models for crystalline phases from 0 K to 6000 K. CALPHAD 19(4):437–447

21. Xiong W, Hedstrom P, Selleby M, Odqvist J, Thuvander M, Chen Q (2011) An improved thermodynamic modeling of the Fe-Cr system down to zero kelvin coupled with key experiments. CALPHAD 35(3):355–366. doi:10.1016/j.calphad.2011.05.002

22. Palumbo M, Fries SG, Pasturel A (2011) Temperature Dependence of Thermodynamic Quantities Calculated from First-Principles. Paper presented at the COST MP0602 HISOLD, Brno, Czech Republic

23. Palumbo M, Fries SG, Hammerschmidt T, Körmann F, Hickel T (2012) SAPIENS Thermophysical Database for Pure Elements: DFT and Experiments. Paper presented at the 18th Symposium on Thermophysical Properties, Boulder, CO USA

24. Vřešťál J, Štrof J, Pavlů J (2012) Extension of SGTE data for pure elements to zero Kelvin temperature—A case study. CALPHAD 37:37–48. doi:10.1016/j.calphad.2012.01.003

25. Rogal J, Divinski SV, Finnis MW, Glensk A, Neugebauer J, Perepezko JH, Schuwalow S, Sluiter MHF, Sundman B (2014) Perspectives on point defect thermodynamics. physica status solidi (b) 251(1):97–129, doi:10.1002/pssb.201350155

26. Palumbo M, Burton B, Costa e Silva A, Fultz B, Grabowski B, Grimvall G, Hallstedt B, Hellman O, Lindahl B, Schneider A, Turchi PEA, Xiong W (2014) Thermodynamic modelling of crystalline unary phases. physica status solidi (b) 251(1):14–32, 10.1002/pssb.201350133

27. Körmann F, Breidi AAH, Dudarev SL, Dupin N, Ghosh G, Hickel T, Korzhavyi P, Muñoz JA, Ohnuma I (2014) Lambda transitions in materials science: Recent advances in CALPHAD and first-principles modelling. physica status solidi (b) 251(1):53–80, doi:10.1002/pssb.201350136

28. Hammerschmidt T, Abrikosov IA, Alfè D, Fries SG, Höglund L, Jacobs MHG, Koßmann J, Lu XG, Paul G (2014) Including the effects of pressure and stress in thermodynamic functions. physica status solidi (b) 251(1):81–96, doi:10.1002/pssb.201350156

29. Becker CA, Ågren J, Baricco M, Chen Q, Decterov SA, Kattner UR, Perepezko JH, Pottlacher GR, Selleby M (2014) Thermodynamic modelling of liquids: CALPHAD approaches and contributions from statistical physics. physica status solidi (b) 251(1):33–52, doi:10.1002/pssb.201350149

30. Sundman B, Ågren J (1981) A Regular Solution Model For Phases With Several Components And Sub-Lattices, Suitable For Computer-Applications. J Phys Chem Solids 42(4):297–301

31. Oates WA, Zhang F, Chen SL, Chang YA (1999) Improved cluster-site approximation for the entropy of mixing in multicomponent solid solutions. Phys Rev B 59(17):11221–11225

32. Andersson JO, Fernández Guillermet A, Hillert M, Jansson B, Sundman B (1986) A Compound-Energy Model of Ordering in a Phase with Sites of Different Coordination Numbers. Acta Metallugrica 34(3):437–445

33. Hillert M (2001) The compound energy formalism. J Alloy Compd 320(2):161–176

34. Hillert M, Jarl M (1978) MODEL FOR ALLOYING EFFECTS IN FERROMAGNETIC METALS. CALPHAD 2(3):227–238

35. Inden G (1975) Determination Of Chemical And Magnetic Interchange Energies In Bcc Alloys.1. General Treatment. Z Metallkd 66(10):577–582

36. Chen Q, Sundman S (2001) Modeling of thermodynamic properties for Bcc, Fcc, liquid, and amorphous iron. J Phase Equil 22(6):631–644. doi:10.1007/s11669-001-0027-9

37. Sundman B, Ohnuma I, Dupin N, Kattner UR, Fries SG (2009) An assessment of the entire Al–Fe system including D03 ordering. Acta Mater 57(10):2896–2908. doi:10.1016/j.actamat.2009.02.046

38. Wang Y, Hector LG, Zhang H, Shang SL, Chen LQ, Liu ZK (2009) A thermodynamic framework for a system with itinerant-electron magnetism. J Physics-Condensed Matter 21(32):326003

39. Shang SL, Wang Y, Liu ZK (2010) Thermodynamic fluctuations between magnetic states from first-principles phonon calculations: The case of bcc Fe. Phys Rev B 82(1):014425, doi:014425 10.1103/PhysRevB.82.014425

40. Shang SL, Saal JE, Mei ZG, Wang Y, Liu ZK (2010) Magnetic thermodynamics of fcc Ni from first-principles partition function approach. J Applied Physics 108(12):123514, doi:123514 10.1063/1.3524480

41. Wang Y, Shang SL, Zhang H, Chen LQ, Liu ZK (2010) Thermodynamic fluctuations in magnetic states: Fe3Pt as a prototype. Philos Mag Lett 90(12):851–859

42. Ågren J, Cheynet B, ClavagueraMora MT, Hack K, Hertz J, Sommer F, Kattner U (1995) GROUP 2: Extrapolation of the heat capacity in liquid and amorphous phases. CALPHAD 19(4):449–480

43. Ågren J (1988) Thermodynamics of Supercooled Liquids and their Glass Transition. Phys Chem Liq 18(2):123–139. doi:10.1080/00319108808078586

44. Han J, Wang WY, Wang C, Wang Y, Liu X, Liu Z-K (2013) Accurate determination of thermodynamic properties for liquid alloys based on ab initio molecular dynamics simulation. Fluid Phase Equilib 360:44–53

45. Pelton AD, Blander M (1986) Thermodynamic Analysis of Ordered Liquid Solutions by a Modified Quasi-Chemical Approach - Application to Silicate Slags. Metallogr Trans B-Process Metallurgy 17(4):805–815

46. Sommer F (1982) Association Model For The Description Of The Thermodynamic Functions Of Liquid Alloys.1. Basic Concepts. Z Metallkd 73(2):72–76

47. Sommer F (1982) Association Model For The Description Of Thermodynamic Functions Of Liquid Alloys.2. Numerical Treatment And Results. Z Metallkd 73(2):77–86

48. Hillert M, Jansson B, Sundman B, Ågren J (1985) A 2-Sublattice Model For Molten Solutions With Different Tendency For Ionization. Metal Trans A 16(2):261–266. doi:10.1007/bf02816052

49. Lu XG, Chen Q (2009) A CALPHAD Helmholtz energy approach to calculate thermodynamic and thermophysical properties of fcc Cu. Philos Mag 89(25):2167–2194. doi:10.1080/14786430903059004

50. Villars P, Calvert LD (1991) Pearson's Handbook of Crystallographic Data for Intermetallic Phases. ASM International, Materials Park

51. FIZ/NIST (2011) Inorganic Crystal Structure Database (ICSD) Version 2011/2. NIST, http://www.nist.gov/srd/nist84.cfm Accessed Jan 9 2012

52. PAULING FILE., http://paulingfile.com/

53. Villars P, Cenzual K (2012) Pearson's Crystal Data., Available via ASM International. http://www.asminternational. org/materials-resources/online-databases/-/journal_content/56/10192/6382084/DATABASE

54. Brown ID, Abrahams SC, Berndt M, Faber J, Karen VL, Motherwell WDS, Villars P, Westbrook JD, McMahon B (2005) Report of the working group on crystal phase identifiers. Acta Crystallogr Sect A 61:575–580. doi:10.1107/s010876730503179x

55. Zhao J-C (2007) Methods for Phase Diagram Determination. Elsevier, Amsterdam

56. Colinet C (1998) Comparison of enthalpies of formation and enthalpies of mixing in transition metal based alloys. Thermochim Acta 314(1–2):229–245

57. Colinet C (1995) High temperature calorimetry: recent developments. J Alloy Compounds 220(1–2):76–87. doi:10.1016/0925-8388(94)06032-0

58. Ipser H, Miikula A, Katayama I (2010) Overview: The emf method as a source of experimental thermodynamic data. CALPHAD 34:271–278, doi:10.1016/j.calphad.2010.05.001

59. Predel B (1982) Recent trends and developments of experimental methods for the determination of thermodynamic quantities of alloys. CALPHAD 6(3):199–216. doi:10.1016/0364-5916(82)90002-5

60. Ferro R, Cacciamani G, Borzone G (2003) Remarks about data reliability in experimental and computational alloy thermochemistry. Intermetallics 11:1081–1094

61. Belova I, Murch GE (2007) Expressions for vacancy-wind factors occurring in interdiffusion in ternary and higher-order alloys. Acta Mater 55:627–634

62. Swoboda B, Van der Ven A, Morgan D (2010) Assessing Concentration Dependence of FCC Metal Alloy Diffusion Coefficients Using Kinetic Monte Carlo. J Phase Equilib Diffus 31(3):250–259. doi:10.1007/s11669-010-9706-8

63. Herzig C, Kohler U, Divinski SV (1999) Tracer diffusion and mechanism of non-Arrhenius diffusion behavior of Zr and Nb in body-centered cubic Zr-Nb alloys. J Appl Phys 85(12):8119–8130. doi:10.1063/1.370650

64. Neumann G, Tuijn C (2009) Self-Diffusion and Impurity Diffusion in Pure Metals: Handbook of Experimental Data, vol 14. Pergamon Materials Series, 1st edn. Elsevier, New York

65. Bakker H, Bonzel HP, Bruff CM, Dayananda MA, Gust W, Horvath J, Kaur I, Kidson GV, LeClaire AD, Mehrer H, Murch GE, Neumann G, Stolica N, Stolwijk NA (eds) (1990) Difusion in Solid Metals and Alloys, vol 26. Landolt-Börnstein. Numerical Data and Functional Relationships in Science and Technology Springer-Verlag, Berlin

66. Gale WF, Totemeier TC (2004) Smithells Metals Reference Book, 8th edn. Elsevier, Amsterdam

67. Cui YW, Kato R, Omori T, Ohnuma I, Oikawa K, Kainuma R, Ishida K (2010) Revisiting diffusion in Fe-Al intermetallics: Experimental determination and phenomenological treatment. Scr Mater 62(4):171–174, http://dx.doi.org/10.1016/j. scriptamat.2009.10.011

68. Cui YW, Jiang M, Ohnuma I, Oikawa K, Kainuma R, Ishida K (2008) Computational study of atomic mobility in Co-Fe-Ni ternary fcc alloys. J Phase Equilib Diffus 29(4):312–321. doi:10.1007/s11669-008-9341-9

69. Engström A, Ågren J (1996) Assessment of diffusional mobilities in face-centered cubic Ni Cr Al alloys. Z Metallkde 87(2):92–97

70. Campbell CE (2009) NIST Diffusion Workshop, http://www.nist.gov/mml/msed/thermodynamics_kinetics/Diffusion-Workshop-Group.cfm

71. Campbell CE, Rukhin AL (2011) Evaluation of self-diffusion data using weighted means statistics. Acta Mater 59(13):5194–5201. doi:10.1016/j.actamat.2011.04.055

72. Brown AM, Ashby MF (1980) Correlations for Diffusion Constants. Acta Metall 28:1085–1101

73. Askill J (1970) Tracer Diffusion Data for Metals. Alloys and Simple Oxides, Plenum, New York

74. Mantina M, Wang Y, Arroyave R, Chen LQ, Liu ZK, Wolverton C (2008) First-principles calculation of self-diffusion coefficients. Phys Rev Lett 100(21):215901

75. Mantina M, Wang Y, Chen LQ, Liu ZK, Wolverton C (2009) First principles impurity diffusion coefficients. Acta Mater 57(14):4102–4108

76. Van Der Ven A, Yu H-C, Ceder G, Thornton K (2010) Vacancy mediated substitutional diffusion in binary crystalline solids. Prog Mater Sci 55(2):61–105, doi:10.1016/j.pmatsci.2009.08.001

77. Mendelev MI, Mishin Y (2009) Molecular dynamics study of self-diffusion in bcc Fe. Phys Rev B 80(14), http://dx.doi.org/10.1103/PhysRevB.80.144111

78. Hallstedt B, Dupin N, Hillert M, Höglund L, Lukas H, Schuster JC, Solak N (2007) Thermodynamic models for crystalline phases. Composition dependent models for volume, bulk modulus and thermal expansion. CALPHAD 31:28–37

79. Lu X-G, Selleby M, Sundman B (2005) Implementation of a new model for pressure dependence of condensed phases in Thermo-Calc. CALPHAD 29(1):49–55. doi:10.1016/j.calphad.2005.04.001

80. Lu X-G, Selleby M, Sundman B (2005) Theoretical modeling of molar volume and thermal expansion. Acta Mater 53(8):2259–2272. doi:10.1016/j.actamat.2005.01.049

81. Fernández Guillermet A (1987) Critical-Evaluation Of The Thermodynamic Properties Of Cobalt. Int J Thermophys 8(4):481–510

82. Kim D, Shang S-L, Liu Z-K (2012) Effects of alloying elements on thermal expansions of γ-Ni and γ'-Ni3Al by first-principles calculations. Acta Mater 60(4):1846–1856. doi:10.1016/j.actamat.2011.12.005

83. Hallstedt B (2007) Molar volumes of Al, Li, Mg and Si. CALPHAD 31(2):292–302. doi:10.1016/j.calphad.2006.10.006

84. Zhang B, Li X, Li D (2013) Assessment of thermal expansion coefficient for pure metals. Calphad 43(1):7–17, http://dx.doi.org/10.1016/j.calphad.2013.08.006

85. Wang Y, Wang JJ, Zhang H, Manga VR, Shang SL, Chen LQ, Liu ZK (2010) A first-principles approach to finite temperature elastic constants. J Phys-Condens Matter 22(22):225404. doi:10.1088/0953-8984/22/22/225404

86. Shang SL, Saengdeejing A, Mei ZG, Kim DE, Zhang H, Ganeshan S, Wang Y, Liu ZK (2010) First-principles calculations of pure elements: Equations of state and elastic stiffness constants. Comput Mater Sci 48(4):813–826. doi:10.1016/j.commatsci.2010.03.041

87. Jacobs MHG, Oonk HAJ (2000) A realistic equation of state for solids. The high pressure and high temperature thermodynamic properties of MgO. CALPHAD 24(2):133–147. doi:10.1016/s0364-5916(00)00019-5

88. Saxena SK (2004) Pressure–volume equation of state for solids. J Phys Chem Solids 65(8–9):1561–1563. doi:10.1016/j.jpcs.2004.02.003

89. Jacobs MHG, van den Berg AP, de Jong BHWS (2006) The derivation of thermo-physical properties and phase equilibria of silicate materials from lattice vibrations: Application to convection in the Earth's mantle. CALPHAD 30(2):131–146. doi:10.1016/j.calphad.2005.10.001

90. Jacobs MHG, de Jong BHWS (2005) An investigation into thermodynamic consistency of data for the olivine, wadsleyite and ringwoodite form of (Mg, Fe)2SiO4. Geochim Cosmochim Acta 69(17):4361–4375. doi:10.1016/j.gca.2005.05.002

91. Brosh E, Makov G, Shneck RZ (2007) Application of CALPHAD to high pressures. CALPHAD 31(2):173–185. doi:10.1016/j.calphad.2006.12.008

92. Grimvall G (1999) Thermophysical Properties of Materials, 2nd edn. Elsevier, Amsterdam

93. Terada Y, Ohkubo K, Mohri T, Suzuki T (2002) Thermal conductivity of intermetallic compounds with metallic bonding. Mater Trans 43(12):3167–3176. doi:10.2320/matertrans.43.3167

94. Nordheim L (1931) Electron theory of metals I. Ann Physik 9:607

95. Terada Y, Ohkubo K, Mohri T, Suzuki T (2000) A comparative study of thermal conductivity in alloys and compounds. Mater Sci Eng A 278(1–2):292–294

96. Schroder K (1983) CRC Handbook of Electrical Resistivitives of Binary Metallic Alloys

97. Gheribi A, Chartrand P (2012) Application of the CALPHAD method to predict the thermal conductivity in dielectric and semiconductor crystals. CALPHAD 39:70–79, http://dx.doi.org/10.1016/j.calphad.2012.06.002

98. Picha R, Vrest IJ, Kroupa A (2004) Prediction of alloy surface tension using a thermodynamic database. CALPHAD 28:141–146

99. Tanaka T, Hack K, Hara S (1999) Use of thermodynamic data to determine surface tension and viscosity of metallic alloys. MRS Bull 24(4):45–50

100. Shi PF, Engström A, Sundman B, Ågren J (2011) Thermodynamic Calculations and Kinetic Simulations of Some Advanced Materials. In: Tan Y, Ju DY (eds) Advanced Material Science and Technology, Pts 1 and 2, vol 675–677. Materials Science Forum. Trans Tech Publications Ltd, Stafa-Zurich, pp 961–974, http://www.scientific.net/MSF.675-677.961

101. Olson GB, Jou HJ, Jung J, Sebastian JT, Misra A, Locci I, Hull D (2008) Precipitation model validation in 3(rd) generation aeroturbine disc alloys. Superalloys 2008. Minerals, Metals & Materials Soc, Warrendale PA; 923-932

102. Wu K, Zhang F, Chen S, Cao W, CY A (2008) A modeling tool for the precipitation simulations of superalloys during heat treatments. Superalloys 2008. Minerals, Metals, & Materials Soc. Warrendale, PA; 933–939

103. Costa e Silva A, Ågren J, Clavaguera-Mora MT, Djurovic D, Gómez-Acebo T, Lee B-J, Liu Z-K, Miodownik P, Seifert HJ (2007) Applications of computational thermodynamics — the extension from phase equilibrium to phase transformations and other properties. CALPHAD 31(1):53–74, doi:10.1016/j.calphad.2006.02.006

104. Turchi PEA, Abrikosov IA, Burton B, Fries SG, Grimvall G, Kaufman L, Korzhavyi P, Rao Manga V, Ohno M, Pisch A, Scott A, Zhang W (2007) Interface between quantum-mechanical-based approaches, experiments, and CALPHAD methodology. CALPHAD 31(1):4–27. doi:10.1016/j.calphad.2006.02.009

105. Shi X, Luo J (2011) Developing grain boundary diagrams as a materials science tool: A case study of nickel-doped molybdenum. Phys Rev B 84(1):014105

106. Ardell AJ (2011) A1-L1$_2$ interfacial free energies from data on coarsening in five binary Ni alloys, informed by thermodynamic phase diagram assessments. J Mater Sci 46(14):4832–4849. doi:10.1007/s10853-011-5395-x

107. Sudbrack CK, Noebe RD, Seidman DN (2007) Compositional pathways and capillary effects during isothermal precipitation in a nondilute Ni–Al–Cr alloy. Acta Mater 55(1):119–130. doi:10.1016/j.actamat.2006.08.009

108. Höglund L, Ågren J (2010) Simulation of Carbon Diffusion in Steel Driven by a Temperature Gradient. J Phase Equil Dif 31(3):212–215. doi:10.1007/s11669-010-9673-0

109. MatNavi (2012) NIMS Materials Database., http://mits.nims.go.jp/index_en.html. Accessed July 2012 2012

110. Landolt-Börnstein Database, Springer Materials, (2014) http://www.springermaterials.com/navigation/
111. Bartolo L, Campbell CE, Kattner UR (2013) CALPHAD File Repositories, http://nist.matdl.org/dspace/
112. American Chemical Society. (2014) CAS REGISTRY, http://www.cas.org/expertise/cascontent/registry/index.html
113. International Union of Pure and Applied Chemistry (IUPAC), (2011) International Chemical Identifier, http://www.iupac.org/home/publications/e-resources/inchi.html
114. Schmid-Fetzer R, Andersson D, Chevailer PY, Eleno L, Fabrichnaya O, Kattner UR, Sundman B, Wang C, Watson A, Zabdyr L, Zinkevich M (2007) Assessment techniques, database design and software facilities for thermodynamics and diffusion. CALPHAD 31:38–52
115. Kaufman JG, Begley EF (2003) MatML: A data interchange markup language. Adv Mater Process 161(11):35–36
116. Chirico RD, Frenkel M, Diky VV, Marsh KN, Wilhoit RC (2003) ThermoML-An XML-based approach for storage and exchange of experimental and critically evaluated thermophysical and thermochemical property data. 2. Uncertainties. J Chem Eng Data 48(5):1344–1359. doi:10.1021/je034088i
117. Frenkel M, Chirico RD, Diky VV, Dong Q, Frenkel S, Franchois PR, Embry DL, Teague TL, Marsh KN, Wilhoit RC (2003) ThermoML - An XML-based approach for storage and exchange of experimental and critically evaluated thermophysical and thermochemical property data. 1. Experimental data. J Chem Eng Data 48(1):2–13. doi:10.1021/je025645o
118. Frenkel M, Chirico RD, Oiky VV, Marsh KN, Dymond JH, Wakeham WA (2004) ThermoML (dagger) - An XML-based approach for storage and exchange of experimental and critically evaluate thermophysical and thermochemical property data. 3. Critically evaluated data, predicted data, and equation representation. J Chem Eng Data 49(3):381–393. doi:10.1021/je049890e
119. Boyce DE, Dawson PR, Miller MP (2009) The Design of a Software Environment for Organizing, Sharing, and Archiving Materials Data. Mater Trans A 40A(10):2301–2318. doi:10.1007/s11661-009-9889-y
120. ECMA International, ECMA-404 Standard, 1st edition (2013), JavaScript Object Notation, http://www.json.org/
121. Binary JSON, (2012) http://bsonspec.org/
122. Shang S, Wang Y, Liu ZKESPEI (2010) Extensible, Self-optimizing Phase Equilibrium Infrastructure for Magnesium Alloys. In: Agnew SR, Neelameggham NR, Nyberg EA, Sillekens WH (eds) Magnesium Technology 2010. Seattle, WA, pp 617–622
123. Wang Y, Curtarolo S, Jiang C, Arroyave R, Wang T, Ceder G, Chen LQ, Liu ZK (2004) Ab initio lattice stability in comparison with CALPHAD lattice stability. Calphad 28(1):79–90, http://dx.doi.org/10.1016/j.calphad.2004.05.002
124. Mei ZG, Shang SL, Wang Y, Liu ZK (2009) Density-functional study of the thermodynamic properties and the pressure-temperature phase diagram of Ti. Phys Rev B 80(10):104116. doi:10.1103/PhysRevB.80.104116
125. Ozolins V (2009) First-Principles Calculations of Free Energies of Unstable Phases: The Case of fcc W. Phys Rev Lett 102(6):065702. doi:10.1103/PhysRevLett.102.065702
126. Zunger A, Wei SH, Ferreira LG, Bernard JE (1990) Special quasirandom structures. Phys Rev Lett 65(3):353–356. doi:10.1103/PhysRevLett.65.353
127. Jiang C, Wolverton C, Sofo J, Chen LQ, Liu ZK (2004) First-principles study of binary bcc alloys using special quasirandom structures. Phys Rev B 69(21):214202. doi:10.1103/PhysRevB.69.214202
128. Shin D, Arroyave R, Liu ZK, Van de Walle A (2006) Thermodynamic properties of binary hcp solution phases from special quasirandom structures. Phys Rev B 74(2):024204. doi:10.1103/PhysRevB.74.024204
129. Wang Y, Zacherl CL, Shang SL, Chen LQ, Liu Z (2011) Phonon dispersions in random alloys: a method based on special quasi-random structure force constants. J Phys-Condens Matter 23(48), 485403.
130. Shang SL, Wang Y, Kim DE, Zacherl CL, Du Y, Liu ZK (2011) Structural, vibrational, and thermodynamic properties of ordered and disordered Ni (1-x) Pt (x) alloys from first-principles calculations. Phys Rev B 83(14), 144204
131. Malakhov DV (1997) Confidence intervals of calculated phase boundaries. CALPHAD 21(3):391–400. doi:10.1016/s0364-5916(97)00039-4
132. Stan M, Reardon B (2003) A Bayesian approach to evaluating the uncertainty of thermodynamic data and phase diagrams. CALPHAD 27(3):319–323. doi:10.1016/j.calphad.2003.11.002
133. National Research Council (2004) Retooling Manufacturing:Bridging Design, Materials, and Production. The National Academies Press, Washington, DC, http://www.nap.edu/catalog.php?record_id=11049
134. Integrated Computational Materials Engineering (2008) A Transformational Discipline for Improved Competitiveness and National Security. The National Academies Press, Washington, DC, http://www.nap.edu/catalog.php?record_id=12199
135. Department of Energy (2003) Human Genome Project information, http://www.ornl.gov/sci/techresources/Human_Genome/home.shtml. Accessed March 3, 2012 2012
136. Obama B (2013) Making Open and Machine Readable the New Default for Government Information. US Government Executive Order, The White House, http://www.whitehouse.gov/the-press-office/2013/05/09/executive-order-making-open-and-machine-readable-new-default-government

9

In situ experimental techniques to study the mechanical behavior of materials using X-ray synchrotron tomography

Sudhanshu S Singh[1], Jason J Williams[1], Peter Hruby[1], Xianghui Xiao[2], Francesco De Carlo[2] and Nikhilesh Chawla[1*]

* Correspondence:
nchawla@asu.edu
[1]Materials Science and Engineering, Arizona State University, Tempe, AZ 85287-6106, USA
Full list of author information is available at the end of the article

Abstract

In situ X-ray synchrotron tomography is an excellent technique for understanding deformation behavior of materials in 4D (the fourth dimension here is time). However, performing *in situ* experiments in synchrotron is challenging, particularly in regard to the design of the mechanical testing stage. Here, we report on several *in situ* testing methods developed by our group in collaboration with Advanced Photon source at Argonne National Laboratory used to study the mechanical behavior of materials. The issues associated with alignment during mechanical testing along with the improvements made to the *in situ* mechanical testing devices, over time, are described. *In situ* experiments involving corrosion-fatigue and stress corrosion cracking in various environments are presented and discussed. These include fatigue loading of metal matrix composites (MMCs), corrosion-fatigue, and stress corrosion cracking of Al 7075 alloys.

Keywords: X-ray tomography; *In situ*; Fatigue; Corrosion; Crack; AA7075

Background

A sound understanding of mechanical behavior requires a thorough understanding of the evolution of deformation with time. Traditionally, the study of material structure has been limited by two dimensional (2D) analyses. This approach is often inaccurate or inadequate for solving many cutting-edge problems. In addition, it is often laborious and time-consuming. We can now use three dimensional (3D) tools and analyses to resolve time-dependent (4D) evolution of a variety of important phenomena [1-4].

X-ray tomography is an extremely promising, non-destructive technique for characterizing microstructures in 3D and 4D. The use of high brilliance and partially coherent synchrotron light allows one to image multi-component materials from the sub-micrometer to nanometer range.

Over the years, many *ex situ* X-ray tomography experiments have been performed to study the behavior of materials under mechanical loading [5-7]. These experiments have consisted of post-mortem characterization after testing. More recently, *in situ* mechanical testing has become more attractive, as a means of visualizing and quantifying microstructural changes as a function of time. Experiments under tension [8-10], cyclic loading [11-14], corrosion-fatigue [15], and creep [16,17] have been conducted.

Several challenges exist when designing an *in situ* mechanical testing stage for synchrotron tomography beamlines. The stage will be on top of several motion stages which have weight limits for precise and reproducible movement. This constrains the size and weight of the motorized linear actuator. Also, since precise rotational movement is necessary for tomography, so the load frame's weight must be distributed as symmetrically as possible about the rotational axis. In addition, the stage requires an X-ray transparent material in the scanning area which should also be able to bear the applied load without excessive deformation. Furthermore, the load train has to be precisely aligned for the smaller samples used in tomography. This becomes even more important for *in situ* experiments on fatigue crack growth, for example. Finally, the distance between the fixture and the detector should be sufficient enough such that there is no contact between the load bearing sleeve and the scintillator during scanning. This can potentially affect the amount of absorption (density-based) contrast, with smaller distances yielding more absorption contrast.

   In this paper, we report on several *in situ* testing methods developed in our group at ASU in collaboration with scientists at the Advanced Photon Source at the Argonne National Laboratory. We begin by describing our initial attempts at *in situ* loading in tension. This is followed by a discussion of the important issues associated with alignment, both in tension and in fatigue loading. Finally, we conclude with a section on *in situ* experiments in corrosion and moisture environments, which carry with them a unique set of challenges.

Experimental details on X-ray synchrotron tomography

X-ray tomography was performed at the 2-BM beamline of the Advanced Photon Source (APS) at Argonne National Laboratory. The details of the tomography system at 2-BM have been described elsewhere [18-20]. In all the experiments, a monochromatic beam with energy of ~ 24 keV (except in the case of corrosion-fatigue experiment, where pink beam was used and will be described later) was focused on the specimen. A scintillator screen (CdWO$_4$ or LuAG:Ce) was used to convert the transmitted X-rays to visible light. This was coupled with an objective lens and a camera (CoolSnap K4 CCD or PCO Dimax CMOS camera) to obtain images. 2D projections were acquired at angular increments of 0.12° over a range of 180°. The tomography at one time step can be completed in about 15–20 minutes when a monochromatic beam is used. These 2D projections were then input into a filtered back-projection reconstruction algorithm (Shepp-Logan filter) to obtain 3D information.

Tensile and fatigue behavior

Our initial *in situ* loading stage used to perform tensile tests is shown in Figure 1. The stage was designed to be less than 2 kg due to the weight limit of the motion stages at the 2-BM beamline. The vertical range where the region of interest on the sample must intersect the X-ray beam is often restrictive and was between 15 mm to 40 mm. Thus, the distance from the bottom of our fixtures to the center of the sample was chosen to be 30 mm — a distance that included the base plate, sub-miniature load cell, bottom sample clamp and half the sample. In the loading stage, the specimen was clamped between the actuator and the load cell. The load cell at the bottom of the loading stage was attached to a fixed lower grip. The upper grip was attached to the stepper motor with linear actuator, which imposes a displacement on the sample to achieve the desire load. A stepper motor was chosen because it could be easily integrated with motor

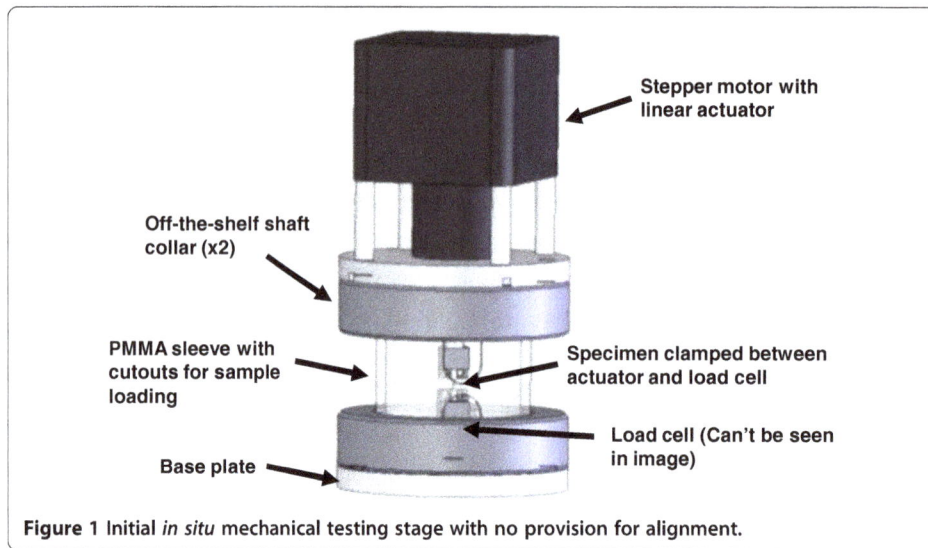

Figure 1 Initial *in situ* mechanical testing stage with no provision for alignment.

drives already present at the beamline. The stepper motor had a captive linear actuator capable of 8 μm per step and a total stroke of 12 mm. The load cell had a capacity of 500 N.

The load was transmitted from the top of the stage to the bottom through a polymer sleeve (Poly methyl methacrylate or PMMA). The PMMA material was chosen due to its transparency to X-rays. Two holes were incorporated in the PMMA sleeve to facilitate the loading of the sample. A simple linear elastic finite element analysis (Cosmos Xpress, Solidworks 2008, Concord, MA) was conducted to determine the stresses and displacements in the sleeve, as shown in Figure 2(a). The modulus of the PMMA was taken as 2.4 GPa and the Poisson's ratio as 0.35. The displacement of the bottom face was constrained and a compressive load of 200 N was applied to the top face. The calculated stress in the region between the holes was only about 0.1 MPa. The displacement in the sleeve was also quite small, on the order of a few micrometers. In addition, experiments at the load of 200 N were conducted to determine the creep rate of the sleeve, if any, during the tomography operation of about 20 minutes (Figure 2b). A negligible displacement rate of 2.5 nm/min was obtained, showing that the displacement in the PMMA did not affect the results significantly.

It is well known that axial alignment is critical during mechanical testing. This is particularly important due to the somewhat smaller sizes of specimens used in *in situ* X-ray tomography, because the bending moments can be exacerbated. Grips can have both concentric and angular (axial) misalignments. Concentric misalignment shifts the vertical axes of the grips laterally away from each other whereas angular misalignment occurs when upper grip's loading axis is at an angle with the lower grip's loading axis. Figure 3 shows a design that incorporates specimen alignment capability. It was designed such that concentric clamp misalignment could be minimized by translating the top clamp in two orthogonal directions with respect to the bottom clamp using the four set screws. Axial misalignment could be minimized by tilting the top clamp about two orthogonal axes with respect to the bottom clamp using the four tilt screws. The top clamp was also allowed to freely rotate about its mounting screw to self-align in the third rotational direction. In addition, a bushing was provided to reduce the lateral deflection of the actuator.

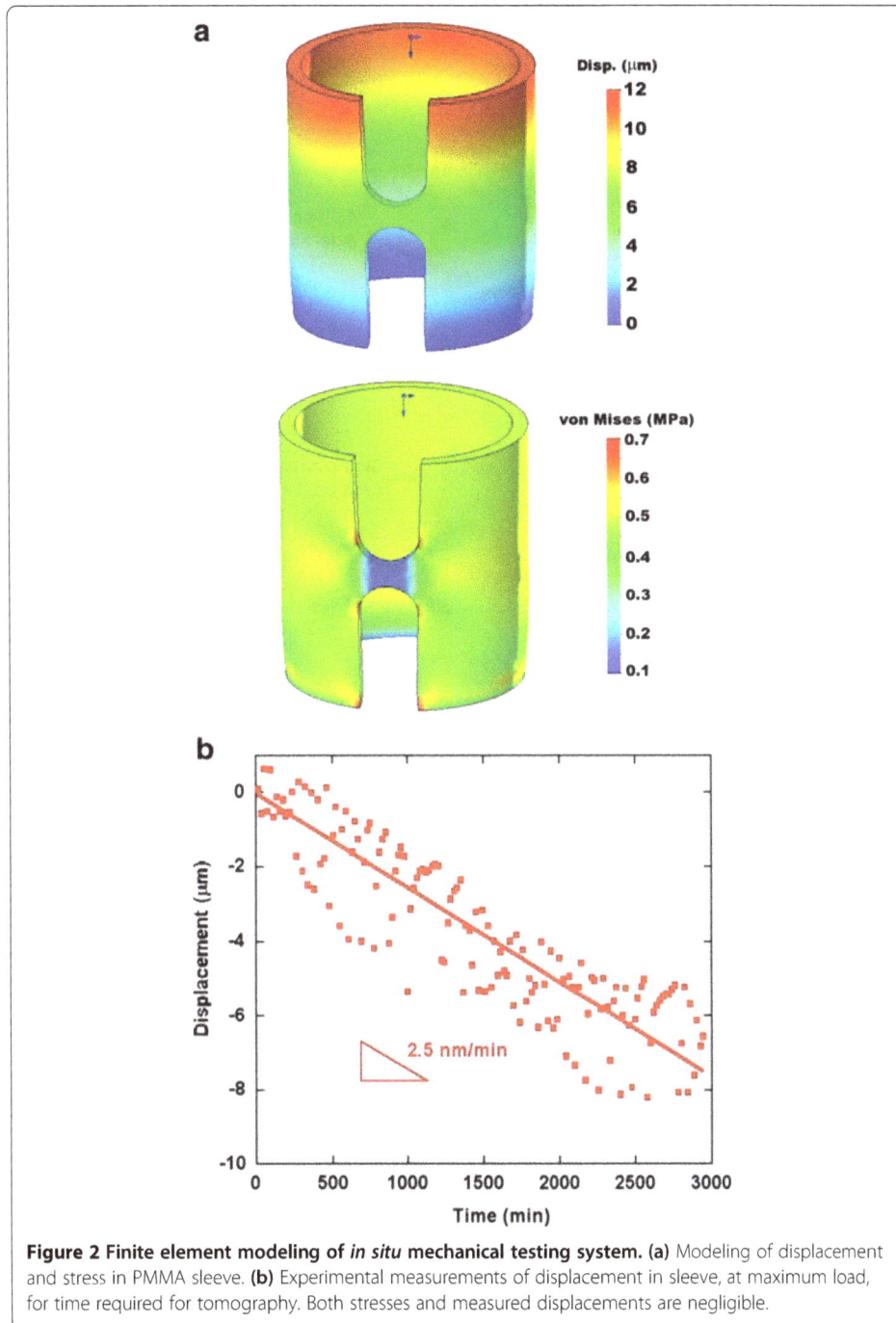

Figure 2 Finite element modeling of *in situ* mechanical testing system. (a) Modeling of displacement and stress in PMMA sleeve. **(b)** Experimental measurements of displacement in sleeve, at maximum load, for time required for tomography. Both stresses and measured displacements are negligible.

The amount of stress in the specimen prior to the self-alignment was small and therefore it did not affect the crack growth behavior.

A straight and parallel alignment specimen was used for initial alignment. It was machined such that it would only fit in the upper and lower grips if they were closely aligned. The benefit of X-ray tomography is the fact that 3D data sets are available, so we used actual images from X-ray tomography for finer alignment, without having to resort to strain gauges. By applying increasing loads to the pre-cracked specimen and measuring crack opening displacements, concentric alignments could be made to

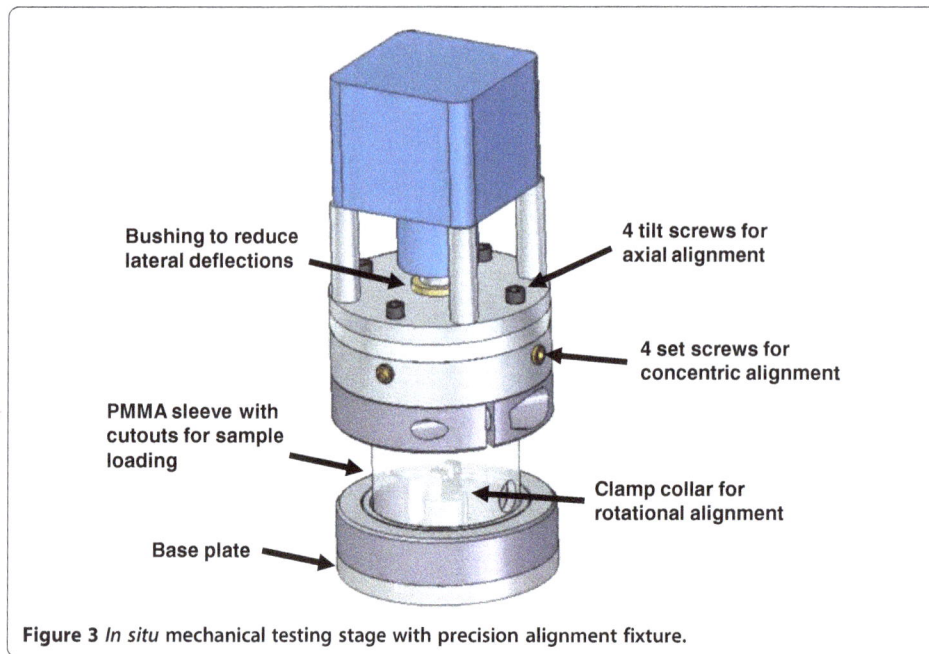

Figure 3 *In situ* mechanical testing stage with precision alignment fixture.

reduce the contributions from cracking in modes II and III. After alignment, the x-axis misalignment was about 1.8 μm, close to zero in the y-axis, and the tilt misalignment about 0.21°, as shown in Figure 4. A symmetric crack growth front was observed when alignment was performed compared to misaligned condition, as shown in Figure 5. It can be seen that with misalignment, the crack grows much further on the right side of the specimen whereas the crack is quite symmetric after alignment.

A few limitations of the loading stage include the maximum cyclic frequency of 2 Hz and the maximum load of only 500 N. Also, given that the fixture must rotate 180 degrees, and that it has a load bearing sleeve, 50 mm was the closest distance our samples could safely be from the scintillator without hitting it - a distance that yields significant phase contrast when using 24 keV X-rays (produced by a bending magnet).

Using the second loading stage shown in Figure 3, *in situ* fatigue crack growth tomography was performed on AA7075-T651 samples [12,13]. The fatigue crack growth rates obtained from these experiments [12,15] were comparable to published data by others [21]. The errors in the ΔK (stress intensity factor range) measurements were up to a maximum of only 7% due to error in measurement of force. The crack lengths can be measured accurately with an error of 1–2 pixels, the effect of which was very small in the crack growth rates (da/dN) and ΔK measurements. To achieve higher resolutions in tomographic reconstructions, the width of our samples was chosen to be about 2.8 mm to achieve a desired resolution of 2 μm. A single edge-notched (SEN) geometry was chosen for these experiments because a CT specimen would be too wide to fit into a 3 mm field of view (necessary to achieve a 2 μm resolution), and a 3 mm wide MT specimen could not be machined precise enough to achieve symmetric crack nucleation on each side of the central hole. The experiments were performed in the Paris law regime because of the limited amount of time available at the beamline (typically three days), and because the highest cycling frequency achieved by our linear actuator was only 2 Hz.

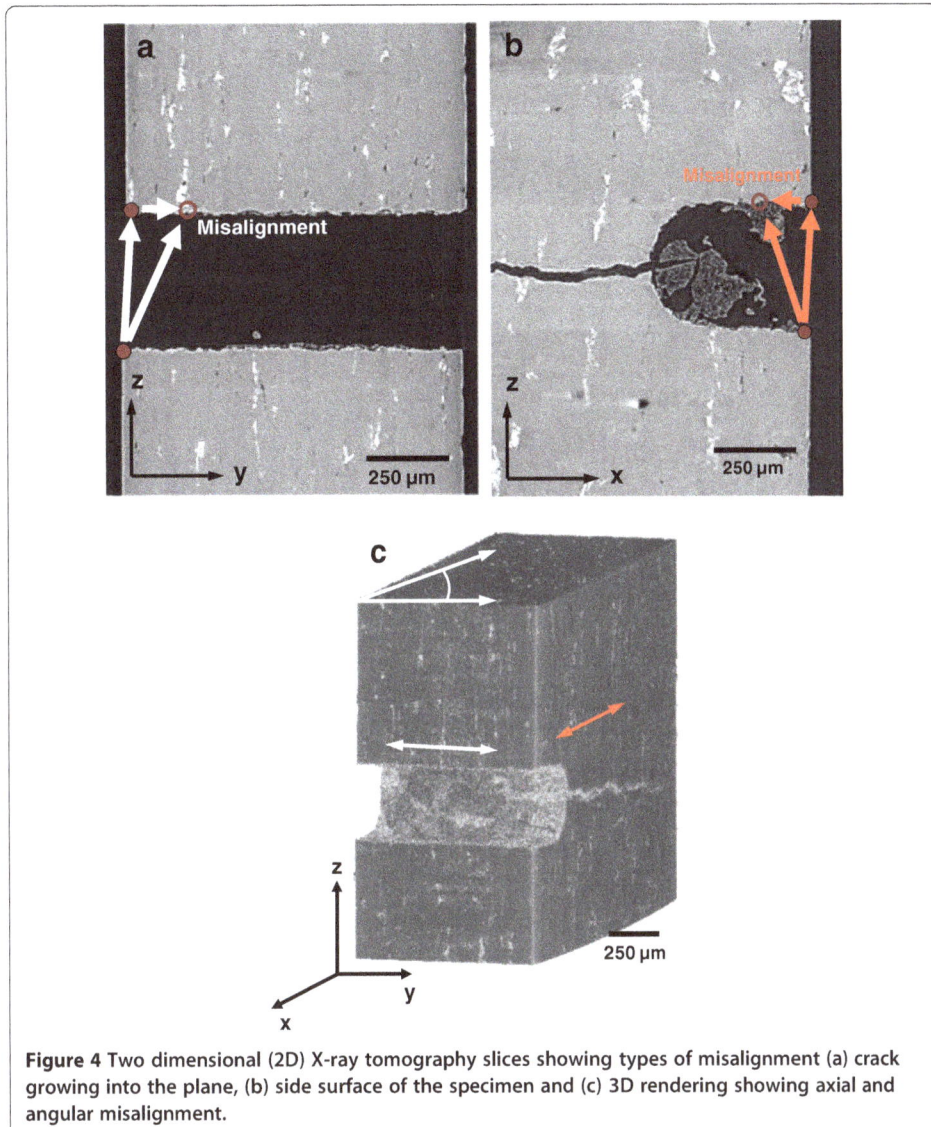

Figure 4 Two dimensional (2D) X-ray tomography slices showing types of misalignment (a) crack growing into the plane, (b) side surface of the specimen and (c) 3D rendering showing axial and angular misalignment.

The third generation *in situ* mechanical testing stage also incorporated load-control capability. Previously, the amount of displacement was manually monitored and was adjusted throughout the crack growth tests to maintain a pseudo-constant load amplitude, ΔP. The design of the third stage (Figure 6) included the ability to automatically control the actuator using feedback from the load cell. Additionally, the sample was inserted from the top of the stage rather than from the cut outs in the sleeve. The reason for this was that small cracks were observed over time near cut outs due to the combination of the applied load and the use of pink beam (used in corrosion-fatigue). Thus, a wider, shorter sleeve was used to facilitate sample loading and to increase load frame stiffness to perform fatigue tests at high R-ratios. The actuator's stroke length was also increased to 25 mm to provide more clearance during sample loading.

We present some preliminary results on the fatigue tests at high R-ratio (R = 0.6) in Al-SiC metal matrix composites (MMCs) using the third loading stage. The effect of high R-ratio on the crack propagation in MMCs has not been well understood. A 2080 aluminum alloy (3.6% Cu, 1.9% Mg, 0.25% Zr) reinforced with 20 volume percent SiC

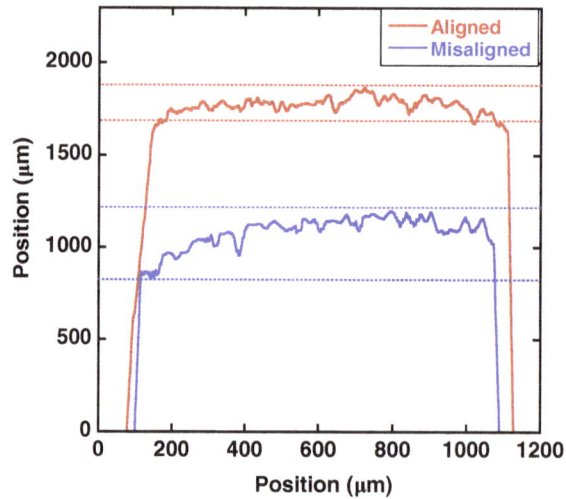

Figure 5 Comparison of crack profile (through thickness) before and after alignment. Note that the crack front is relatively symmetrical after alignment.

particles (average particle size of 25 µm) was used. The composite was processed by blending SiC and Al powders, compacting the powder mixture, hot pressing, and hot extrusion (Alcoa Inc., Alcoa, PA). Details of the powder metallurgy process for fabrication of these composite materials can be found elsewhere [22]. SEN specimens were machined by EDM for the fatigue tests.

2D X-ray synchrotron tomography images showing the progression of fatigue cracking have been shown in Figure 7. The fracture of a particle ahead of the crack tip during high load ratio fatigue can be clearly seen. Figure 7a shows that the particle (circled) was not fractured when the experiment was paused for tomography after 7,000 fatigue cycles. However the particle fractured at 8000 cycles (Figure 7b) and then the crack passed through the same fractured particle as shown in Figure 7c. This provides insight into the fundamental understanding on how the fatigue crack interacts with SiC particles, and the

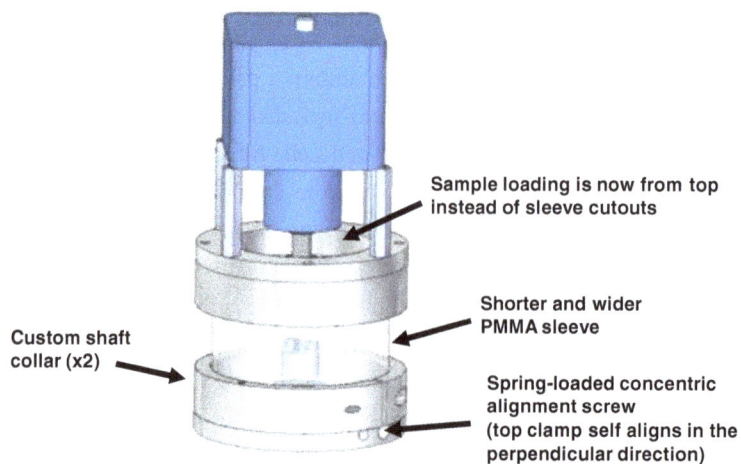

Figure 6 Third generation *in situ* loading stage to perform mechanical testing with load-control capability and additional stiffness for higher R-ratio fatigue crack growth experiments.

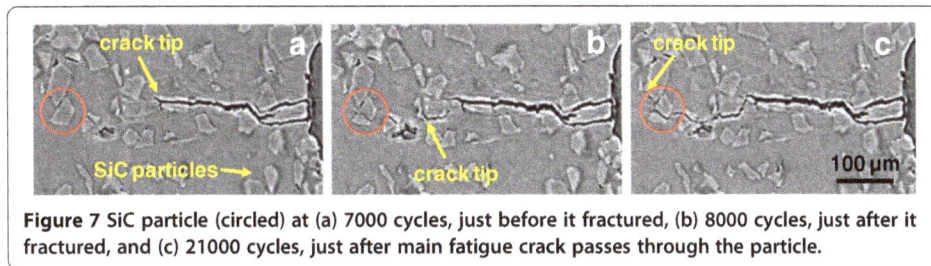

Figure 7 SiC particle (circled) at (a) 7000 cycles, just before it fractured, (b) 8000 cycles, just after it fractured, and (c) 21000 cycles, just after main fatigue crack passes through the particle.

role of particle fracture, both ahead and right at the crack tip, in controlling fatigue crack propagation at high R-ratios.

Corrosion-fatigue and stress corrosion cracking (SCC)

In order to study the effects of corrosion-fatigue as well as stress corrosion cracking, new designs were developed. Corrosion-fatigue experiments were performed in a liquid environment, as shown in Figure 8. The PMMA sleeve has not been shown here in order to clearly show the arrangements inside the stage. Corrosion-fatigue experiments were performed on AA7075-T651 in EXCO solution (4 M NaCl, 0.5 M KNO_3 and 0.1 M HNO_3). EXCO was chosen to ensure a significant amount of corrosion in the

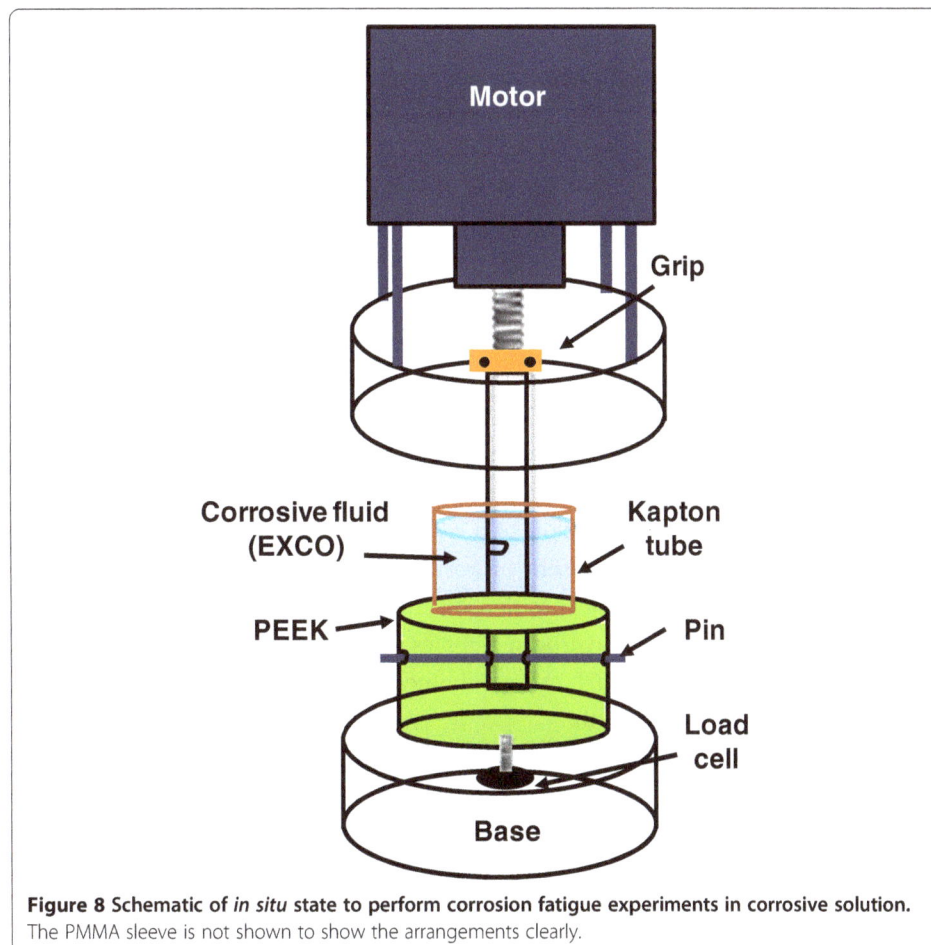

Figure 8 Schematic of *in situ* state to perform corrosion fatigue experiments in corrosive solution. The PMMA sleeve is not shown to show the arrangements clearly.

limited time available at the synchrotron beamline. SEN specimens were machined along LT orientation by EDM for the corrosion-fatigue tests. The biggest challenge in the experiment was to avoid the reaction of corrosive fluid with any part of the loading stage, especially the grips. Therefore, the material used for the bottom grip must be chemically inert to the solution while also sustaining the applied load. A polymeric PEEK (Polyether ether ketone) cylindrical grip was chosen to replace the bottom steel grip since it is chemically resistant to the solution and has good mechanical strength [23]. The bottom grip was fitted with the load cell. A rectangular hole was made at the top part of the PEEK grip to accommodate the sample. A hole was made on the side wall of the cylindrical grip to accommodate the stainless steel pin which passes through the hole made in the sample. Epoxy was added to make this a permanent and strong connection. The top part of the sample was clamped as previously described. Kapton tubing, which is also chemically inert to EXCO solution, was attached to the PEEK cylinder with wax to hold the solution around the sample during the test. The height of the Kapton tube was chosen such that the notch of the specimen was immersed in the corrosive fluid.

"Pink beam", which is a polychromatic beam with low and high energies removed from the white beam spectrum, was used during corrosion experiments, because the higher photon flux compared to a monochromatic beam allowed for significantly faster data acquisition. Thus the progression of corrosion could be documented with a higher frequency. While a typical scan with a double-multi-layer monochromatic (DMM) beam takes twenty minutes to complete, a pink beam scan can take only 0.5–1 second. Although much faster scans are technically possible using a pink beam, significant X-ray attenuation by the corrosive fluid required higher exposure times (2–3 minutes) to achieve a suitable signal-to-noise ratio.

Using the above design, we showed that the fatigue crack growth rate of AA7075-T651 was much higher in EXCO solution than in ambient air [15]. Figure 9 (a) shows a 2D X-ray tomography image of the fatigue crack along with corrosion product and the hydrogen bubbles inside the crack formed due to reaction between AA7075 and the EXCO solution. As shown in Figure 9(b), the 3D reconstruction of the crack, bubbles, corrosive fluid and corrosion products was performed using commercially available software (MIMICS, Ann Arbor, MI). Corrosion products were segmented on a few slices (from the part of crack), as shown in Figure 9(b). Figure 10 shows the changes in the shape of the bubbles and the formation of a new bubble in a fatigue cycle. Formation of a new hydrogen bubble can be clearly seen at position 3 which was not present in position 2. All bubbles were squeezed as the crack closes during unloading (position 4). These results provided insights to the fundamentals of evolution of hydrogen bubbles inside a growing fatigue crack. In particular, local variations in pH cause inhomogeneous formation of reaction products and will affect crack growth.

The presence of moisture in the environment is known to deteriorate the mechanical properties of high strength alloys. In order to understand the effect of moisture, *in situ* stress corrosion cracking (SCC) experiments were conducted in moisture, as shown in Figure 11(a). The PMMA sleeve has not been shown here for clear visualization of the arrangements inside the stage. The sample was loaded on both sides using steel grips attached to the actuator on one side and the load cell on other. To introduce moisture

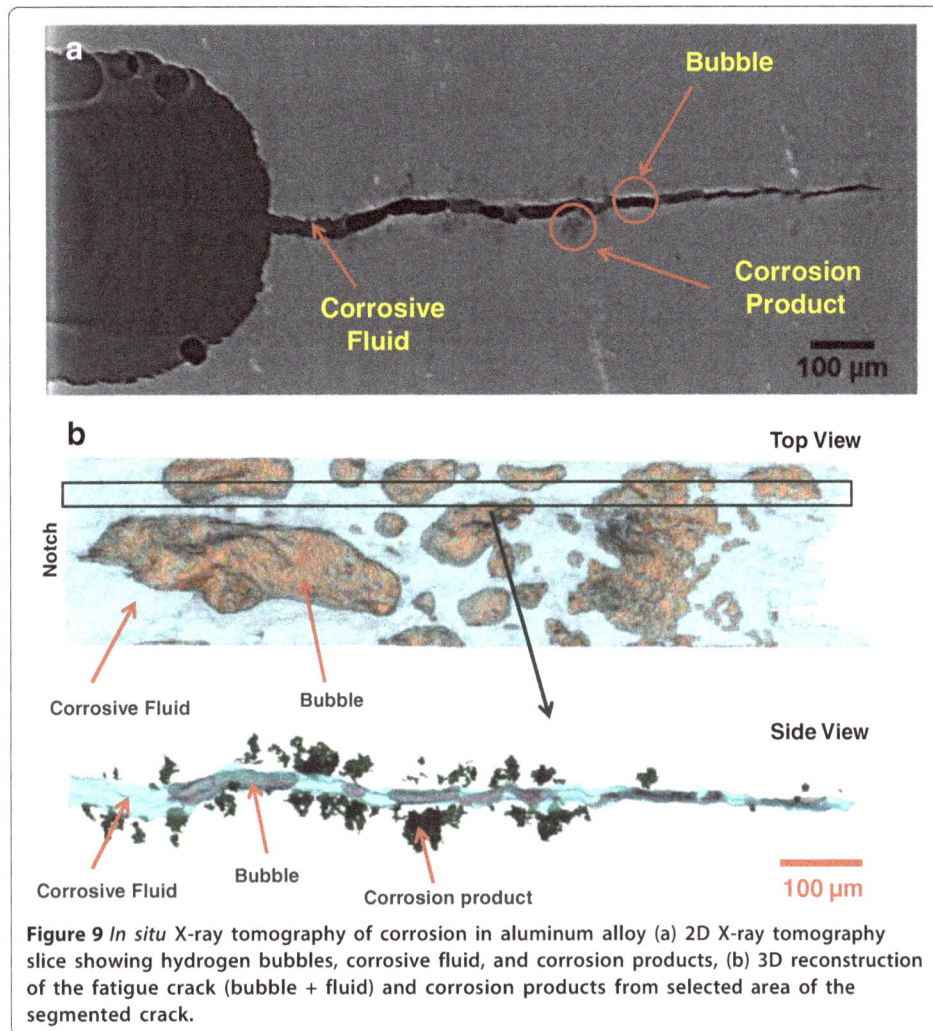

Figure 9 *In situ* X-ray tomography of corrosion in aluminum alloy (a) 2D X-ray tomography slice showing hydrogen bubbles, corrosive fluid, and corrosion products, (b) 3D reconstruction of the fatigue crack (bubble + fluid) and corrosion products from selected area of the segmented crack.

in the cell, an annular (ring shape) wet sponge was placed around the load cell at the base of the loading stage (a gap was provided between load cell and the sponge). Since stress corrosion cracking experiments require long time exposure of moisture, maintaining the constant humidity inside the chamber is a big challenge and is also critical to minimize the variability in results. In order to minimize the loss of moisture and to maintain constant relative humidity, the top part of the loading stage was covered with a plastic wrap as shown in Figure 11a. A humidity sensor was placed in front of the notch to measure the relative humidity throughout the test. These arrangements led to the constant humidity of 95–96% inside the system throughout the test, as shown in Figure 11b.

The *in situ* stress corrosion cracking experiment was performed on AA7075 in under-aged (UA) condition under constant load. The under-aged AA7075 was used due to the limited time available at the synchrotron beamline as it has already been established that the under-aged alloy is most susceptible to stress corrosion cracking in moisture [24]. Small SEN specimens were machined along the ST orientation by EDM for the SCC tests. Preliminary results obtained from these arrangements are shown in Figure 12, which contains 2-D X-ray tomography images of a SCC crack over time at

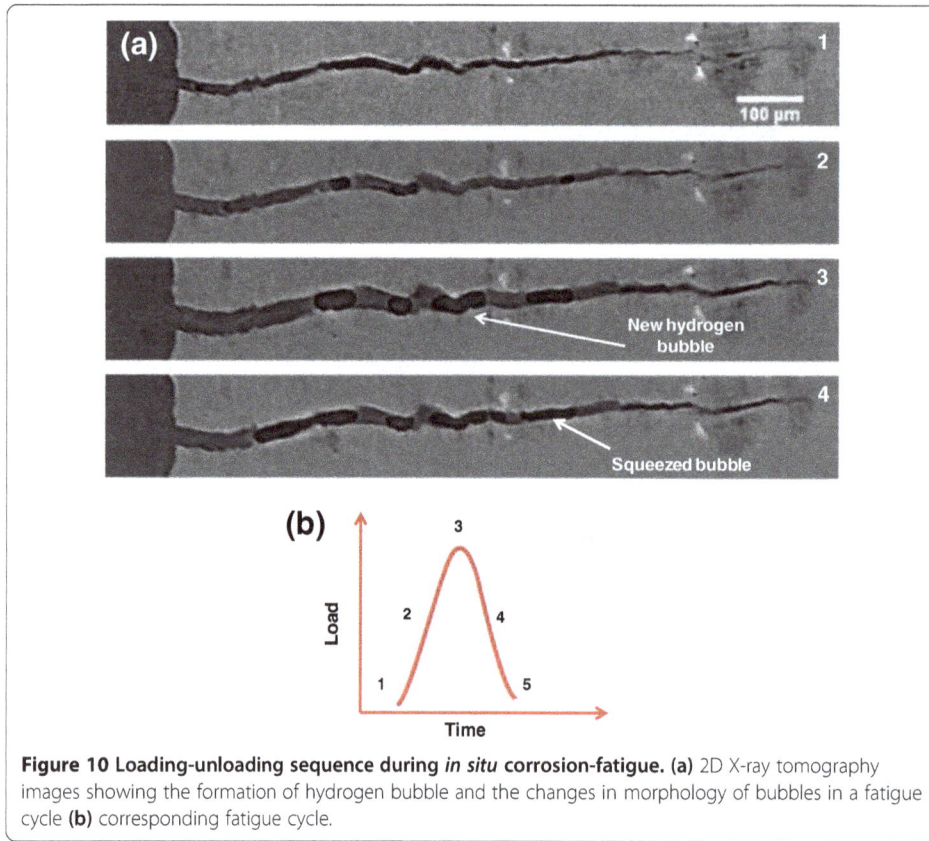

Figure 10 Loading-unloading sequence during *in situ* corrosion-fatigue. (a) 2D X-ray tomography images showing the formation of hydrogen bubble and the changes in morphology of bubbles in a fatigue cycle **(b)** corresponding fatigue cycle.

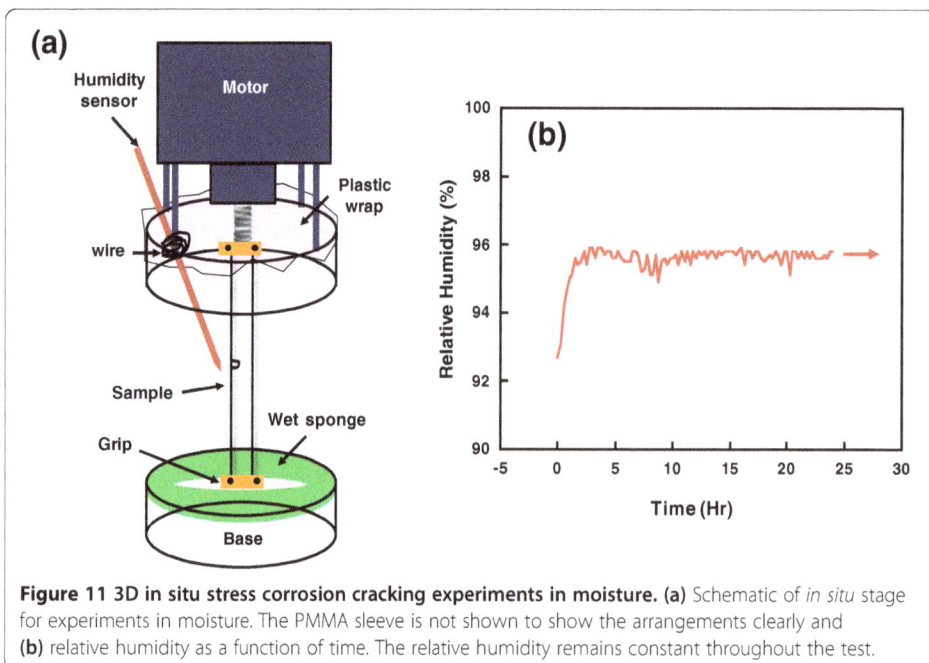

Figure 11 3D in situ stress corrosion cracking experiments in moisture. (a) Schematic of *in situ* stage for experiments in moisture. The PMMA sleeve is not shown to show the arrangements clearly and **(b)** relative humidity as a function of time. The relative humidity remains constant throughout the test.

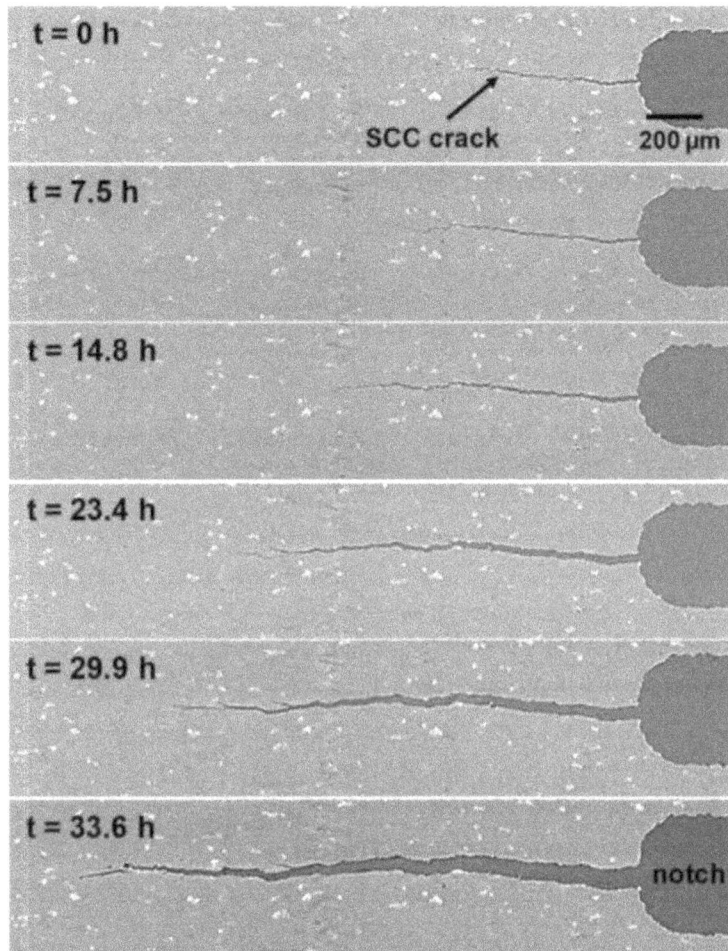

Figure 12 X-ray synchrotron tomography images of the center of the specimen with time. The crack is growing at constant load (SCC).

constant load at the center of the specimen. These results show that the SCC tests in moisture can be performed by these arrangements. It should be noted that the same arrangements can be used to perform fatigue test in moisture by changing the constant load to cyclic load.

Summary

In situ techniques using X-ray synchrotron tomography to understand the mechanical behavior of materials under variety of conditions have been explored. We have described several *in situ* loading methodologies, challenges, and solutions. The provisions for alignment led to symmetrical crack front in the second generation *in situ* mechanical testing stage. The third generation *in situ* loading stage provided load-control capability and additional stiffness to perform high R-ratio fatigue crack growth experiments. These loading stages were used for mechanical testing (monotonic and cyclic loading) in ambient air as well as in corrosive environments such as EXCO solution or moisture.

Competing interests

The authors declare that they have no competing interests.

Authors' contribution

SS carried out experiments, analysis of data, and wrote the manuscript. JJW designed the testing jigs, carried out experiments, and analysis. PH helped with experimental analysis and image segmentation. XX and FDC provided experimental support at APS. NC helped design the experiments, ideas for crack growth and corrosion, and helped in writing the manuscript. All authors read and approved the final manuscript.

Acknowledgements

We acknowledge financial support from the Office of Naval Research, under contract number N000141010350 (Dr. Asuri Vasudevan, Program Manager). Use of the Advanced Photon Source was supported by the U.S. Department of Energy, Office of Science, Office of Basic Energy Sciences, under Contract No. DE-AC02-06CH11357. We also thank Carl Mayer at Arizona State University for useful discussions.

Author details

[1]Materials Science and Engineering, Arizona State University, Tempe, AZ 85287-6106, USA. [2]Advanced Photon Source, Argonne National Laboratory, Argonne, IL, USA.

References

1. Buffiere JY, Maire E, Adrien J, Masse JP, Boller E (2010) *In situ* experiments with x ray tomography: an attractive tool for experimental mechanics. Exp Mech 50:289–305
2. Maire E, Buffiere JY, Salvo L, Blandin JJ, Ludwig W, Letang JM (2001) On the application of x-ray microtomography in the field of materials science. Adv Eng Mater 3:539–546
3. Beckmann F, Grupp R, Haibel A, Huppmann M, Nothe M, Pyzalla A, Reimers W, Schreyer A, Zettler R (2007) *In situ* synchrotron x-ray microtomography studies of microstructure and damage evolution in engineering materials. Adv Eng Mater 9:939–950
4. Buffiere JY, Cloetens P, Ludwig W, Maire E, Salvo L (2008) *In situ* x-ray tomography studies of microstructural evolution combined with 3D modeling. MRS Bull 33:611–619
5. Williams JJ, Flom Z, Amell A, Chawla N, Xiao X, De Carlo F (2010) Damage evolution in SiC particle reinforced Al alloy matrix composites by X-ray synchrotron tomography. Acta Mater 58:6194–6205
6. Link T, Zabler S, Epishin A, Haibel A, Bansal M, Thibault X (2006) Synchrotron tomography of porosity in single-crystal nickel-base superalloys. Mater Sci Eng A 425:47–54
7. Caty O, Maire E, Bouchet R (2008) Fatigue of metal hollow spheres structures. Adv Eng Mater 10:179–184
8. Williams JJ, Chapman NC, Jakkali V, Tanna VA, Chawla N, Xiao X, De Carlo F (2011) Characterization of damage evolution in SiC particle reinforced Al alloy matrix composites by in-situ X-ray synchrotron tomography. Metall Mater Trans A 42:2999–3005
9. Maire E, Zhou S, Adrien J, Dimichiel M (2011) Damage quantification in aluminium alloys using *in situ* tensile tests in X-ray tomography. Eng Fract Mech 78:2679–2690
10. Withers PJ, Preuss M (2012) Fatigue and damage in structural materials studied by X-ray tomography. Annu Rev Mater Res 42:81–103
11. Guvenilir A, Breunig TM, Kinney JH, Stock SR (1997) Direct observation of crack opening as a function of applied load in the interior of a notched tensile sample of Al-Li 2090. Acta Mater 45(5):1977–1987
12. Williams JJ, Yazzie KE, Padilla E, Chawla N, Xiao X, De Carlo F (2013) Understanding fatigue crack growth in aluminum alloys by *in situ* X-ray synchrotron tomography. Int J Fatigue 57:79–85
13. Williams JJ, Yazzie KE, Phillips NC, Chawla N, Xiao X, Carlo FD, Iyyer N, Kittur M (2011) On the correlation between fatigue striation spacing and crack growth rate: a three-dimensional (3-D) X-ray synchrotron tomography study. Metall Mater Trans A 42A:3845–3847
14. Khor KH, Buffiere JY, Ludwig W, Toda H, Ubhi HS, Gregson PJ, Sinclair I (2004) *In situ* high resolution synchrotron x-ray tomography of fatigue crack closure micromechanisms. J Phys Condens Matter 16:S3511–S3515
15. Singh SS, Williams JJ, Xiao X (2012) Carlo FDe, Chawla N (2012) *In situ* three dimensional (3D) X-ray synchrotron tomography of corrosion fatigue in Al7075 alloy. In: Srivatsan TS, Imam AM, Srinivasan R (eds) Fatigue of Materials II: advances and emergences in understanding. Materials Science and Technology, Pittsburgh
16. Pyzalla A, Camin B, Lehrer B, Wichert M, Koch A, Zimnik K, Boller E, Reimers W (2006) *In-situ* observation of creep damage in Al-Al$_2$O$_3$ MMCs by synchrotron X-ray tomography. Int Centre Diffraction Data:1097–2102
17. Issac A, Sketa F, Reimers W, Caminb B, Sauthoff G, Pyzalla AR (2008) *In situ* 3D quantification of the evolution of creep cavity size, shape, and spatial orientation using synchrotron X-ray tomography. Mater Sci Eng A 478:108–118
18. De Carlo F, Tieman B (2004) High-throughput X-ray microtomography system at the advanced photon source beamline 2-BM. SPIE 5535:644–651, Ulrich Bonse, editors
19. Peele G, De Carlo F, McMahon PJ, Dhal BB, Nugent KA (2005) X-ray phase contrast tomography with a bending magnet source. Rev Sci Instrum 76:083707-1-5
20. De Carlo F, Albee P, Chu YS, Mancini DC, Tieman B, Wang SY (2002) High-throughput real-time X-ray microtomography at the advanced photon source. Proc SPIE 4503:1–13, U. Bonse, ed
21. Mikheevskiy S (2009) Elastic–plastic fatigue crack growth analysis under variable amplitude loading spectra. Dissertation, University of Waterloo, Ph.D

22. Chawla N, Andres C, Jones JW, Allison JE (1998) Effect of SiC volume fraction and particle size on the fatigue resistance of a 2080 Al/SiCp composite. Metall Mater Trans A 29:2843–2854
23. Plastic chemical resistance chart, plastic international., http://www.plasticsintl.com/plastics_chemical_resistence_chart.html
24. Holroyd NJH, Scamansc GM (2011) Crack propagation during sustained-load cracking of Al-Zn-Mg-Cu aluminum alloys exposed to moist air or distilled water. Metall Mater Trans A 42:3979–3998

10

Three-dimensional sampling of material structure for property modeling and design

McLean P Echlin[*], William C Lenthe and Tresa M Pollock

*Correspondence:
mechlin@engineering.ucsb.edu
University of California Santa
Barbara, Materials Dept. Building
503 Santa Barbara, CA, 93106-5050,
USA

Abstract

Newly developed 3-D tomographic techniques permit acquisition of quantitative materials data for input to structure-property models. At the mesoscale, techniques that enable sampling of larger material volumes provide information such as grain size and morphology, 3-D interfacial character, and chemical gradients. However, systematic approaches for determining the characteristic material volume for 3-D analysis have yet to be established. In this work, the variability in properties due to microstructure is discussed in the context of a methodology for defining volume elements that link microstructure, properties, and design. As such, we propose a 3-D sampling methodology based on convergence of microstructural parameters and associated properties and design considerations.

Keywords: Representative volume element; Microstructure volume elements; Property volume elements; Femtosecond laser; Tomography; Serial sectioning

Background

With the dramatic increases in capability of 3-D tomographic techniques in the past decade [1-11], it is now possible to acquire quantitative information on material structure for higher fidelity property models. Tomographic data can be acquired across many lengthscales, using techniques such as atom probe tomography, transmission electron microscope tomography, focused ion beam serial sectioning, femtosecond laser tomography, microtomes, and both benchtop and synchrotron x-ray techniques. However, protocols for gathering 3-D data, in terms of defining representative volume elements and statistical sampling approaches, remain poorly defined for most engineering materials and their corresponding properties.

Simplified volumetric representations of materials are often made in an attempt to reduce the amount of data being passed into a component design process. Examples of these volumetric reductions include representative volume elements (RVEs) [12,13], statistical volume elements (SVEs) [14,15], and statistically equivalent representative volume elements (SERVE) [16-18]. Criteria for the degree of volumetric reduction are often linked to continuum modeling assumptions, convergence of a given property, or statistical representation of specific microstructural features. Existing approaches have defined specialized RVEs that are augmented sets of sampled volume elements which are statically selected to be representative, in aggregate. Statistical representative volume element sampling has been proposed and applied [15,19-21] to materials such as titanium and fiber composites.

Engineering materials possess microstructural features across a wide spectrum of lengthscales that determine the property responses. Figure 1 shows the approximate lengthscales at which material properties (e.g., elastic modulus, creep, and fatigue) are controlled and the corresponding volume element size necessary to characterize them. The range in requisite volume element sizes is rooted in the types, sizes, and variation in the microstructures that dictate the material property responses. The property responses can be categorized as one of the following: (1) structure insensitive - properties mainly being described by the atomistic characteristic of the material, (2) dependent on the 'mean' structure - due the fact that they are dependent on the aggregate microstructural features and are relatively 'flaw insensitive', and (3) having dependence on the 'extremes' of microstructures - these are properties that depend on microstructures that are statistically rare and require very large material volume element sizes, sophisticated combinations of volume element sampling, or may require volumes entirely too large to quantify. In the case of properties such as fatigue, which may rely on very rare microstructural features, extreme value statics can be employed [22,23].

Even when employing existing structure-property models, it is still very difficult to precisely select the correct tomographic volume size for collection of experimental information that is statistically representative, can satisfy the requirements of homogenization theory, and is appropriate for the design of engineering components. We propose a method to quantitatively define volume elements that can be linked directly to microstructures/properties/design elements of interest with a finite range over which the volume element is valid.

The representative volume element

To properly describe material properties and associated constitutive response, it is necessary to select a representative volume element for analysis. Typically, RVEs are defined as

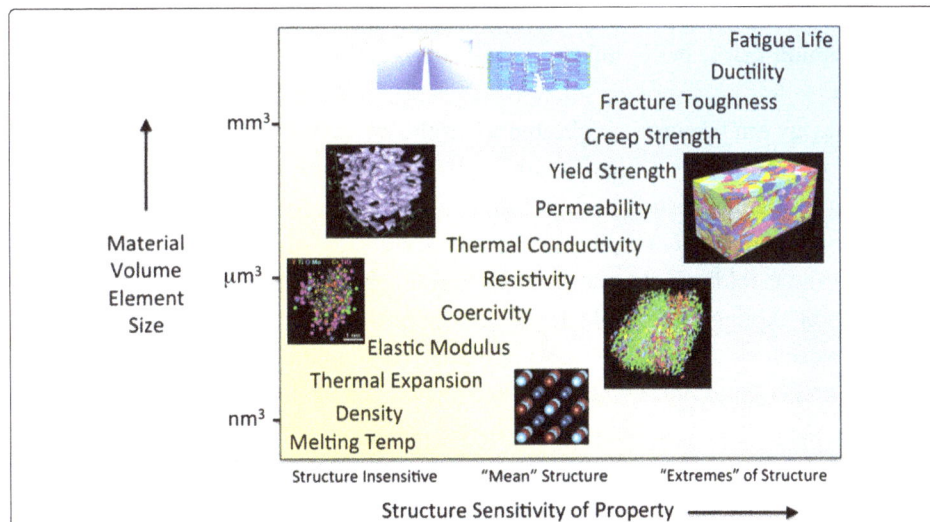

Figure 1 Material properties require differently sized material volume elements to accurately describe them. Properties can be categorized as one of three types: (1) structure insensitive - properties mainly being described by the atomistic characteristic of the material, (2) dependent on the mean structure - due the fact that they are dependent on the aggregate microstructural features and are relatively flaw insensitive, and (3) having dependence on the extremes of microstructures - these are properties that depend on microstructures that are statistically rare and require very large material volume element sizes.

a conveniently sized volume which is smaller than the macroscopic sample of interest, but large enough to be representative of the material as a whole and therefore valid for continuum homogenization assumptions. The breadth of these definitions are well summarized by Gitman [12], with most falling into one of two categories: (1) microstructure-based descriptions and (2) mechanics-based descriptors. RVEs often are defined to be of a volume that is large enough to have constitutive material properties, often only elastic properties, but small enough to be computationally tractable.

For mechanics problems, often RVEs are sized so that the Hill conditions [13,14] are met. These conditions are met when the property of interest becomes independent of RVE size, according to one of the following boundary conditions: uniform displacement, uniform traction, displacement-traction, or periodic. As described by Qidwai [19], RVEs are commonly estimated by tracking property convergence as a function of increasing volume element size. They also note that the RVE dataset size often is collected and then (erroneously) used for many or all material properties. For specific applications, these definitions are useful, but not in design cases which require many material properties to be considered.

Limitations of the RVE
Continuum mechanics RVE definitions require that a volume element size randomly sampled from the bulk will have uniform material property response. Different properties have a dependence on microstructures which exist at specific length scales. For example, the elastic modulus has a strong dependence on interatomic bonding (atomic scale) and grain structure/texture (the mean structure at the mesoscale), whereas fatigue life can have a dependence on distributions of pores or other extrinsic material flaws (the extremes of structure at the mesoscale). Fatigue life would therefore require a much larger volume element to be appropriately modeled, as shown in Figure 1. It would then follow that an RVE size for the property of elastic modulus should not be used for modeling fatigue life. The discrepancies in types of RVE definitions illustrates the need for a volume element description that is better linked to the material parameters of microstructures and properties.

The RVE strategy can become complicated when the microstructures that govern the property or properties of interest span across lengthscales. In such a case, the RVE would need to be large to define mesoscale-sized features such as dendrites, grain texture, and shrinkage pores, while simultaneously having finer-scale microstructural features, such as precipitates and carbides. This strategy is problematic because it pushes the limits of the capabilities of existing tomography tools and computational methods. As such, one would prefer to decouple the fine and coarse resolution systems into separate tomography experiments. Similar types of decoupled methods have become popular with hierarchical modeling systems [24,25], although these models often only take average parameter inputs from each of the RVEs. A distinct benefit of using 3-D datasets for inputs to microstructure modeling is the ability to directly input microstructure descriptors that contain more information than single value parameters; examples of this include morphological parameters [26-28], shape parameters [29], and microstructure distribution functions [30].

New tools for gathering larger volumes of material *in situ* in a SEM have been developed recently [11,31,32]. One such tool, the TriBeam system [11], uses ultrashort femtosecond laser pulses to remove material through a layer-by-layer ablation process. This tool allows for the gathering of 3-D datasets either using established FIB serial sectioning [3] or by femtosecond laser ablation, which produces datasets 3 orders of magnitude volumetrically larger in time periods ranging from a few hours to a few days. The access to these substantially larger 3-D datasets motivates new approaches for probing and analyzing material volume elements. Here we use datasets generated by this new technique to address the volume element challenge.

Methods

First, we describe new volume element definitions that connect microstructure and property level descriptors with standard modes of component design. Then two sampling methods are described which (1) bound convergence criteria and the corresponding estimates of variability and (2) compare sampled volume elements from different component locations. Examples of the application of these methods will be described in 'Results' section. The ability of the new TriBeam technique to gather mesoscale-sized datasets appropriate for a range of mechanical properties is discussed.

Volume element definitions

The representative volume element is not typically clearly defined for microstructural representation, compared to property representation. As such, we define an infrastructure that connects the volume element concepts from the materials realm to the design domain, with emphasis given to building hierarchically on materials descriptors. These volume element definitions will be described presently, starting with the most fundamental microscopic descriptors and then moving up to the macroscopic scale. The most basic volume elements are defined as *microstructural volume elements* (MVEs), which have volumes that scale with the microstructural features of relevance. Examples of structural features that may constrain the size of the MVE include grain size, precipitate volume fraction, dendrite spacing, texture, and precipitate size. They can be defined as average quantities, distributions, or scalar quantities depending on the requirements of the structure-property models being used. Often, MVEs can be one, two, or *n*-point descriptors, which are covered in detail elsewhere [21,33]. Next, there are *property volume elements* (PVEs), which are linked to MVEs by existing or yet to be developed structure-property models and therefore have sizes that scale with the microstructure volume elements on which they depend. Examples of properties that define the PVE are yield strength, elastic modulus, thermal conductivity, and permeability, which will be discussed as example cases in 'Results' section. Contrary to intuition, PVEs are not simply defined as the maximum size of their dependent MVEs; this will be more rigorously addressed later in the paper. The *design volume element* (DVE) is composed of the volume of an engineering component being designed or alternatively as a sub-region of the component of interest. For example, an engineering component may be designed to remain elastic over its entire volume, except for a small volume of material located adjacent to a stress-concentrating notch. In this event, the MVE and PVE for the elastic modulus would apply to the bulk of the component and, for example, the yield strength PVE would then be used to size the notch. Therefore, the DVE is specific to the component and its anticipated application. A schematic of the relations of the MVE/PVE/DVEs is shown in Figure 2, with example structure-property models inset within the connectivity between MVE and PVE, and example MVE and PVE definitions labeled within the boxes.

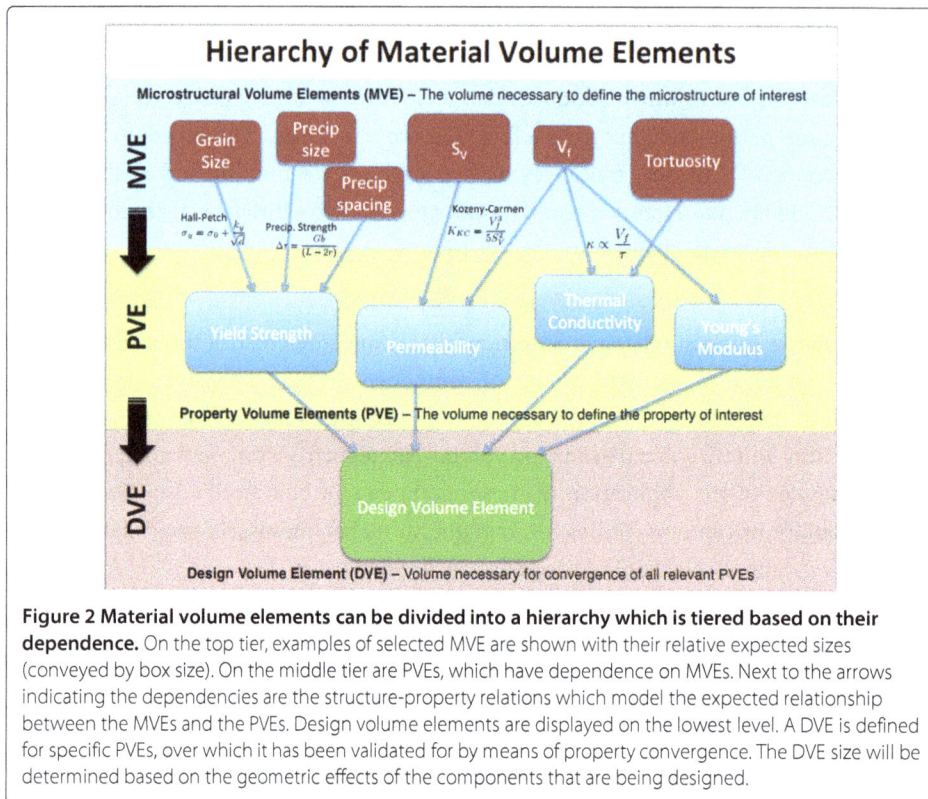

Figure 2 Material volume elements can be divided into a hierarchy which is tiered based on their dependence. On the top tier, examples of selected MVE are shown with their relative expected sizes (conveyed by box size). On the middle tier are PVEs, which have dependence on MVEs. Next to the arrows indicating the dependencies are the structure-property relations which model the expected relationship between the MVEs and the PVEs. Design volume elements are displayed on the lowest level. A DVE is defined for specific PVEs, over which it has been validated for by means of property convergence. The DVE size will be determined based on the geometric effects of the components that are being designed.

The volume element definitions presented in this section can be used in the following two ways: (1) a DVE volume can be defined as the size at which point all MVEs and PVEs converge; (2) in the event that the DVE is instead limited by the physical design constrains of the part, then alternately the variability of the MVEs and PVEs applicable to the design problem can be assessed to provide information for the minimum material property design limits. In other words, the DVE size limits the problem and therefore the variability in PVEs at the prescribed size can be evaluated. Examples of these methods will be given in the 'Results' section.

Sampling for convergence size

Two distinct materials sampling methods are presented to define volume elements for MVEs or PVEs. The first method is used to randomly sample n volumes of equal size across a range of increasing volume sizes (V_1, V_2, ..., V_i) in order to determine the volume V_c at which microstructure or property convergence occurs, shown in Figure 3. This method has been applied to materials such as tungsten copper (WCu) composite, using the MVEs of volume fraction (V_f) and surface area to volume ratio (S_v) for the PVEs of permeability (K), and polycrystalline Young's modulus (E) which is shown in more detail in 'Results' section and in [34]. Convergence of these MVEs or PVEs was determined using a standard 99% confidence interval bound (to be within 5% of the mean), shown in Figure 4, and applying statistical hypothesis tests, such as the t test and z test. For example, the t test confidence interval (CI) around the sampled volume fraction average, $\overline{V_f}$, is defined as

$$CI = \overline{V_f} \pm t^* \frac{\sigma_{V_f}}{\sqrt{n}} \tag{1}$$

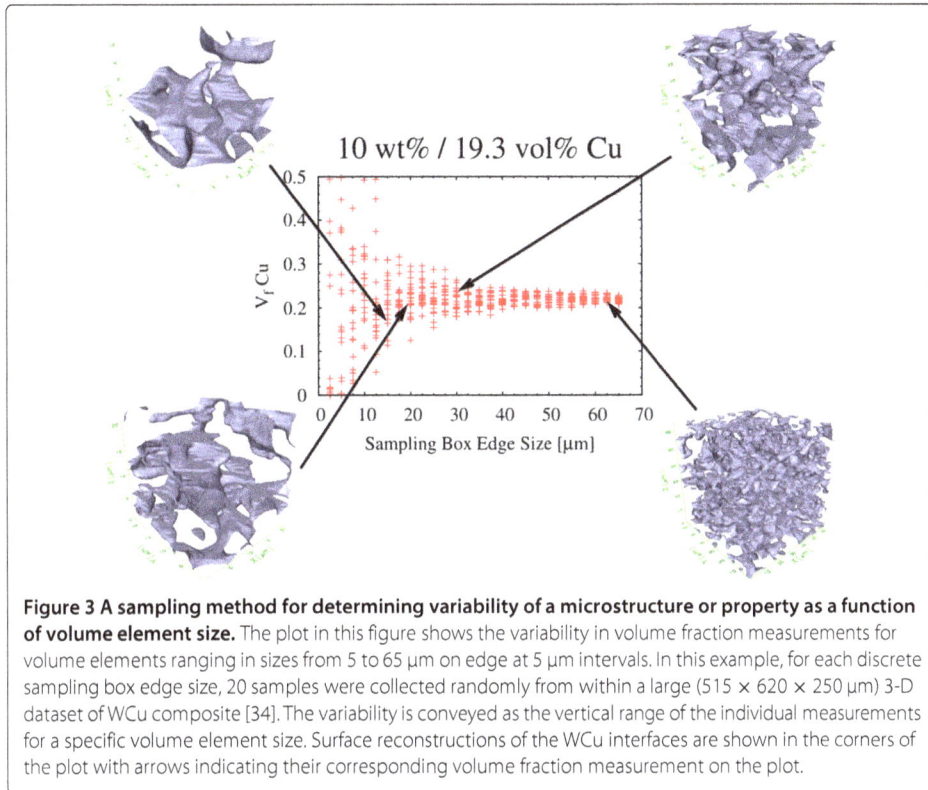

Figure 3 A sampling method for determining variability of a microstructure or property as a function of volume element size. The plot in this figure shows the variability in volume fraction measurements for volume elements ranging in sizes from 5 to 65 µm on edge at 5 µm intervals. In this example, for each discrete sampling box edge size, 20 samples were collected randomly from within a large (515 × 620 × 250 µm) 3-D dataset of WCu composite [34]. The variability is conveyed as the vertical range of the individual measurements for a specific volume element size. Surface reconstructions of the WCu interfaces are shown in the corners of the plot with arrows indicating their corresponding volume fraction measurement on the plot.

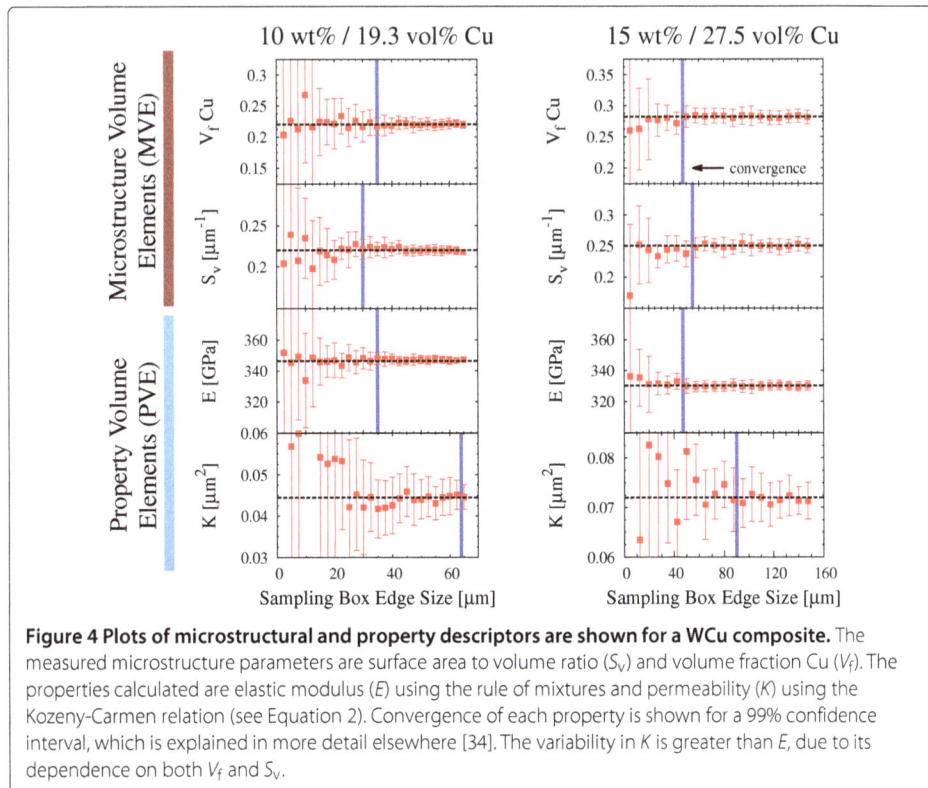

Figure 4 Plots of microstructural and property descriptors are shown for a WCu composite. The measured microstructure parameters are surface area to volume ratio (S_v) and volume fraction Cu (V_f). The properties calculated are elastic modulus (E) using the rule of mixtures and permeability (K) using the Kozeny-Carmen relation (see Equation 2). Convergence of each property is shown for a 99% confidence interval, which is explained in more detail elsewhere [34]. The variability in K is greater than E, due to its dependence on both V_f and S_v.

where σ_{V_f} is the sampled standard deviation in V_f, t^* is evaluated from the t distribution for the desired confidence index, and n is the number of randomly sampled volumes. Furthermore, the confidence interval can be expressed in terms of the coefficient of variation $\left(C_v = \sigma_{V_f}/\overline{V_f}\right)$. This general methodology of tracking convergence has been used regularly in problems such as random composites [35], ice cream [36], hydrided Zircaloy cladding [37], and the WCu composite discussed in this research.

Sampling for rare or site-specific structural features

The second method for sampling is more relevant for properties that require volume element sizes which are inaccessible due to a mismatch between the capabilities of tomography and the volume to be sampled or require sampling from site-specific locations. Examples of rare features that can be interrogated include interconnect defects in electronics components and porosity in cast metals. Examples of site-specific microstructural features include grain boundaries or specially oriented crystals. Using a tool such as the TriBeam [11] or a dual-beam focused ion beam (FIB), targeted dataset acquisitions can be made from multiple locations within a sample, shown schematically in Figure 5.

It is often of interest to collect datasets which are spatially located near design features which are known to accelerate events leading to failure or be deleterious to local material properties. Stress concentrators such as notches or cracks are particularly detrimental under fatigue and static loads [38] where the local structure and properties are of strong interest. In such cases, one would like to measure the microstructure features and local properties nearby the geometric irregularity that will be preferentially sampled by the design feature. It is also important to consider cases where the design-imposed geometric constraints interrogate a volume that is much smaller than the microstructure or property volume elements.

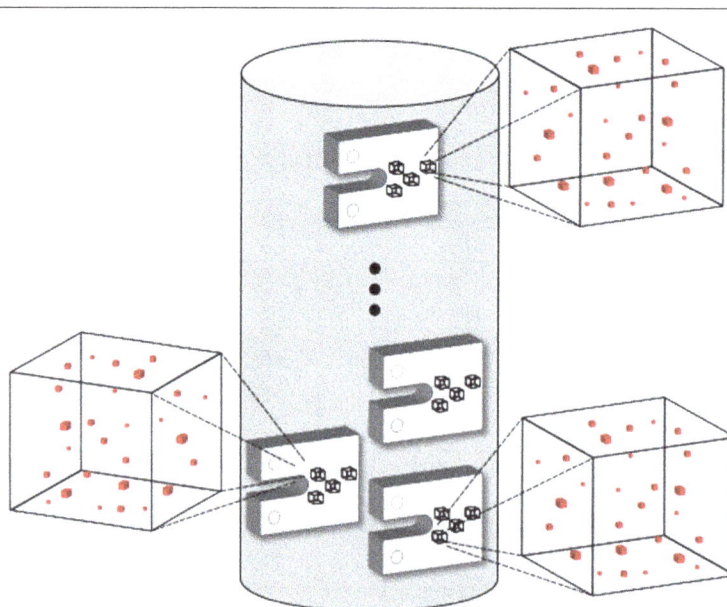

Figure 5 Sampling method for use with extremely large property or microstructure volume elements (MVEs or PVEs). A component can be randomly (or strategically) sampled to find long range variation in microstructure or to capture 'rare' features. In this methodology, a volume element size may be chosen that is not fully converged but sampled often enough to define the rare events.

This method of sampling for rare structures was employed to gather large 3-D datasets from a high-strength steel containing widely spaced titanium nitride (TiN) inclusion phases that have sizes ranging from 1 to 10 µm at volume fractions of 0.01% to 0.05%. During crack advance, the TiN inclusions are deleterious to fracture toughness, a material property which has a large PVE. TiN 3-D datasets were gathered from many locations on a compact tension sample, collecting different plastic zone-sized volumes that a crack tip would sample during crack propagation [39]. These datasets were then used, in aggregate, to measure the various spatial distributions of TiN inclusions in a high-strength steel and assess how the TiN inclusions contribute to the variability in fracture toughness.

Results

The MVE/PVE/DVE hierarchy has been designed to connect the existing structure-property relations (models) with component design architecture. In the following sections, we demonstrate its use for several sample systems with 3-D data gathered by femtosecond laser-aided tomography in both the vacuum chamber (TriBeam) and in ambient laboratory air with optical imaging.

MVE and PVE variability

WCu composite datasets were collected using the TriBeam [11] in less than 48 h with volumes as large as $615 \times 525 \times 250$ µm with a 250-nm slice thickness [34]. The TriBeam uses a femtosecond laser to ablate sections of material at rates that are 4 to 5 orders of magnitude faster than the standard focused ion beam source available in many dual-beam FIB microscope system. The resulting image stacks from the tomography experiments are composed of 100s to 1000s of secondary electron images, which were segmented, registered, and reconstructed in 3-D.

Two different composition WCu datasets were collected using the TriBeam system [34] and sampled to analyze the convergence of MVEs and PVEs, shown in Figure 4. Samples were collected by randomly selecting volumes at 5 µm intervals between 5 and 65 µm on the edge for a W-10 wt.% Cu composite dataset and 5 to 160 µm on edge for a W-15 wt.% Cu composite dataset. A total of 20 random samples were taken for each volume in order to calculate the variability in two microstructural parameters and two material properties. A sensitivity study was performed to determine the number of samples necessary for variability analysis and the results are shown in Figure 6. These data show that for n (number of randomly sampled volume elements) greater than 5 to 10, variability plateaus; therefore, all analyses were performed at 20 samples per volume. Figure 4 shows the average value of each of 20 samples plotted as solid squares, for both W-10 wt.% Cu and W-15 wt.% Cu composites, with the standard deviations in the sample sets indicated with bar lines for volume fraction (V_f), surface area to volume ratio (S_v), permeability (K), and polycrystalline effective elastic modulus (E). The microstructural or property average values that the data are converging toward are shown as a horizontal dotted line, while the converged volume element size (with 99% confidence interval to be within 5% of the mean) is shown as a vertical solid line. The MVEs (V_f and S_v) converge faster than the PVEs (K and E), as illustrated by the positions of the vertical lines in Figure 4. Also, the aggregate-converged PVE size is non-intuitively larger than the largest dependent MVE convergence size. This result demonstrates the compounding variability that accrues with

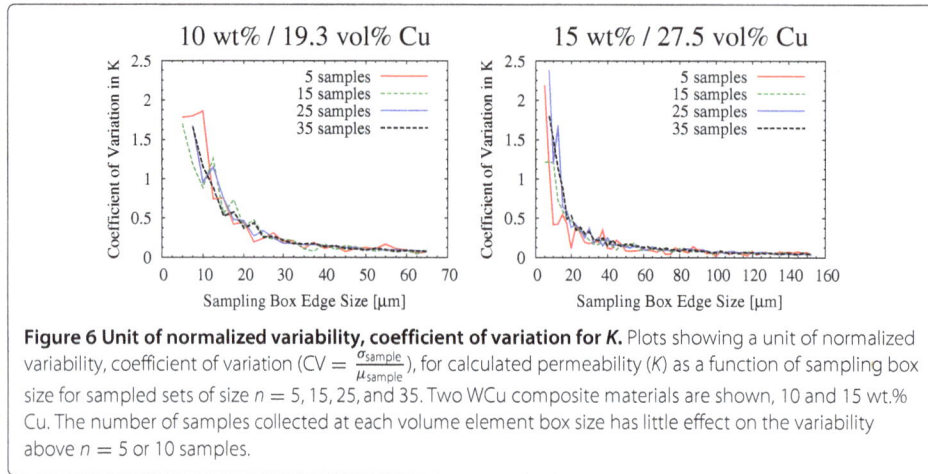

Figure 6 Unit of normalized variability, coefficient of variation for K. Plots showing a unit of normalized variability, coefficient of variation (CV $= \frac{\sigma_{sample}}{\mu_{sample}}$), for calculated permeability (K) as a function of sampling box size for sampled sets of size $n = 5, 15, 25,$ and 35. Two WCu composite materials are shown, 10 and 15 wt.% Cu. The number of samples collected at each volume element box size has little effect on the variability above $n = 5$ or 10 samples.

multiple MVE dependence of a PVE. Linking the predicted MVE variability at a specified PVE size will be shown to be a valuable tool in 'Discussion' section.

Error analysis, which is well detailed elsewhere [40], can be also be performed for many existing analytical structure-property models to determine the uncertainty in a sampled property average. For example, the variability in permeability (K), as defined by the Kozeny-Carmen relation,

$$K = \frac{V_f^3}{5S_V^2} \tag{2}$$

where V_f is the volume fraction Cu, and S_V is the surface area to volume ratio which can be represented as

$$\sigma_K = \sqrt{\left(\frac{\partial K}{\partial V_f}\right)^2 \sigma_{V_f}^2 + \left(\frac{\partial K}{\partial S_V}\right)^2 \sigma_{S_V}^2 + 2 \frac{\partial K}{\partial S_V} \frac{\partial K}{\partial V_f} \sigma_{S_V V_f}} \tag{3}$$

where σ terms represent standard deviation and $\sigma_{S_V V_f}$ is the covariance of the two MVEs. In Figure 7, the uncertainty has been plotted with and without the covariance term, $\sigma_{S_V V_f}$, which is only necessary for a PVE (e.g., K) that has MVEs that are codependent (e.g., V_f and S_V). Notably, the uncertainty calculation with the covariance term included has a better fit with the sampled data volume, suggesting that the MVEs of V_f and S_V are both dependent on similar microstructural features and lengthscales, which is more clearly shown in the standard deviation plots of the same three calculations in Figure 8. Therefore, error analysis can be used to estimate PVE sizes from MVEs where the structure-property relation is analytically defined, whereas the PVE size must be determined by direct measurements of variability in the property from sampled volumes in all other cases. For example, error analysis requires a different approach in more complicated structure-property modeling relations such as numerical simulations of fluid flow or plasticity models that rely on non-analytical solutions or have stochastic components.

Design volume elements

Design volume elements (DVEs) are component specific. Consider for example the use of the WCu composite in a non-structural thermal protection application where ablative cooling via vaporization of Cu is required. Assuming stresses are thermal and well below

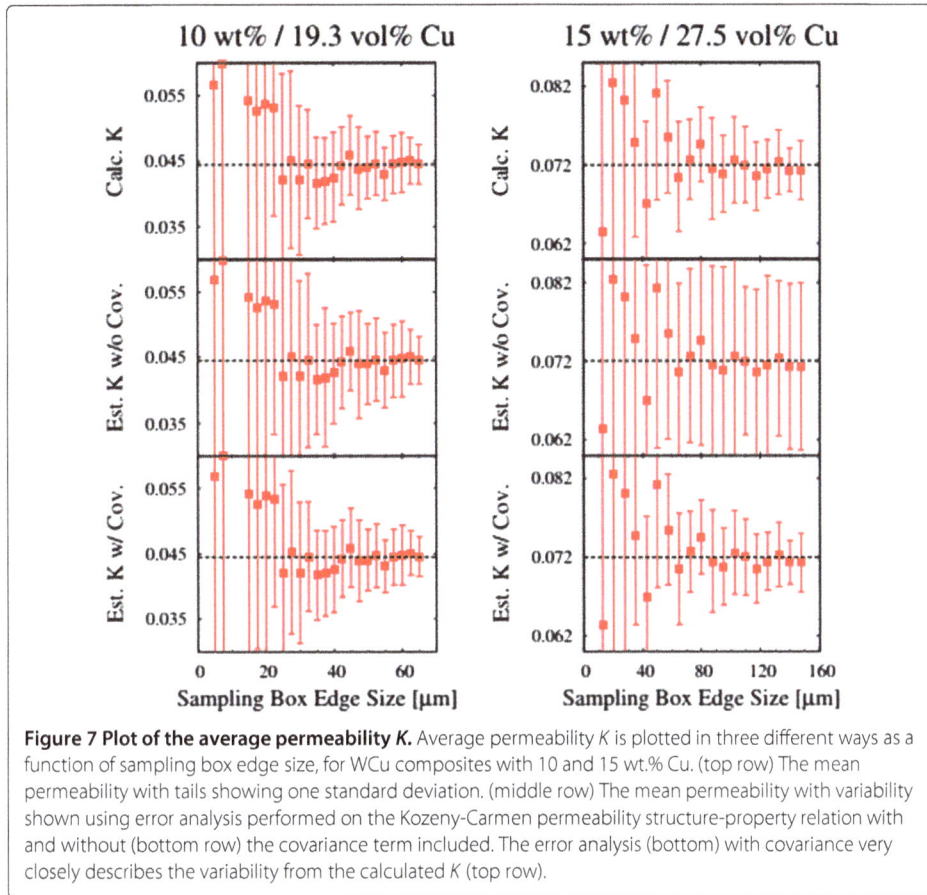

Figure 7 Plot of the average permeability K. Average permeability K is plotted in three different ways as a function of sampling box edge size, for WCu composites with 10 and 15 wt.% Cu. (top row) The mean permeability with tails showing one standard deviation. (middle row) The mean permeability with variability shown using error analysis performed on the Kozeny-Carmen permeability structure-property relation with and without (bottom row) the covariance term included. The error analysis (bottom) with covariance very closely describes the variability from the calculated K (top row).

the yield strength of the tungsten phase, the two primary properties of interest would have elastic modulus and permeability.

A volume element dependency chart similar to Figure 2 can be constructed for the DVE for the described thermal protection application, shown in Figure 9. This example shows the magnitude dependence in the amount of variability in the elastic modulus as

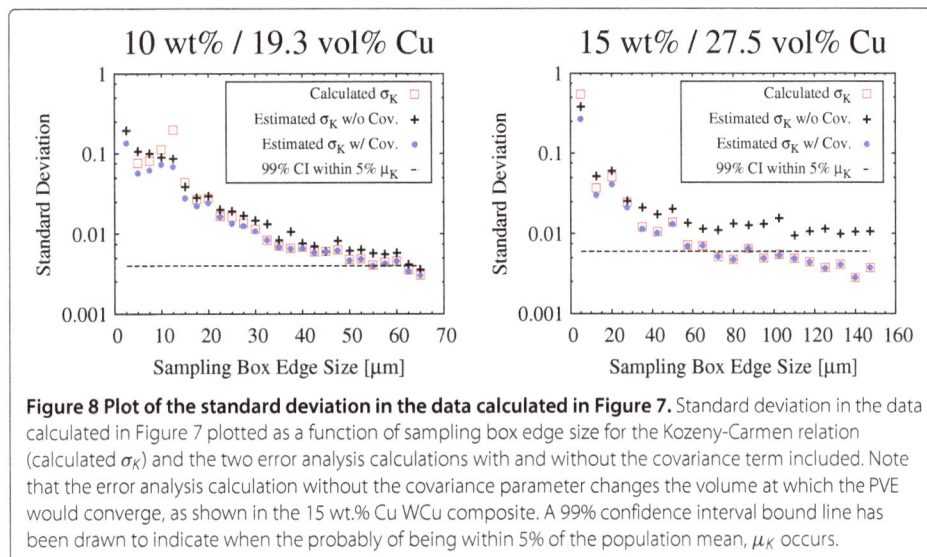

Figure 8 Plot of the standard deviation in the data calculated in Figure 7. Standard deviation in the data calculated in Figure 7 plotted as a function of sampling box edge size for the Kozeny-Carmen relation (calculated σ_K) and the two error analysis calculations with and without the covariance term included. Note that the error analysis calculation without the covariance parameter changes the volume at which the PVE would converge, as shown in the 15 wt.% Cu WCu composite. A 99% confidence interval bound line has been drawn to indicate when the probably of being within 5% of the population mean, μ_K occurs.

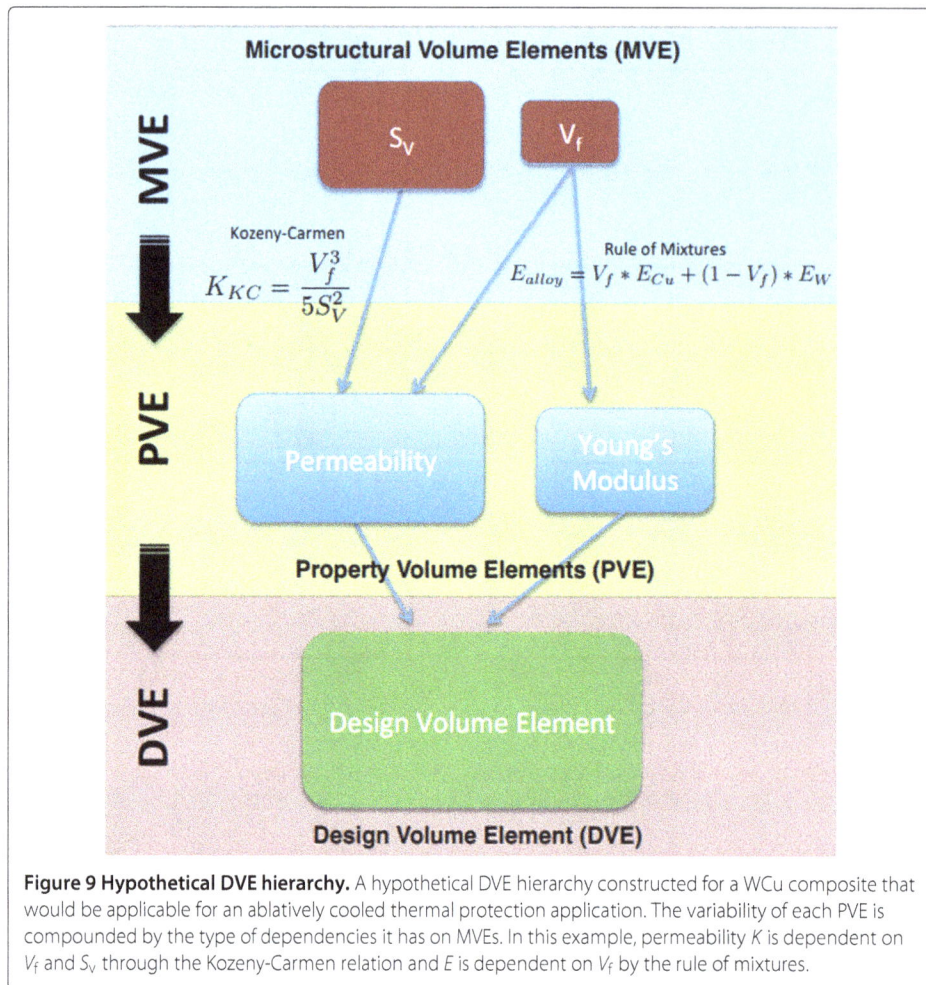

Figure 9 Hypothetical DVE hierarchy. A hypothetical DVE hierarchy constructed for a WCu composite that would be applicable for an ablatively cooled thermal protection application. The variability of each PVE is compounded by the type of dependencies it has on MVEs. In this example, permeability K is dependent on V_f and S_v through the Kozeny-Carmen relation and E is dependent on V_f by the rule of mixtures.

a function of V_f and the variability in permeability with both V_f and S_v. In this application, a DVE of size greater than the combined variability of both the elastic modulus and permeability will be necessary. Using the convergence data shown in Figure 4, then the DVE volume required for the W-10 wt.% Cu composite would be $> 65\,\mu m$ on edge and the W-15 wt.% Cu composite volume would be $> 90\,\mu m$ on edge for convergence in both properties. However, if a smaller volume is selected, the variability of the properties can be determined for the specified volume as indicated by the bar widths in Figures 3 and 4. For example, if a DVE was determined to be $60\,\mu m$ on edge in the W-15 wt.% Cu dataset, then 1 standard deviation from the mean ranges between 0.064 and 0.078, with a minimum permeability of 0.05. One can imagine that a larger set of properties would be required in other applications, i.e., yield strength, elastic modulus, fatigue strength, and permeability in this event the DVE would likely be even larger.

Discussion

RVEs used in design often assume uniform material properties and typically have sizes based only on mechanics considerations. However, to predict properties with a prescribed degree of confidence, it is important to account for the distribution of each of the microstructural features that influence the property of interest. Therefore, microstructural variability, shown in Figure 10, must be included in the property calculation in

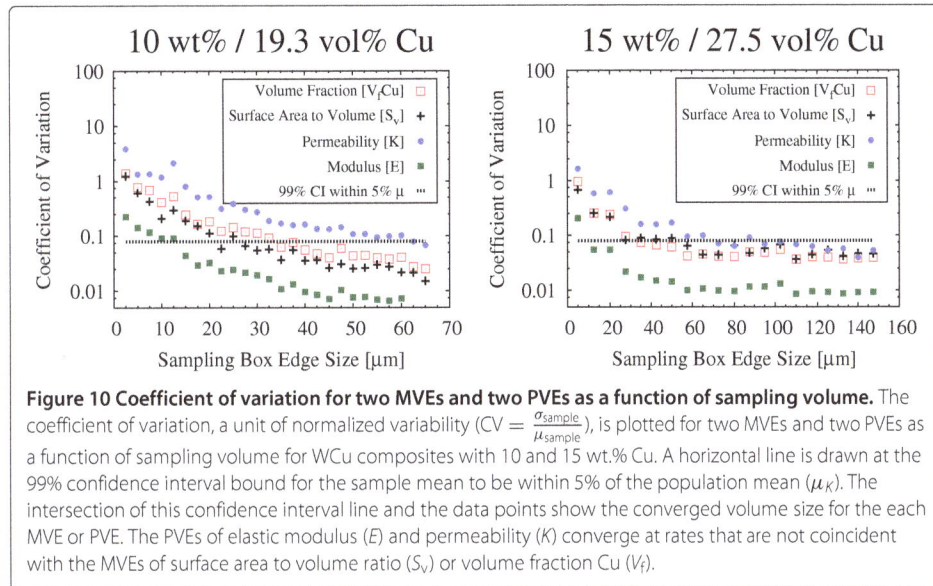

Figure 10 Coefficient of variation for two MVEs and two PVEs as a function of sampling volume. The coefficient of variation, a unit of normalized variability (CV $= \frac{\sigma_{sample}}{\mu_{sample}}$), is plotted for two MVEs and two PVEs as a function of sampling volume for WCu composites with 10 and 15 wt.% Cu. A horizontal line is drawn at the 99% confidence interval bound for the sample mean to be within 5% of the population mean (μ_K). The intersection of this confidence interval line and the data points show the converged volume size for the each MVE or PVE. The PVEs of elastic modulus (E) and permeability (K) converge at rates that are not coincident with the MVEs of surface area to volume ratio (S_v) or volume fraction Cu (V_f).

order to correctly predict the variability in material response and the lower bounds of the property. For analytically defined structure-property models such as the permeability PVE, shown in 'Results' section, convergence can be predicted using error analysis (see Equation 3). Furthermore, variability in microstructure must also be considered when specific design geometries are introduced, such as notches, which will create higher localized stresses. Mechanics calculations which have included material properties have been applied in works such as those by Lazzarin and Berto [38], where the material toughness was incorporated with the standard notched crack tip strain field calculation in order to resolve fatigue life calculations. We assert that models such as this could be enhanced with the use of converged material and property volume element calculations, which can define a range of variability in material parameters that can be directly input to mechanics calculations such as these.

Large 3-D datasets are often either not computationally tractable or collection is not feasible experimentally. In either of these situations where the microstructure or property volume element (MVE or PVE) may not have converged, statistical analysis can be applied. For average properties, one can apply distribution assumptions (typically Gaussian) and predict the variability expected for the non-converged dataset. For example, if a dataset was gathered at a volume smaller than the converged sizes shown in Figure 4, then the analysis shown in 'Sampling for convergence size' section can be applied with an assumption of a 99% (or other) confidence interval to predict the approximate volume necessary for convergence. Furthermore, the expected variability in the measured microstructure or property parameters at the collected dataset volume element size can be inferred from the confidence interval bounds. However, this methodology will not be predictive for properties that rely on extreme value microstructure descriptors, such as fatigue, except in cases where the microstructural distribution has been well characterized [20,23,41-43]. Finally, we note that the rapid advancement of 3-D tomography techniques will increasingly enable the collection of microstructural datasets of large enough volumes to properly bound material properties as well as enable the development of improved property models for a range of materials systems.

Conclusions

Based on the above results, we have derived the following:

- A method for categorizing and quantifying volume elements based on microstructure, properties, and design has been presented (MVE, PVE, and DVE).
- Sampling methods for determining convergence of MVEs and PVEs are presented for the case of a WCu 3-D dataset.
- PVEs converge (variability decreases to a specified confidence interval) at rates that are greater than those of their MVE dependencies.
- MVEs and PVEs converge at different rates and sizes; therefore, a volume element should only be used to model properties for which it has been validated.
- Microstructural volume elements that are smaller than the converged size can be useful to calculate the expected variability for that volume.

Competing interests
The authors declare that they have no competing interests.

Author's contributions
MPE performed the datasets acquisitions and reconstructions and analysis. WCL provided statistical analysis and convergence criteria. TMP participated in the design and coordination of this work through its entirety. All authors read and approved the final manuscript.

Acknowledgements
The authors acknowledge the Office of Naval Research ONR-DURIP grant no. N000141010783 for the support of the development of the TriBeam system and Air Force grant FA9550-12-1-0445 for the support of this research. We also thank FEI Corporation for supporting the development of the TriBeam system. The authors acknowledge David Rowenhorst for visualization code in IDL and Michael Uchic for useful discussions on tomography. The thoughtful comments from the reviewers are also greatly appreciated.

References

1. Midgley PA, Weyland M (2003) 3D electron microscopy in the physical sciences: the development of z-contrast and {EFTEM} tomography In: Proceedings of the international workshop on strategies and advances in atomic level spectroscopy and analysis. Ultramicroscopy. (3–4):413–431. doi:10.1016/S0304-3991(03)00105-0
2. Holzer L, Indutnyi F, Gasser P, Munch B, Wegmann M (2004) Three-dimensional analysis of porous BaTiO$_3$ ceramics using FIB, nanotomography. J Microsc 216(1):84–95. doi:10.1111/j.0022-2720.2004.01397.x
3. Uchic MD, Groeber MA, Dimiduk DM, Simmons JP (2006) 3D microstructural characterization of nickel superalloys via serial-sectioning using a dual beam FIB-SEM In: Viewpoint set no. 4: 3D characterization and analysis of materials. Organized by G. Spanos. Scripta Mater 55(1):23–28. doi:10.1016/j.scriptamat.2006.02.039
4. Jensen DJ, Poulsen HF (2012) The three dimensional x-ray diffraction technique. Mater Char 72(0):1–7. doi:10.1016/j.matchar.2012.07.012
5. Suter RM, Hennessy D, Xiao C, Lienert U (2006) Forward modeling method for microstructure reconstruction using x-ray diffraction microscopy: single-crystal verification. Rev Sci Instrum 77(12). doi:10.1063/1.2400017
6. Landis EN, Keane DT (2010) X-ray microtomography. Mater Char 61(12):1305–1316. doi:10.1016/j.matchar.2010.09.012
7. Alkemper J, Voorhees PW (2001) Quantitative serial sectioning analysis. J Microsc 201(3):388–394. doi:10.1046/j.1365-2818.2001.00832.x
8. Spowart J, Mullens H, Puchala B (2003) Collecting and analyzing microstructures in three dimensions: a fully automated approach 55(10):35–37. doi:10.1007/s11837-003-0173-0
9. Rowenhorst DJ, Lewis AC, Spanos G (2010) Three-dimensional analysis of grain topology and interface curvature in a beta-titanium alloy. Acta Mater 58(16):5511–5519. doi:10.1016/j.actamat.2010.06.030
10. Uchic M, Groeber M, Spowart J, Shah M, Scott M, Callahan P, Shiveley A, Chapman M (2012) An automated multi-modal serial sectioning system for characterization of grain-scale microstructures in engineering materials (preprint). Technical report, DTIC Document
11. Echlin MP, Mottura A, Torbet CJ, Pollock TM (2012) A new TriBeam system for three-dimensional multimodal analysis. Rev Sci Instrum 83(2):doi:10.1063/1.3680111
12. Gitman IM, Askes H, Sluys LJ (2007) Representative volume: existence and size determination. Eng Fract Mech 74(16):2518–2534. doi:10.1016/j.engfracmech.2006.12.021
13. Hill R (1963) Elastic properties of reinforced solids: some theoretical principles. J Mech Phys Solid 11(5):357–372. doi:10.1016/0022-5096(63)90036-X
14. Ostoja-Starzewski M (2006) Material spatial randomness: from statistical to representative volume element. Probabilist Eng Mech 21(2):112–132. doi:10.1016/j.probengmech.2005.07.007

15. Niezgoda SR, Turner DM, Fullwood DT, Kalidindi SR (2010) Optimized structure based representative volume element sets reflecting the ensemble-averaged 2-point statistics. Acta Mater 13:4432–4445. doi:10.1016/j.actamat.2010.04.041

16. Groeber M, Haley BK, Uchic MD, Dimiduk DM, Ghosh S (2006) 3D reconstruction and characterization of polycrystalline microstructures using a FIB-SEM system. Mater Char 57:259–273. doi:10.1016/j.matchar.2006.01.019

17. Swaminathan S, Ghosh S, Pagano N (2006) Statistically equivalent representative volume elements for unidirectional composite microstructures: part i - without damage. J Compos Mater 40(7):583–604

18. Shan Z, Gokhale AM (2002) Representative volume element for non-uniform micro-structure. Comput Mater Sci 24(3):361–379. doi:10.1016/S0927-0256(01)00257-9

19. Qidwai SM, Turner DM, Niezgoda SR, Lewis AC, Geltmacher AB, Rowenhorst DJ, Kalidindi SR (2012) Estimating the response of polycrystalline materials using sets of weighted statistical volume elements. Acta Mater 60:5284–5299. doi:10.1016/j.actamat.2012.06.026

20. Groeber M, Ghosh S, Uchic MD, Dimiduk DM (2008) A framework for automated analysis and simulation of 3d polycrystalline microstructures. part 2: synthetic structure generation. Acta Mater 56(6):1274–1287

21. McDowell DL, Ghosh S, Kalidindi SR (2011) Representation and computational structure-property relations of random media. JOM 63(3):45–51. doi:10.1007/s11837-011-0045-y

22. Brundidge CL, Pollock TM (2012) Processing to fatigue properties: benefits of high gradient casting for single crystal airfoils. In: Huron ES, Reed RC, Hardy MC, Mills MJ, Montero RE, Portella PD, Telesman J (eds) Superalloys 2012. Wiley, Hoboken, pp 379–385. doi:10.1002/9781118516430.ch41. http://dx.doi.org/10.1002/9781118516430.ch41

23. Przybyla CP, McDowell DL (2010) Microstructure-sensitive extreme value probabilities for high cycle fatigue of Ni-base superalloy {IN100}. Int J Plast 26(3):372–394. doi:10.1016/j.ijplas.2009.08.001

24. Vernerey F, Liu WK, Moran B, Olson G (2009) Multi-length scale micromorphic process zone model. Comput Mech 44(3):433–445

25. McDowell DL (2010) A perspective on trends in multiscale plasticity. Int J Plast 26(9):1280–1309

26. Rowenhorst D, Kuang J, Thornton K, Voorhees P (2006) Three-dimensional analysis of particle coarsening in high volume fraction solid–liquid mixtures. Acta mater 54(8):2027–2039

27. Madison J, Spowart J, Rowenhorst D, Pollock T (2008) The three-dimensional reconstruction of the dendritic structure at the solid-liquid interface of a Ni-based single crystal. JOM 60(7):26–30

28. Rowenhorst D, Voorhees P (2012) Measurement of interfacial evolution in three dimensions. Annu Rev Mater Res 42:105–124

29. MacSleyne J, Uchic MD, Simmons JP, Graef MD (2009) Three-dimensional analysis of secondary $\gamma\prime$ precipitates in Rene-88 {DT} and UMF-20 superalloys. Acta Mater 57(20):6251–6267. doi:10.1016/j.actamat.2009.08.053

30. Groeber M, Ghosh S, Uchic MD, Dimiduk DM (2008) A framework for automated analysis and simulation of 3D polycrystalline microstructures.: part 1: statistical characterization. Acta Mater 56(6):1257–1273. doi:10.1016/j.actamat.2007.11.041

31. Taklo MM, Klumpp A, Ramm P, Kwakman L, Franz G (2011) Bonding and TSV in 3D IC integration: physical analysis with plasma FIB. Microsc Anal 25(7):9–12

32. Altmann F, Beyersdorfer J, Schischka J, Krause M, Franz G, Kwakman L (2012) Cross section analysis of Cu filled TSVS based on high throughput plasma-FIB milling In: ISTFA, 2012: ASM international conference proceedings of the 38th international symposium for testing and failure analysis. ASM International, Geauga County, pp 39–43

33. Torquato S (2002) Random heterogeneous materials: microstructure and macroscopic properties. Interdisciplinary applied mathematics: mechanics and materials, vol. 16.. Springer, New York. http://books.google.com/books?id=PhG_X4-8DPAC

34. Echlin MP, Mottura A, Wang M, Mignone PJ, Riley DP, Franks GV, Pollock TM (2013) Three-dimensional characterization of the permeability of W-Cu composites using a new "Tribeam" technique. Acta Materi 64:307–315. doi:10.1016/j.actamat.2013.10.043

35. Kanit T, Forest S, Galliet I, Mounoury V, Jeulin D (2003) Determination of the size of the representative volume element for random composites: statistical and numerical approach. Int J Solid Struct 40:3647–3679. doi:10.1016/S0020-7683(03)00143-4

36. Kanit T, N'Guyen F, Forest S, Jeulin D, Reed M, Singleton S (2006) Apparent and effective physical properties of heterogeneous materialsrepresentativity of samples of two materials from food industry. Comput Meth Appl Mech Eng 195(33–36):3960–3982. doi:10.1016/j.cma.2005.07.022

37. Pelissou C, Baccou J, Monerie Y, Perales F (2009) Determination of the size of the representative volume element for random quasi-brittle composites. Int J Solid Struct 46(14–15):2842–2855. doi:10.1016/j.ijsolstr.2009.03.015

38. Lazzarin P, Berto F (2005) Some expressions for the strain energy in a finite volume surrounding the root of blunt v-notches. Int J Fract 135(1–4):161–185

39. Echlin MP, Pollock TM (2013) A statistical sampling approach for measurement of fracture toughness parameters in a 4330 steel by 3-D femtosecond laser-based tomography. Acta Mater 61(15):5791–5799. doi:10.1016/j.actamat.2013.06.023

40. Taylor JR (1997) An Introduction to error analysis: the study of uncertainties in physical measurements. A series of books in physics. University Science Books, Herndon. http://books.google.com/books?id=giFQcZub80oC

41. Groeber M, Ghosh S, Uchic MD, Dimiduk DM (2007) Developing a robust 3-D characterization-representation framework for modeling polycrystalline materials. JOM 59(9):32–36

42. Przybyla C, McDowell D (2012) Microstructure-sensitive extreme-value probabilities of high-cycle fatigue for surface vs. subsurface crack formation in duplex ti–6al–4v. Acta Mater 60(1):293–305

43. Jha S, Caton M, Larsen J (2007) A new paradigm of fatigue variability behavior and implications for life prediction. Mater Sci Eng 468:23–32

11

Anisotropy in plastic deformation of extruded magnesium alloy sheet during tensile straining at high temperature

David E Cipoletti[1,3*], Allan F Bower[1] and Paul E Krajewski[2]

* Correspondence:
david.cipoletti@bucknell.edu
[1]School of Engineering, Brown
University, Providence, RI 02912, USA
[3]Present Address: College of
Engineering, Bucknell University,
Lewisburg, PA 17837, USA
Full list of author information is
available at the end of the article

Abstract

Experimental measurements are used to characterize the anisotropy of flow stress in extruded magnesium alloy AZ31 sheet during uniaxial tension tests at temperatures between 350°C and 450°C, and strain rates ranging from 10^{-5} to 10^{-2} s^{-1}. The sheet exhibits lower flow stress and higher tensile ductility when loaded with the tensile axis perpendicular to the extrusion direction compared to when it is loaded parallel to the extrusion direction. This anisotropy is found to be grain size, strain rate, and temperature dependent, but is only weakly dependent on texture. A microstructure based model (D. E. Cipoletti, A. F. Bower, P. E. Krajewski, Scr. Mater., 64 (2011) 931–934) is used to explain the origin of the anisotropic behavior. In contrast to room temperature behavior, where anisotropy is principally a consequence of the low resistance to slip on the basal slip system, elevated temperature anisotropy is found to be caused by the grain structure of extruded sheet. The grains are elongated parallel to the extrusion direction, leading to a lower effective grain size perpendicular to the extrusion direction. As a result, grain boundary sliding occurs more readily if the material is loaded perpendicular to the extrusion direction.

Keywords: Magnesium alloys; Grain boundary sliding; Creep; Finite element method; Crystal plasticity

Background

Wrought magnesium alloys commonly display anisotropic behavior during deformation, which has a profound influence on their ductility and forming limits. This anisotropy results both from the production of the sheet or extruded materials, and from deformation induced during forming of the desired shape [1]. Texture has been studied in a wide variety of the hexagonal metals and has been shown to influence plastic deformation for a majority of them, including titanium, zinc, and zirconium [2-9]. The influence of texture on deformation in magnesium has been studied, however, its effect is not completely understood [10,11]. During room temperature deformation of magnesium alloys, plastic deformation occurs mostly by $\langle a \rangle$ slip on the basal plane or by $\{10\bar{1}2\}\langle\bar{1}011\rangle$ twinning. The limited number of systems for plastic deformation, together with the strong deformation induced texture, result in low ductility [12]. For example, the orientation of the tensile axis with respect to the extrusion direction was shown to affect the strength, ductility, and work hardening in tensile specimens of magnesium alloy AZ31 [13,14]. The anisotropy observed during room temperature deformation of extruded sheet persists at elevated temperature [15]. However, the origin of the anisotropy at elevated temperatures is less certain. As temperature is increased, $\langle a \rangle$ slip on the prismatic plane becomes active, as well as $\langle c + a \rangle$ slip on the pyramidal plane [16-18]. At the same time, additional deformation mechanisms such as grain boundary sliding and grain boundary diffusion become active, significantly increasing ductility. However, there is still debate on whether the increase in ductility is due to the activation of additional slip systems or the additional deformation mechanisms [19]. Specifically, the importance of $\langle c + a \rangle$ slip to increase ductility during elevated temperature deformation is discussed in the work of Hutchinson [12]. Barnett et al. show that anisotropy is affected by an increase in temperature through the activation of additional deformation mechanisms specifically grain boundary sliding (GBS). Their work suggests that an increase in the fraction of strain due to GBS is responsible for the decrease in the Lankford r-value that occurs with increasing temperature during tensile straining of magnesium AZ31 sheet [20]. Stanford et al. [21] argues against this conclusion based on their measurements of observable shear at grain boundaries during tensile tests between 25 and 250°C, and supports the conclusion that increased prismatic and pyramidal slip activity is responsible. Recently, a study was performed on compression of samples made from equal channel angular extrusion (ECAE) at midrange temperatures by Foley et al. [22]. They found that in compression tests between 115°C and 200°C, texture effects dominated the mechanical properties of the material, and furthermore, that strength and ductility could be enhanced by fine tuning the ECAE process to achieve grain refinement and specific texture. These observations suggest that the key to developing processing maps to obtain highly formable magnesium alloys is to understand the relationship between crystallographic texture evolution and the prevalent deformation mechanisms. Thus, based on the conclusions discussed here, research into the interplay between GBS and dislocation creep, specifically slip on the $\langle c + a \rangle$ system is necessary.

With this in mind, the goal of this study is to determine the nature and causes of plastic anisotropy during elevated temperature straining of extruded magnesium alloy AZ31 sheets, using a combination of experiments and numerical simulations. Successful implantation of this combined approach to influence industrial processing application has previously been carried out by Krajewski et al. with Al-Mg alloys [23]. They utilize a multiscale modeling framework such that a microstructure-based finite element model is incorporated into a computational model of dome forming experiments which in turn is incorporated into a macroscale model of a license plate pocket forming operation. This multiscale modeling framework provides a direct link from the influence of deformation mechanisms at the microstructural level to large scale forming operations in an industrial application specifically within the automotive industry [23]. Our investigation focuses on deformation mechanisms that occur at temperatures and strain rates typical of the quick plastic forming process (QPF). Quick plastic forming is a hot blow forming process developed by General Motors for forming sheet material into complex shapes at a strain rate and temperature acceptable for production volumes [24]. The process is adapted from the superplastic forming process (SPF) that allows large tensile elongations without failure by straining at slow rates and elevated temperatures [25]. Elevated temperature processes such as QPF are necessary to produce complex 3D parts from magnesium alloy sheet.

Uniaxial tensile tests and strain rate jump tests were used to determine the flow stress of extruded magnesium alloy AZ31 sheet as a function of strain rate and temperature. Tests were conducted for specimens with tensile axis both parallel and perpendicular to the extrusion direction. A significant anisotropy was observed at low strain rates, which decreases as the strain rate is increased. There are two possible causes for this anisotropy: firstly the sheet is textured, and loading orientations which favor basal slip tend to have faster rates of dislocation creep and consequently lower flow stresses. Secondly, extruded sheet has a non equi-axed grain structure, with the grains elongated parallel to the extrusion direction. As a result, the grain boundary structure is anisotropic, and tends to promote grain boundary sliding when the loading axis is transverse to the extrusion direction.

To distinguish between these two possible sources of anisotropy, numerical simulations were used to quantify the relative contributions of dislocation creep and grain boundary sliding to plastic flow in the magnesium alloy sheet specimens. In an earlier paper by Cipoletti et al. [26], the microstructure based model developed by Bower and Wininger [27] was extended and calibrated for magnesium alloy AZ31. In this work, the previously developed model is utilized to specifically investigate the root of the anisotropic behavior of extruded sheet material. The computations account for dislocation creep within the grains using a crystal plasticity model appropriate for hcp metals, while using a sharp-interface model to account for grain boundary diffusion and sliding. The model was used to compute the flow stress of extruded AZ31 alloy specimens, with microstructures taken directly from experimental micrographs, as functions of temperature and strain rate. The predicted flow stresses were found to be in excellent agreement with experiment. The computations show that anisotropy in grain shape is the principal cause of flow stress anisotropy in extruded AZ31 sheet, with texture induced plastic anisotropy playing a secondary role.

The remainder of this paper will be organized as follows. First an experimental analysis of the extruded magnesium AZ31 sheet material will be explained along with conclusions and implications of the results. Then the finite element model will be discussed in greater detail along with the each of the deformation mechanisms that are modeled. Finally, the model will be employed to investigate the experimental results and the implications of its findings will be discussed along with the conclusions of the study.

Methods

Sample preparation and observation

As received material, annealed samples, and deformed samples were analyzed using optical microscopy to determine grain size and general topography, and were analyzed using the method of electron backscatter diffraction (EBSD) on the scanning electron microscope (SEM) to determine texture and grain orientation. The samples were prepared in careful adherence to the following procedure for observing the microstructure characteristics. Each specimen was first cut using a low speed diamond saw to minimize any chance of twinning during the cutting process. The specimens were then mounted in epoxy cold mounts to avoid the possibility of a change in the microstructure due to the heat and pressure associated with a hot mount. Conductive filler was used at the specified ratio for each mount that housed specimens destined to be analyzed using EBSD to ensure electrical conductivity. The samples were mechanically polished with fine grain SiC paper, followed by finer polishing using suspended alumina solution with particle size $1\,\mu$m, $0.3\,\mu$m, and $0.05\,\mu$m in order of decreasing particle size. Specimens to be observed optically were etched using an acetic-picric acid etchant. Grain size was measured on the optically observes samples using the linear intercept method in accordance with ASTM standard E112-96. Samples to be observed using EBSD were electro-polished in a solution of 10 percent nitric acid in methanol at 2 volts for one second with an aluminum cathode and a copper anode. All EBSD imaging was done on a JEOL 845 scanning electron microscope, and all EBSD patterns were obtained using an Oxford Instruments HKL Nordlys detector and the Oxford Instruments INCA software package.

Material specifications

Magnesium alloy AZ31 with composition Mg – Al 3.5 – Zn 1.5 – Mn 0.2 – Cu 0.05 – Ca 0.04 – Fe 0.01 – Ni 0.01 (in weight percent) was extruded into sheet material 175 mm wide and xx mm thick. Elevated temperature tensile test specimens were cut from the extruded sheet so that the tensile axis was rotated 0° or 90° from the extruded direction (ED). The specimens cut 0° to the ED were removed from sheet with a final thickness of 1.30 mm, and the specimens cut 90° to the ED were removed from sheet with a final thickness of 1.67 mm.

Optical microscopy and EBSD were used to characterize the initial microstructure. Due to the extrusion process, the grain size measured on the transverse face of the specimens loaded 0° to the ED was approximately $18\,\mu$m whereas on the specimens loaded 90° to the ED the initial grain size was measured as approximately $9\,\mu$m (Figure 1 describes with more clarity the exact face that was observed). This suggests that during the extrusion process, the grains were elongated in the extrusion direction. Thus, although the same grains are being measured, they have an apparent grain size based on the orientation. Figure 2 shows EBSD texture maps for the face oriented 0° and 90° to the ED where this effect can be observed. The extrusion process also imparts a strong texture in the sheet,

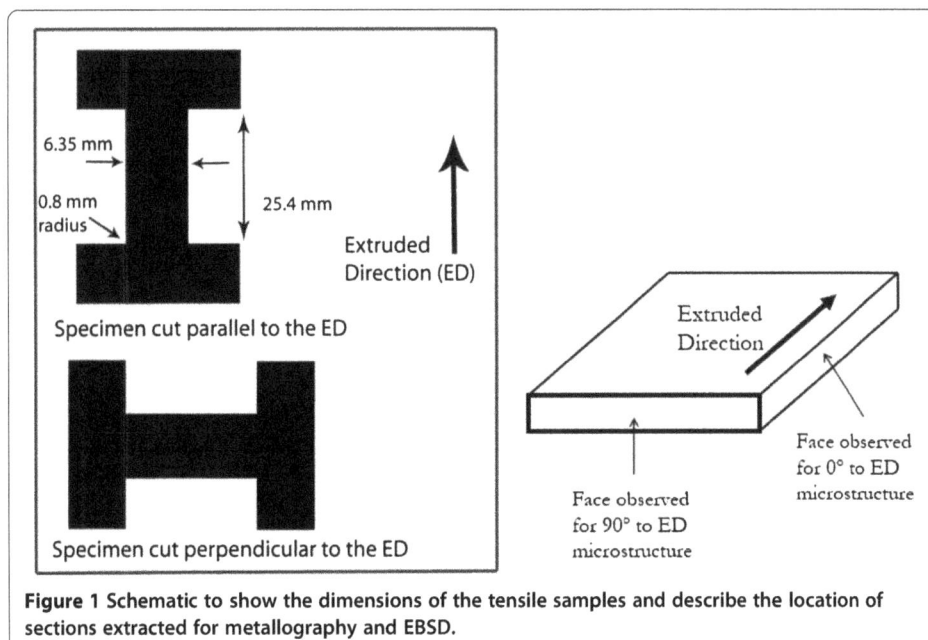

Figure 1 Schematic to show the dimensions of the tensile samples and describe the location of sections extracted for metallography and EBSD.

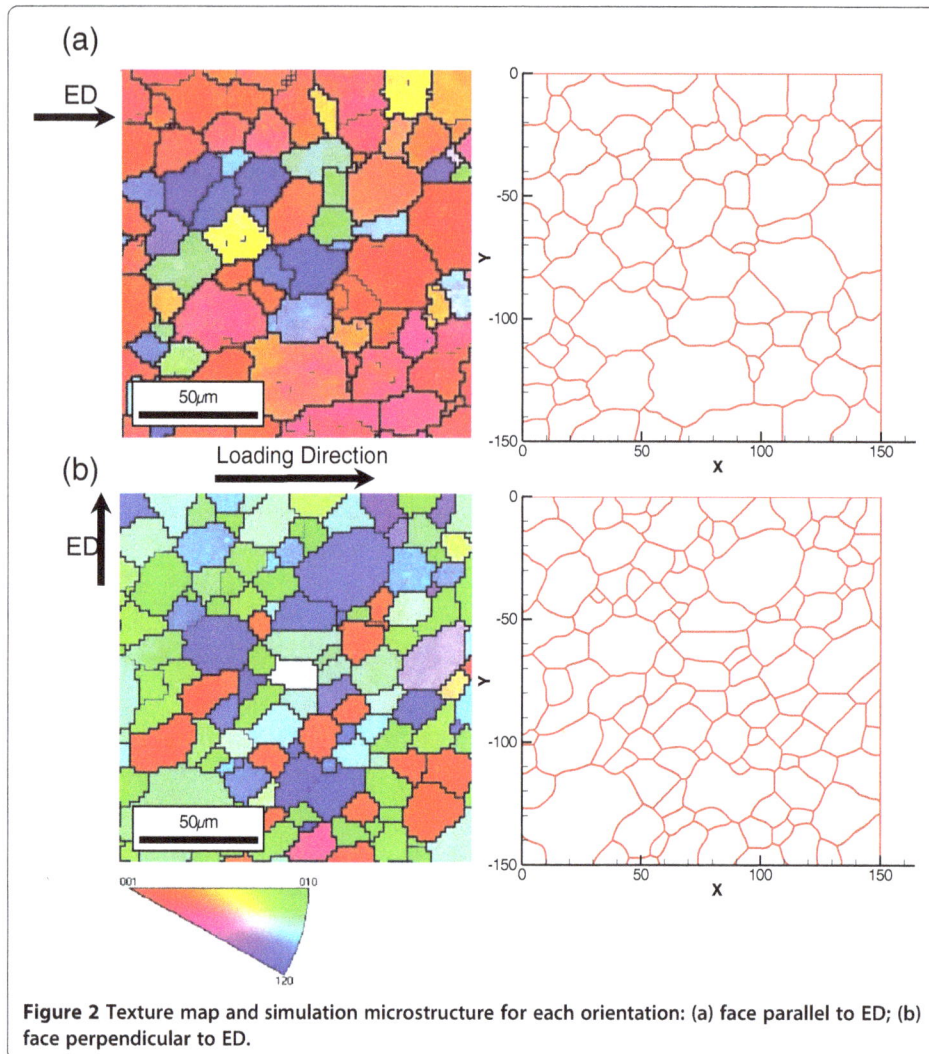

Figure 2 Texture map and simulation microstructure for each orientation: (a) face parallel to ED; (b) face perpendicular to ED.

which can also be observed in EBSD texture maps. As observed in Figure 2a, most grains are oriented with the basal plane normal to the surface for the specimen oriented 0° to the ED, whereas the grains of the specimen oriented 90° to the ED are mostly oriented in the $\langle 10\bar{1}0 \rangle$ and $\langle 11\bar{2}0 \rangle$ directions (Figure 2b). This suggests that in addition to elongating the grains, the extrusion process tends to rotate the grains so that the basal plane is aligned along the extrusion direction. Thus, by adjusting the loading direction to be 0° to the ED, the tensile axis is aligned with the long axis of the elongated grains and the basal plane of a majority of the grains. In contrast, if the loading direction is 90° to the ED, the tensile axis is perpendicular to the long axis of the elongated grains and perpendicular to the basal plane of most grains. It is important to note that the texture of the specimens that these microstructures are meant to model does not change but the orientation of the tensile axis rotates against the strong texture of the specimen.

Elevated temperature tensile tests

Elevated temperature tensile tests were performed to determine the response of the material to an applied strain rate at temperature. A standard dog bone style elevated temperature tensile specimen was used for all tests. The specimen dimensions are shown in Figure 1. Each experiment was performed on an Instron 5568. The tensile specimens were loaded into a pre-heated furnace, and testing was begun 2.5 minutes after the specimen reached testing temperature to ensure even temperature distribution in the specimen. At the completion of the test, which was determined by fracture in all cases, the specimen was immediately removed and quenched in water to preserve the microstructure. Two different types of tensile tests were performed on the specimens, a constant strain rate test and a step strain rate test.

Constant strain rate to failure tests were performed at two different strain rates at 450°C for tensile specimens loaded parallel and perpendicular to the extrusion direction, the results of which are seen in Figure 3. The crosshead speed of the test was varied to account for uniform thinning of the sample during deformation. This was done using an algorithm through the test frame, but does not guarantee constant strain rate under conditions of extreme necking. A greater flow stress was reached in the material when the specimen was pulled at a faster strain rate and the material exhibited more ductility at the lower strain rate.

Step strain rate tests were performed to obtain the steady state flow stress as a function of strain, strain rate, and temperature. In these tests, the specimen was first subjected to an initial strain rate of $3.0 \times 10^{-3} s^{-1}$ to ensure that the specimen was firmly positioned in the testing apparatus and to stabilize the microstructure. Then the specimen was subjected to multiple strain rates sequentially in decreasing order from $3.0 \times 10^{-2} s^{-1}$ to $1.0 \times 10^{-4} s^{-1}$. Each strain rate was held constant for a minimum of 2 percent engineering strain. When the stress had ceased to decay after each change in strain rate, reaching essentially a steady state value, the stress was recorded in the specimen. Each specimen was subjected to multiple strain rate cycles so that the flow stress at multiple strains could be determined based on a linear interpolation of the coordinates for the steady state stress and strain at each strain rate. An example stress vs. strain curve that shows two complete cycles of a step strain rate test is shown in Figure 4. The approximately horizontal lines show how the

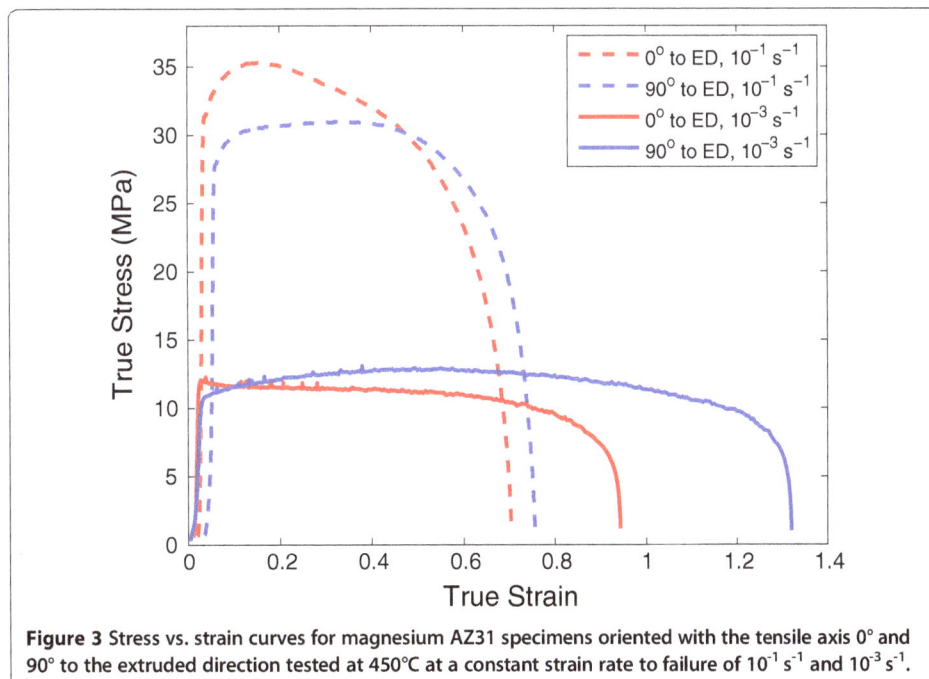

Figure 3 Stress vs. strain curves for magnesium AZ31 specimens oriented with the tensile axis 0° and 90° to the extruded direction tested at 450°C at a constant strain rate to failure of $10^{-1} s^{-1}$ and $10^{-3} s^{-1}$.

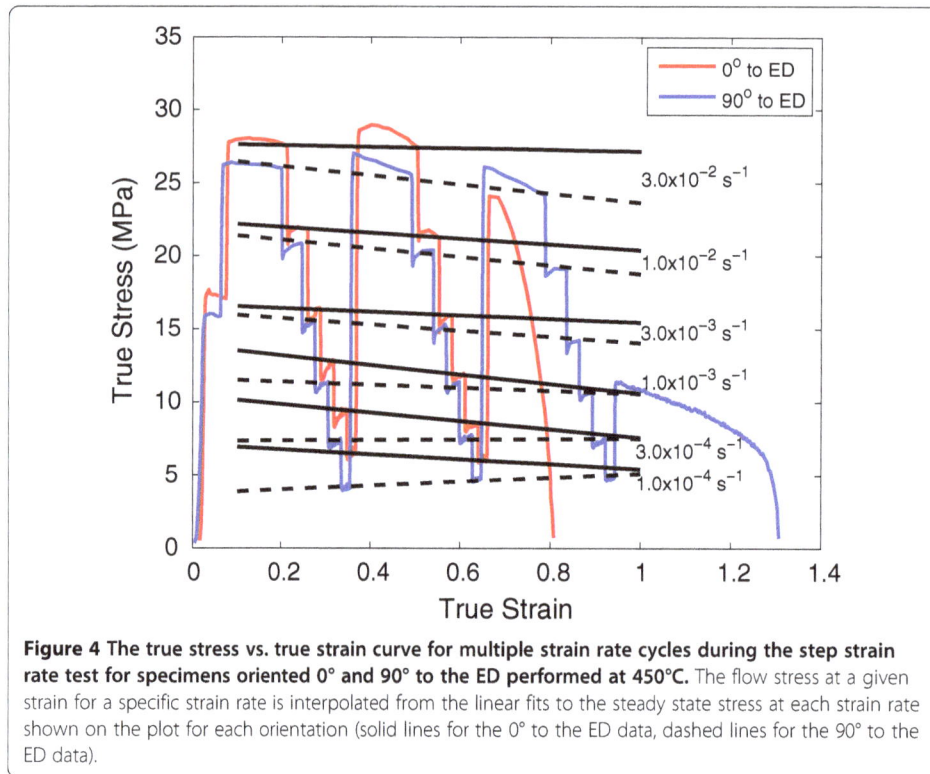

Figure 4 The true stress vs. true strain curve for multiple strain rate cycles during the step strain rate test for specimens oriented 0° and 90° to the ED performed at 450°C. The flow stress at a given strain for a specific strain rate is interpolated from the linear fits to the steady state stress at each strain rate shown on the plot for each orientation (solid lines for the 0° to the ED data, dashed lines for the 90° to the ED data).

stress was interpolated to give the steady state flow stress at various strain rates. It is assumed that the microstructure remains relatively constant during testing at each strain rate and that minimal hardening or softening occurs in the sample during the test. This assumption is supported by observing that the stress at each strain rate recorded at low and high levels of strain was approximately equal; the interpolation lines in Figure 4 have close to no slope. However, there is a change in the stress during each strain rate step, thus the range of error possible from the tests was recorded and error bars were placed on the plots compiled from the step strain rate test results.

Model description

Recently, Cipoletti et al. [26] developed a microstructure based finite element model that predicts elevated temperature deformation mechanisms in magnesium alloys. Details of the model as well as how it was calibrated and validated are discussed in ref. [25]. In the present study, the model is utilized to understand the relationship between anisotropy and the active deformation mechanisms in magnesium AZ31 extruded sheet. Thus, a brief description of it is included here for clarity and to highlight how it was adapted to study the effect of anisotropy in this work.

Two representative two-dimensional microstructures have been assembled directly from micrographs of the specimens cut 0° and 90° to the ED (Figure 2), one of which is shown in Figure 5. The orientation of each grain in the simulation microstructure has been taken directly from the actual material using data obtained from the EBSD analysis. Each microstructure is modeled as a collection of single crystal grains separated by sharp grain boundaries. A constant strain rate is applied in the x direction to induce a state of plane strain uniaxial extension in the polycrystal. Three mechanisms of deformation are accounted for

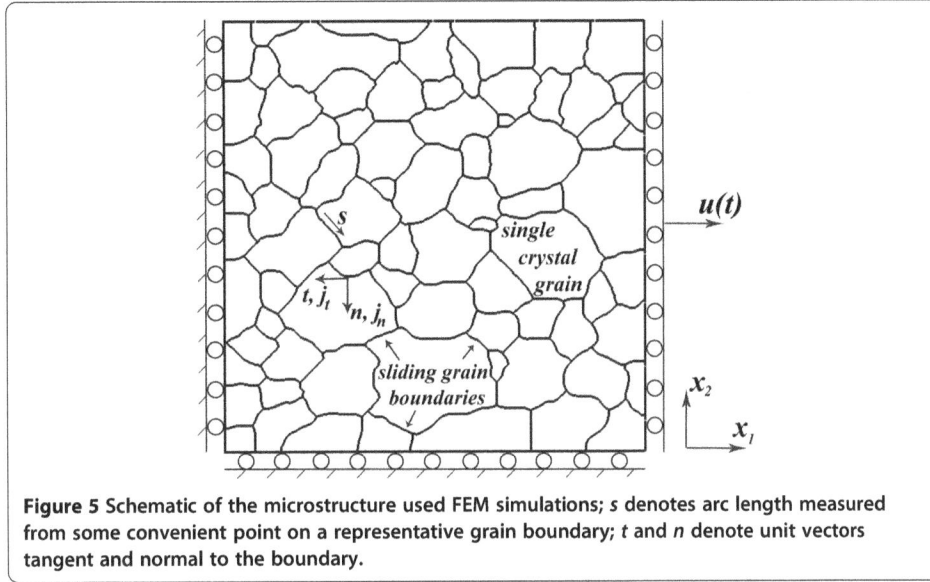

Figure 5 Schematic of the microstructure used FEM simulations; s denotes arc length measured from some convenient point on a representative grain boundary; t and n denote unit vectors tangent and normal to the boundary.

in the polycrystal: (i) dislocation creep (DC) within the grains; (ii) grain boundary diffusion (GBD); and (iii) grain boundary sliding (GBS). The assumptions and constitutive equations that are used to model these processes will be described in more detail below.

The main objective of the finite element simulations is to calculate the stress and displacement distributions induced in the microstructure. The results are used to deduce the volume average uniaxial stress σ as a function of the applied uniaxial strain and strain rate (calculated from the displacements applied to the boundary of the microstructure). After a brief elastic transient, the uniaxial stress settles to a constant value, which depends on the applied strain rate. Consequently, our focus is placed on the variation of steady-state flow stress, σ as a function of uniaxial strain rate, $\dot{\varepsilon}$.

Dislocation creep

A classical single crystal model is used to describe dislocation creep within the grains, all of which are modeled as hexagonal close packed crystals. The strain rate $\dot{\varepsilon}_{ij} = \left(\partial \dot{u}_i / \partial x_j + \partial \dot{u}_j / \partial x_i\right)/2$ in each grain is broken down into an elastic and plastic strain rate as.

$$\dot{\varepsilon}_{ij} = \dot{\varepsilon}^e_{ij} + \dot{\varepsilon}^p_{ij} \tag{1}$$

Linear elastic constitutive equations for a transversely isotropic material with Young's modulus E_p and E_t, shear modulus μ_p and μ_t, and Poisson's ratio v_p, v_{tp}, and v_p [28] are used to relate the elastic strain rate to the stress rate. Plastically, each grain deforms by shearing on the basal (0001) [11-20], pyramidal (1–100) [11-20], and prismatic (1–101) [11-20] slip systems. The shear rate on a slip system $\dot{\gamma}^\alpha$ is determined from the model for solute drag creep proposed by Frost and Ashby [29] as

$$\dot{\gamma}^\alpha = \dot{\gamma}\left(\frac{q^\alpha}{\tau^\alpha_0}\right)^n \tag{2}$$

where $n \approx 5.5$ is the stress exponent, $q^\alpha = s^\alpha_i \sigma_{ij} m^\alpha_j$ denotes the resolved shear stress on the slip system α, τ_0 is the critical resolved shear stress for the slip system α, and $\dot{\gamma}$ is the characteristic slip rate. The characteristic slip rate is temperature dependent as defined by the following Arrhenius relationship

$$\dot{\gamma} = \dot{\gamma}_o \exp\left(-Q_{DC}/kT\right) \tag{3}$$

where $\dot{\gamma}_o$ is a pre-exponential factor, Q_{DC} is the activation energy for dislocations to escape pinning points, k is the Boltzmann constant, and T is temperature. The strength of the different slip systems in magnesium was taken into account by adjusting τ_0. This data was taken from experimental work on single crystal magnesium [16,18,30-32]. The resolved shear stress on the HCP slip systems are of the following ratio, Basal: Prismatic: Pyramidal = 4:10:11 (a more in depth discussion of this choice of ratio is included in ref. [26]). Finally, to determine the plastic strain rate, the shear rate calculated on each of the active slip systems is summed over all slip systems,

$$\dot{\varepsilon}^p_{ij} = \sum_{\alpha=1}^{N} \dot{\gamma}^\alpha(q^\alpha)\left(s_i^\alpha m_j^\alpha + s_j^\alpha m_i^\alpha\right)/2 \tag{4}$$

where s_i^α and m_i^α denote the components of unit vectors parallel to the slip direction and slip plane normal, respectively.

Grain boundary sliding and diffusion

In addition to DC, the model allows for displacement with respect to adjacent grains along their common grain boundary in response to normal stress, defined as $\sigma_n = \sigma_{ij}n_i n_j$, and shear stress, defined as $\sigma_t = \sigma_{ij}t_i n_j$, acting on a representative boundary. Due to this displacement, discontinuities develop in the velocity field across each grain boundary, which are defined in terms of their normal and tangential component as follows

$$\begin{aligned}[v_n] &= \left(\dot{u}_i^+ - \dot{u}_i^-\right)n_i \\ [v_n] &= \left(\dot{u}_i^+ - \dot{u}_i^-\right)t_i\end{aligned} \tag{5}$$

where $u_i^\pm$ denotes the displacement of each grain immediately adjacent to a point on the grain boundary.

   GBS is a result of tangential displacement between adjacent grains along their common grain boundary due to shear stress acting on their interface. The GBS velocity is calculated relative to the adjacent grains using the following linear-viscous constitutive equation

$$[v_t] = \eta\sigma_t \tag{6}$$

where η is the fluidity of the grain boundary. The fluidity of the grain boundary is defined by the following relationship,

$$\eta = \frac{\Omega\eta_o \exp\left(-Q_{GB}/kT\right)}{kT} \tag{7}$$

where Ω is the atomic volume and Q_{GB} is the corresponding activation energy for GBD given by Frost and Ashby [29]. Although there has been substantial work defining the relationship between GBS and grain boundary migration in magnesium alloys [33], specifically in relation to their misorientation, grain boundary migration has been neglected in our computations, for simplicity. Note that our computations naturally enforce compatibility between grain boundary sliding, grain boundary diffusion, and dislocation plasticity within the grains.

GBD is a result of displacement between adjacent grains perpendicular to their common grain boundary due to normal stress acting on their interface. Thinking in terms of energy, the total free energy of the system is increased due to the normal stress acting on the grain boundary. By allowing atoms to diffuse from parts of the grain boundary in compression to the parts of the boundary in tension, the free energy is reduced. This atomic migration has the effect of displacing material points within neighboring grains towards each other in areas of compression and away from each other in areas of tension. This displacement results in a velocity discontinuity across the common grain boundary, defined by the following linear diffusion law

$$[v_n] = \frac{\Omega D_{GB} \delta_{GB} \exp(-Q_{GB}/kT)}{kT} \frac{\partial^2}{\partial s^2} [-\sigma_n] \tag{8}$$

where $D_{GB} \exp(-Q_{GB}/kT)$ is the grain-boundary diffusivity and δ_{GB} is the thickness of the interface where the diffusion is occurring.

All the grain boundaries in our microstructure terminate either at a triple junction, or at the edge of the specimen. The constraints which control behavior at these points are explained in detail in ref. [27].

Finite element simulations

The finite element method is used to solve the equations of mechanical equilibrium, the constitutive equations (1, 2, 3, 4) for the grains, and the stress-velocity discontinuity equations (5, 6, 7, 8). Deformation within the grains is modeled using a standard procedure. Line elements are introduced along the grain boundaries to interpolate the velocity discontinuity as well as the distribution of the normal and tangential stress along the boundary to solve the equations governing GBS and GBD. Details of these procedures are provided in ref. [27].

The stress and displacement fields in the microstructure are provided as output from the finite element computations, from which the contribution to the total strain rate from each deformation mechanism is determined as follows,

$$\begin{aligned}
\dot{\varepsilon}_{ij}^{dislocation} &= \frac{1}{V} \int_V \sum_\alpha \dot{\gamma}^\alpha \frac{1}{2} \left(s_i^\alpha m_j^\alpha + s_j^\alpha m_i^\alpha \right) dV \\
\dot{\varepsilon}_{ij}^{sliding} &= \frac{1}{V} \int_\Gamma [v_t] \frac{1}{2} \left(n_i t_j + t_i n_j \right) ds \\
\dot{\varepsilon}_{ij}^{diffusion} &= \frac{1}{V} \int_\Gamma [v_n] n_i n_j ds
\end{aligned} \tag{9}$$

where V denotes the area of the entire 2D microstructure and Γ denotes the assembly of grain boundaries within V. The rate of shear on each slip system is used to compute the plastic strain rate resulting from dislocation creep, while the velocity discontinuity due to the shear and normal stress acting on the grain boundary is used to compute the plastic strain rate resulting from grain boundary sliding and diffusion, respectively.

Setup of initial conditions

The model was validated for deformation of magnesium alloy AZ31 at 450°C in [26]. The material model parameters from this process are shown in Table 1. The model can be shown to account for changes in temperature as long as testing remains at temperature

Table 1 Values for material parameters used in the FEM simulations

Parameter	Value
Grain size L	16 μm
Melting Temperature T_M	923 K
Atomic volume Ω	2.32×10^{-29} m^3
Elastic constant, c_{11}	59.7 GN m^{-2}
Elastic constant, c_{33}	61.7 GN m^{-2}
Elastic constant, c_{44}	16.4 GN m^{-2}
Elastic constant, c_{12}	26.2 GN m^{-2}
Elastic constant, c_{13}	21.7 GN m^{-2}
Initial yield stress, basal slip system τ_0	4 MN m^{-2}
Initial yield stress, prismatic slip system τ_0	10 MN m^{-2}
Initial yield stress, pyramidal slip system τ_0	11 MN m^{-2}
Characteristic strain rate pre-exponential $\dot{\gamma}_0$	1.76×10^{-8} s^{-1}
Solute drag creep stress exponent n	5.5
Activation energy for dislocation creep Q_{DC}	2.39×10^{-19} J
Grain boundary diffusion pre-exponential $\delta_{GB} D_{GBt}$	4.74×10^{-8} m^3 s^{-1}
Grain boundary sliding pre-exponential η_0	712.6 m s^{-1}
Grain boundary diffusion activation energy Q_{GB}	1.53×10^{-19} J

values where the assumptions made for elevated temperature deformation are still valid. The temperature cannot be reduced to a point where twinning becomes integral to deformation or to where the critical resolved shear stress on the prismatic and pyramidal slip systems has increased substantially over that of the basal slip system.

A change in temperature has a substantial effect on deformation. When the testing temperature was decreased from 450°C to 350°C, the flow stress increased over the range of strain rates tested. The stress exponent was equal to approximately 5 at fast strain rates for both testing temperatures and decreased to between 2 and 4 at decreased strain rates; for the 450°C tests the stress exponent decreased more substantially for both orientations. This suggests that grain boundary sliding and diffusion are more prevalent at higher temperatures.

The model accounts for the change in deformation due to a change in forming temperature in the constitutive equations for DC, GBS, and GBD. This change is based on the activation energy for diffusion, which was taken from Frost and Ashby's Deformation Maps [29]. The activation energy for GBS was assumed to be equal to that for diffusion. The pre-exponential terms in the constitutive laws for GBD and GBS were determined previously by Cipoletti et al. [26]. The temperature dependence of DC is similarly accounted for using an Arrhenius relationship for the characteristic slip rate. The activation energy for DC is adjusted slightly from the activation energy for power law creep defined by Frost and Ashby. The pre-exponential term is fit to experiment at 350°C and 450°C.

The starting microstructure used by the finite element simulations was taken directly from an actual micrograph of magnesium AZ31. In this way, the topography of the actual material could be captured as input to the model including the grain shape, the relative size of the grains, the path of the grain boundaries, and the orientation of the grains. This was done by taking the micrograph obtained using the EBSD method on the SEM (Figure 2) and mapping the grain and grain boundary information into a format that could be understood by the finite element code (Figure 5). The grains show good agreement with the actual data and although the grain orientation is not shown on the simulation micrograph it has been mapped directly from the EBSD data.

Results and discussion

Both experiments and computations were used to determine the variation of flow stress with applied strain rate. Representative results are shown in Figure 6. The relationship between stress and strain rate can be described by a relationship of the form

$$\dot{\varepsilon} = A\sigma^n \tag{10}$$

where A is a constant and n is the stress exponent. Thus, the stress exponent could be extrapolated from the results. The strain rate sensitivity m, which is defined as

$$m = \frac{d(\log(\sigma))}{d(\log(\dot{\varepsilon}))} \tag{11}$$

and is equal to the inverse of the stress exponent, n, can also be interpreted from the results. A change in the stress exponent (and in the strain rate sensitivity) is often used to gain insight on the prominent deformation mechanisms in the microstructure.

The results of experimental step strain rate tests are plotted in Figure 6, for two orientations of the tensile axis with respect to the extrusion direction, and for two temperatures. For all cases, as the strain rate is increased to faster rates the flow stress increases. The stress exponent can be extracted from the curves using equation (10). At 450°C the stress exponent is equal to approximately 5 for high strain rates. When the strain rate is decreased, the stress exponent for the specimen loaded 90° to the ED decreases below 2, whereas for the specimen loaded 0° to the ED, it decreases to

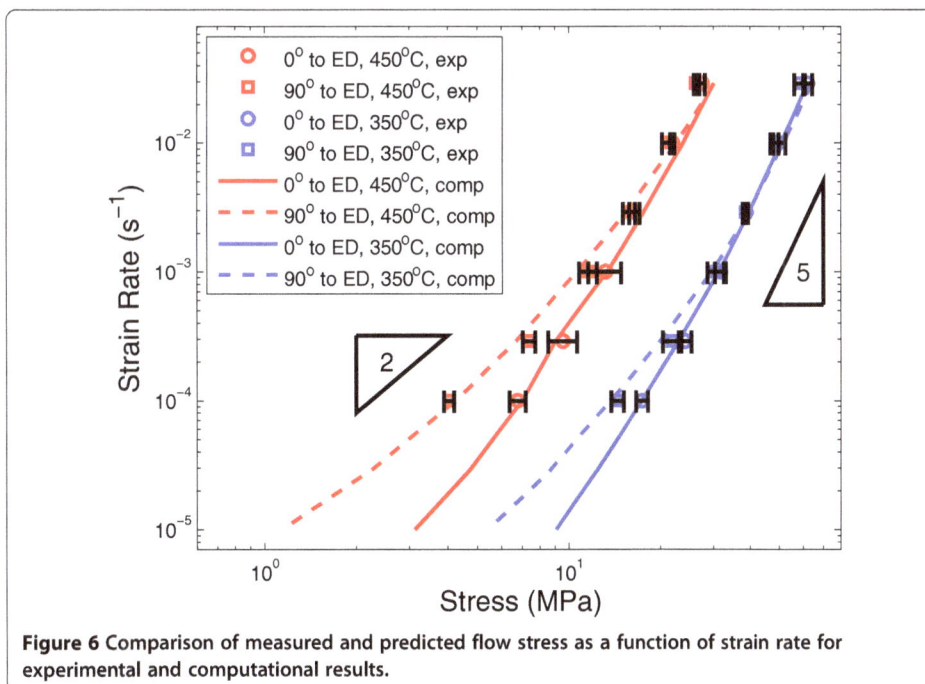

Figure 6 Comparison of measured and predicted flow stress as a function of strain rate for experimental and computational results.

approximately 3. For a decrease in testing temperature to 350°C there is an increase in flow stress across the range of strain rates tested. The stress exponent remains at approximately 5 at fast strain rates, but there is a smaller transition to the lower strain rates, to approximately 4 for the specimen oriented with the tensile axis 0° to the ED and approximately 3 for the specimen oriented with the tensile axis 90° to the ED.

The finite element predictions show good agreement with the experimental curves of flow stress vs. strain rate for when the tensile axis is oriented 0° and 90° to the extrusion direction at a testing temperature of 450°C and 350°C, as shown in Figure 6. In the high strain rate regime the stress exponent, n, approaches 5 for each curve (a strain rate sensitivity of 0.2). The model is able to capture the transition in the stress exponent from the high strain rate regime to the low strain rate. The stress exponent decreases to between 2 and 4 in the low strain rate regime. The stress exponent for the specimen loaded perpendicular to the ED is smaller at low strain rates, which suggests that anisotropy affects the stress exponent.

The variation of the stress exponent suggests a change in the deformation mechanism from fast to slow strain rates. A stress exponent of approximately 2 is characteristic of deformation dominated by GBS, whereas for a higher stress exponent, on the order of 5, it is likely that DC is dominant. Within the lower range of strain rates at each temperature, there is a greater discrepancy between the two orientations. The change in stress exponent is greater for the data from the sample oriented 90° to the ED, and furthermore the flow stress extends to smaller values, suggesting that GBS is more prevalent for this orientation.

Our FEA computations are able to partition the total strain rate in the polycrystal into contributions from each deformation mechanism as shown in Figure 7. In general, at slower strain rates the greatest contribution to the total strain rate is GBS, while at

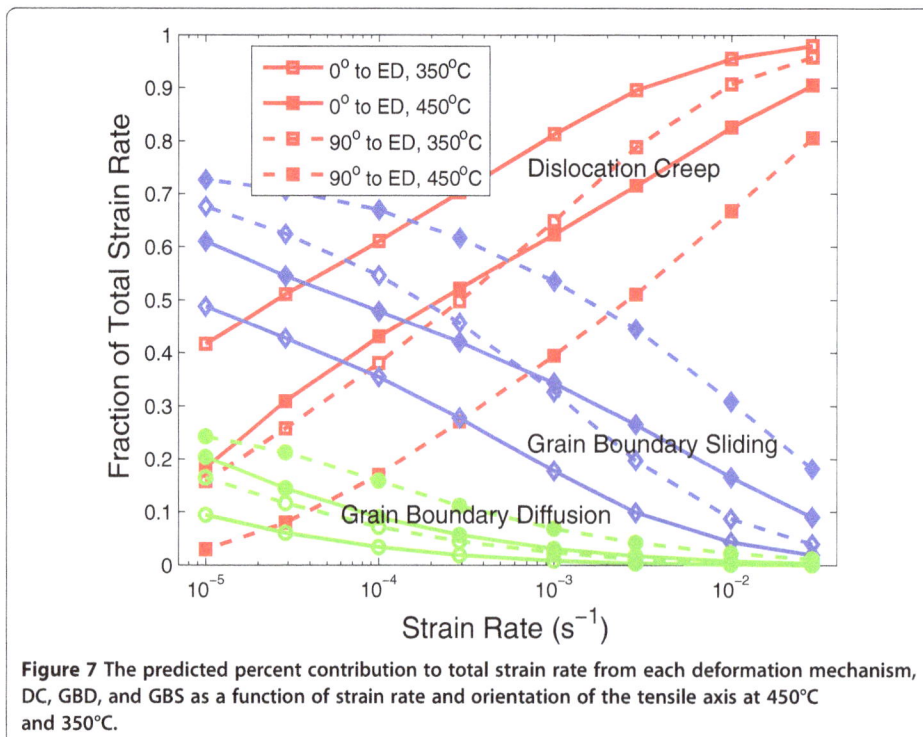

Figure 7 The predicted percent contribution to total strain rate from each deformation mechanism, DC, GBD, and GBS as a function of strain rate and orientation of the tensile axis at 450°C and 350°C.

faster strain rates DC emerges as the dominant deformation mechanism. The model confirms that the change in stress exponent observed experimentally is associated with a transition of the dominant deformation mechanism from DC to GBS. In addition, the model elucidates the origin of the anisotropy in flow stress in extruded AZ31 sheet. In the simulation in which the tensile axis is oriented 90° to the ED, the contribution to the strain rate from GBS as a percentage of total strain rate than the data for the specimen oriented 0° to the ED at all strain rates, and correspondingly a smaller amount of DC across the range of strain rates tested as shown for the results at 450°C and 350°C in Figure 7. Thus, changing the loading direction from parallel to perpendicular to the ED drives the transition point of the dominant deformation mechanism from GBS to dislocation creep to occur at higher strain rates. A corresponding transition in strain rate sensitivity is observed when the direction of the loading axis is oriented perpendicular to the extrusion direction.

Figure 7 also shows the effects of temperature on the deformation mechanism. Although the resistance for each deformation mechanism was increased in the model for a decrease in temperature from 450°C to 350°C, DC is shown to be dominant across the range of strain rates for the lower temperature tests at each orientation. The decrease in temperature led to an increase in the amount of plasticity due to DC and a decrease in the amount of GBS and GBD relative to the total strain rate in the microstructure. The transition point between deformation dominated by DC and deformation due to GBS migrated to slower strain rates. In summary for a decrease in temperature from 450°C to 350°C, the strain rate of the transition point decreased one order of magnitude. For the specimen oriented 0° to the ED, the transition point decreased from $1.5 \times 10^{-4} s^{-1}$ at 450°C to $1.5 \times 10^{-5} s^{-1}$ at 350°C, and for the specimen oriented 90° to the ED the transition point migrated from $2.0 \times 10^{-3} s^{-1}$ at 450°C to $2.3 \times 10^{-4} s^{-1}$ at 350°C.

It is instructive to plot the model's prediction of the contribution from each slip system to the total DC in the simulation, which is investigated in Figure 8 as a function of strain rate at 350°C and 450°C. First, consider the effect of temperature on distribution of strain rate due to DC. Although the flow stress increased substantially in the microstructure during the simulation at 350°C there was little change in the distribution of slip from the test at 450°C. These results are not unexpected due to the fact that the model assumes that the ratio between the slip system strength does not change in this range of temperature, and for each orientation there was no change in the starting microstructure for the simulations at 350°C and 450°C.

However, there is an effect on the orientation on the contribution from each slip system to the total DC in the simulation. It should be noted that the strain rate on each system is normalized by total plastic strain rate due to plastic shearing within the grains, which is significantly smaller than the total strain rate at low strain rates. When the loading is parallel to the ED, slip on the prismatic system is dominant followed by the basal and pyramidal system respectively for the entire range of strain rates. However, when loading is 90° to the ED, slip on the basal system is dominant followed by slip on the prismatic and pyramidal, which each contribute approximately equal amounts to the total slip. At low strain rates for the sample oriented 90° to the ED there is a transition of basal dominated slip with slip on the prismatic and pyramidal planes. However, at these low strain rates the strain rate associated with dislocation creep in the microstructure is small, and the flow stress is substantially decreased.

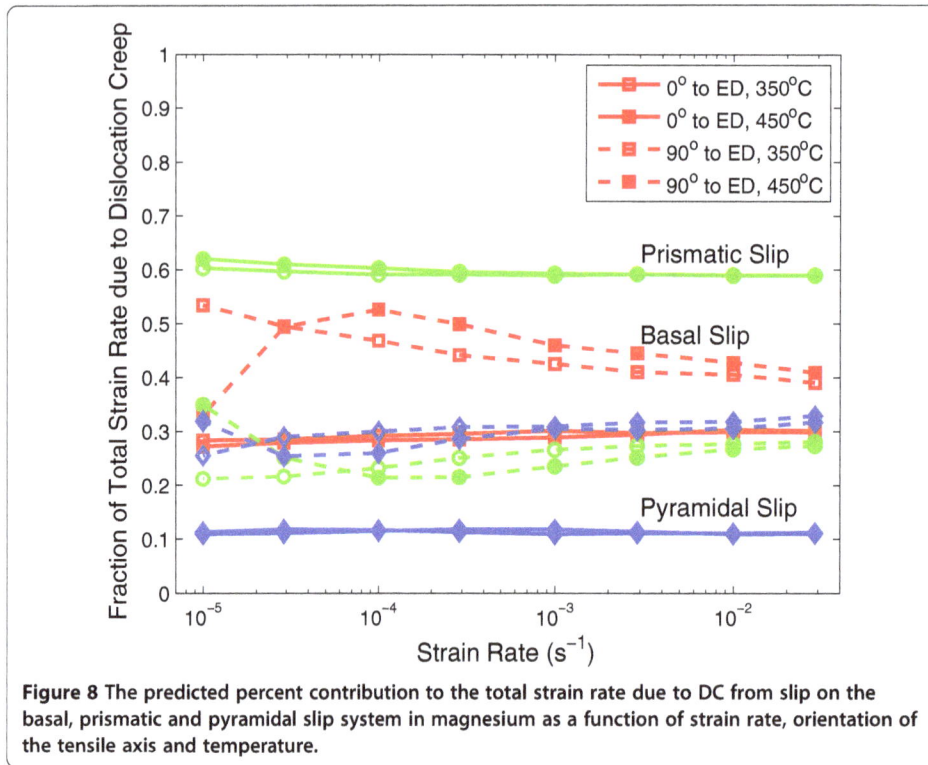

Figure 8 The predicted percent contribution to the total strain rate due to DC from slip on the basal, prismatic and pyramidal slip system in magnesium as a function of strain rate, orientation of the tensile axis and temperature.

These results are worth discussing in comparison to the results of the work introduced previously by Gehrmann et al. [1]. Their group focused on the effect of texture during compression tests from 100°C to 200°C and thus the boundary conditions are different than those considered here, although the effect of their implementation is similar. They determined that when basal slip was suppressed due to orientation of the basal planes parallel to the compression direction, large fracture strains were made possible by significant slip on the prismatic system that was activated as low as 100°C [1]. In the work presented here for the sample oriented so that the tensile axis is parallel to the ED, as seen in Figure 2, the basal planes are also parallel to the tensile axis. Due to their orientation, slip on the basal planes is suppressed for a majority of the grains, and thus as seen in Figure 8, DC is accommodated by slip on the prismatic system.

Our results are also worth comparing to the work of Agnew & Duygulu [19] who by using a combination of experimental and computational techniques performed a comprehensive study of deformation in magnesium AZ31 from room temperature to 250°C and strain rates of 10^{-5} s^{-1} to 0.1 s^{-1}. They concentrated on the effect of strain hardening, strain rate sensitivity, anisotropy, and the stress and strain at fracture. In conclusion they found that the most important factor to increase formability, at temperatures on the high end of the range they investigated, was an increase in the strain rate sensitivity. This conclusion supports the results of the present work where it was shown that the material oriented such that the tensile axis was 90° to the ED exhibited greater strain rate sensitivity (a lower stress exponent, Figure 6) and also exhibited greater strain prior to failure, Figure 3.

However, Agnew and Duygulu reason that an increase in the activity of the ⟨c + a⟩ slip system, which is activated at elevated temperatures, is what leads to the increase in strain rate sensitivity [19]. In contrast, the work of Hutchinson et al. comes to a different conclusion than that of Agnew and Duygulu. Their work used a similar material and experimental procedure to Agnew and Duygulu with additional analysis using optical and transmission electron microscopy to specifically look at the contribution of $\langle c + a \rangle$ dislocations [12]. They observed multiple grains after deformation at elevated temperatures and found that two-thirds of those observed did not contain $\langle c + a \rangle$ dislocations. Thus, they concluded that although slip on the pyramidal system is important to increasing the strain rate sensitivity, and thus increasing ductility, it cannot be the only factor. Their results suggested that increased ductility is more dependent on an increase in GBS in the material, which is in turn dependent on the grain size of the material [12]. This conclusion is supported by the work presented here, which shows that GBS is significant to increasing the strain rate sensitivity (decreasing the stress exponent) and that GBS is strongly influenced by the grain size of the material. The contribution from GBS increases for smaller grain size. Furthermore, pyramidal slip is shown to be influential but not dominant. The contribution of slip on the pyramidal plane is significant and is shown to be greater for the specimen loaded 90° to the ED, which is the specimen that also shows greater strain rate sensitivity at low strain rates. It should be noted that the results presented in this work are at higher temperatures, and as expected, there is a greater contribution of GBS observed. Thus, our conclusions are consistent with results [19,21] that suggest GBS is not responsible for the increase in strain rate sensitivity at temperatures below 250°C.

Experiments and the finite element model both show anisotropy of the magnesium AZ31 extruded sheet during elevated temperature deformation. The stress exponent and flow stress curves along with ductility and yield strength are shown to be dependent on temperature and strain rate. When the tensile axis is parallel to the extrusion direction, the specimen exhibits greater yield strength and decreased ductility compared to when the tensile axis is rotated 90°. It is assumed that the origin of this behavior lies in the extrusion process, which elongates the grains and aligns the basal plane parallel to the extrusion direction. The apparent grain size of the microstructure for the specimen with the tensile axis aligned 0° to the ED is 18 μm and the grains on this face are strongly oriented in the basal direction, whereas for the specimen with the tensile axis aligned 90° to the ED the apparent grain size is 9 μm and the grains are oriented in the directions perpendicular the basal direction along this face. The orientation of the textured sheet along with the discrepancy in the apparent grain size due to the elongation of grains by extrusion is assumed to lead to the anisotropy shown during deformation.

There are two possible explanations for the anisotropy observed in both experimental tensile tests and in predictions by the finite element model. The first explanation is that the anisotropy results from the texture of the grains, which will effect deformation because of the orientation of the various slip planes with respect to the tensile axis. The second possibility is that the anisotropic behavior results from differences in the GBS mechanism caused by the elongation of the grains parallel to the ED. GBS has been shown to be more prevalent for smaller grain sized material, thus when the tensile axis is parallel to the ED, the apparent grain size is smaller and a greater contribution from GBS would be expected. The model has been used to investigate the relative contribution from each possible explanation.

Our computations allow us to distinguish between these two possible mechanisms. In the simulations described above, two simulation microstructures were constructed from micrographs of the AZ31 sheet material. The material representing the microstructure with tensile axis parallel to the extrusion direction had an apparent grain size with $18\,\mu m$, and the texture was such that the basal planes are oriented perpendicular to the loading direction. The microstructure with tensile axis perpendicular to the ED had grain size $9\,\mu m$ and basal planes parallel to the loading direction. Our simulations enable us to create 'virtual' materials with the identical texture, but with the grain size reversed (i.e. with $9\,\mu m$ parallel to the ED and $18\,\mu m$ perpendicular to the ED, representing grains elongated transverse to the ED). The resulting strain rate vs. stress curves for these additional simulations along with the original results from the simulations with the experimentally correct microstructure characteristics are plotted in Figure 9. The flow stress clearly correlates with grain size, not texture. Similarly, Figure 10 compares the contribution to the total strain rate from each deformation mechanism for the actual materials and the two 'virtual materials'. Again the strain rates correlate with grain size rather than texture. However, a noticeable separation between the plots of corresponding grain size can be observed in Figure 10. For each grain size when the orientation is 0° to the ED, the contribution from DC is slightly greater, and when the orientation is 90° to the ED the contribution from GBS is greater. This suggests that GBS is more likely for loading oriented 90° to the ED, which may be due to the more even distribution of slip on the various systems for this orientation, which would promote a more even distribution of stress, allowing for greater sliding. Thus, the orientation does have an effect on the response of the microstructure to deformation; however, the root of the anisotropy is a result of the difference in the effective grain size rather than orientation.

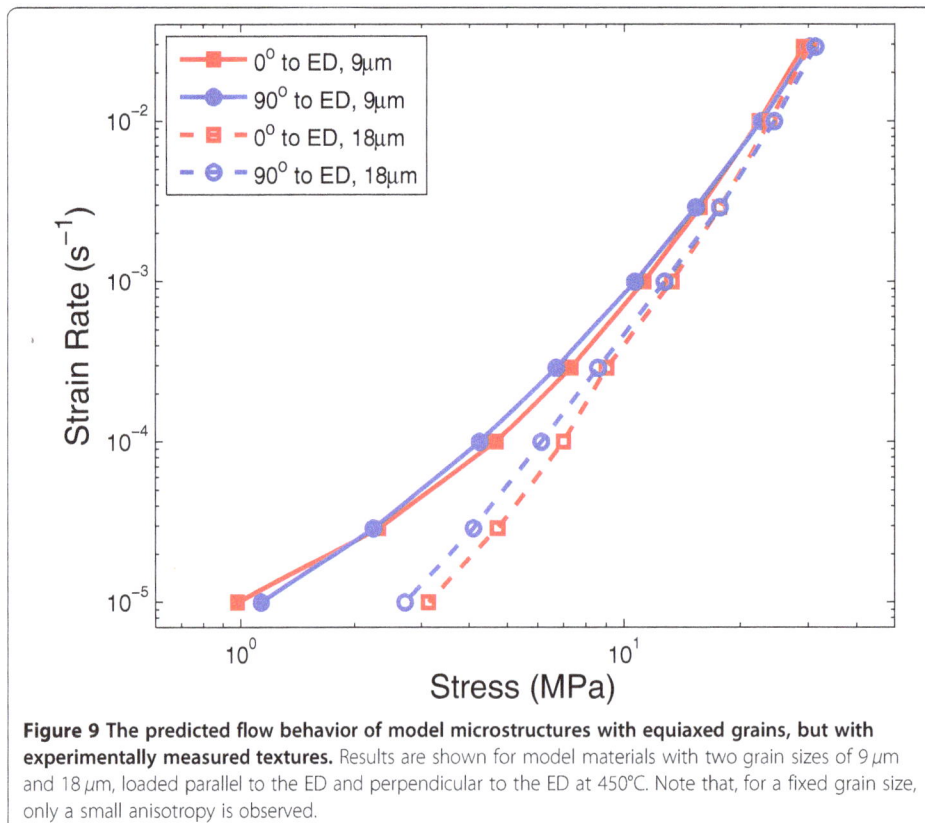

Figure 9 The predicted flow behavior of model microstructures with equiaxed grains, but with experimentally measured textures. Results are shown for model materials with two grain sizes of $9\,\mu m$ and $18\,\mu m$, loaded parallel to the ED and perpendicular to the ED at 450°C. Note that, for a fixed grain size, only a small anisotropy is observed.

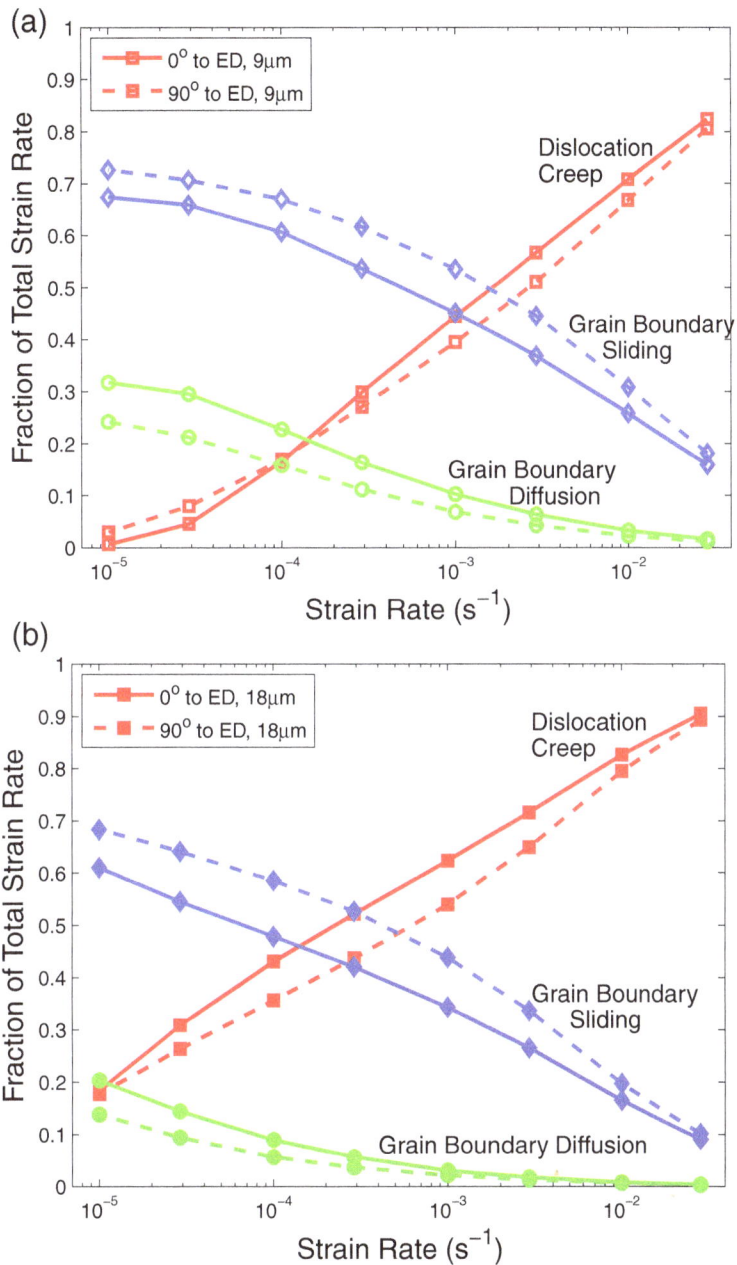

Figure 10 The predicted percent contribution to total strain rate from each deformation mechanism, DC, GBD, and GBS as a function of strain rate and orientation of the tensile axis at 450°C with constant grain size of (a) 9 μm and (b) 18 μm. The difference in deformation mechanisms at a fixed grain size is caused by texture.

Conclusions

A combination of experiment and numerical simulation was used to determine the flow stress and deformation mechanisms in extruded magnesium alloy AZ31 sheet at temperatures between 350°C and 450°C, and strain rates ranging from 10^{-5} to 10^{-2} s^{-1}. Our results show that

1. The flow stress in extruded magnesium alloy sheet is anisotropic, with a higher flow stress for tensile deformation parallel to the extrusion direction than transverse to

the ED. Material loaded with the tensile axis perpendicular to the ED also has a higher tensile ductility than material loaded parallel to the ED.

2. For both loading orientations, the flow stress has a power-law relationship to strain rate $\dot{\varepsilon} = A\sigma^n$. The stress exponent n decreases with strain rate, consistent with a transition in deformation mechanism from grain boundary sliding at low strain rates to dislocation creep at high strain rates. The transition from one mechanism to the other varies with temperature and the orientation of the loading axis.

3. At 450°C, the critical strain rate where the dominant deformation mechanism transitions from DC to GBS shifts from $1.5 \times 10^{-4} s^{-1}$ when the tensile axis is 0° to the ED to $2.0 \times 10^{-3} s^{-1}$ when the tensile axis is 90° to the ED.

4. At 350°C, the critical strain rate where the dominant deformation mechanism transitions from DC to GBS shifts from $1.5 \times 10^{-5} s^{-1}$ when the tensile axis is 0° to the ED to $2.3 \times 10^{-4} s^{-1}$ when the tensile axis is 90° to the ED.

5. For deformation with the tensile axis 0° to the ED, slip on the prismatic system is dominant followed by slip on the basal and pyramidal system respectively.

6. For deformation with the tensile axis 90° to the ED, slip on the basal system is dominant followed by slip on the pyramidal and prismatic system respectively.

7. The contribution of each slip system to the total strain rate due to DC is not dependent on strain rate or temperature over the ranges considered.

8. Two microstructural features contribute to the plastic anisotropy of extruded magnesium alloy sheet. The material has an anisotropic texture, with basal planes lying predominantly perpendicular to the extrusion direction. In addition, the grain structure is anisotropic, with grains elongated parallel to the ED. The numerical simulations indicate that the anisotropy in grain shape leads to an increase in grain boundary sliding for material loaded transverse to the ED, and provides the dominant contribution to the plastic anisotropy at elevated temperature. The texture plays a lesser role.

Availability of supporting data

The data supporting the results of this article are included within the article in graphical form.

Competing interests
The authors declare that they have no competing interests.

Authors' contributions
DC performed both the experimental and computational work, and drafted the manuscript. AB developed the microstructure based finite element model, which was adapted by DC adapted for use with magnesium alloys. PK planned the experiments. AB and PK helped to manage the project and were instrumental in composing the final manuscript. All authors read, discussed, and approved the final manuscript.

Acknowledgements
This work was supported by the General Motors Collaborative Research Laboratory in Computational Materials Research at Brown University.

Author details
[1]School of Engineering, Brown University, Providence, RI 02912, USA. [2]General Motors Company, 30001 Mound Rd, Warren, MI 48090, USA. [3]Present Address: College of Engineering, Bucknell University, Lewisburg, PA 17837, USA.

References
1. Gehrmann R, Frommert M, Gottstein G (2005) Texture effects on plastic deformation of magnesium. Mater Sci Eng A 395:338–349. 10.1016/j.msea.2005.01.002

2. Ecob N, Ralph B (1983) The effect of grain size on deformation twinning in a textured zinc alloy. J Mater Sci 18:2419–2429. 10.1007/BF00541848

3. Lebensohn R, Tome C (1993) A self-consistent anisotropic approach for the simulation of plastic deformation and texture development of polycrystals: application to zirconium alloys. Acta Metallurgica et Materialia 41:2611–2624. 10.1016/0956-7151(93)90130-K

4. Lee HP, Esling C, Bunge HJ (1988) Development of the rolling texture in titanium. Textures and Microstructures 7:317–337. 10.1155/TSM.7.317

5. Philippe M, Beaujean I, Bouzy E et al (1994) Effect of texture and microstructure on the mechanical properties of Zn alloys. MSF 157–162:1671–1674. 10.4028/www.scientific.net/MSF.157-162.1671

6. Philippe M, Bouzy E, Fundenberger J-J (1998) Textures and anisotropy of titanium alloys. MSF 273–275:511–522. 10.4028/www.scientific.net/MSF.273-275.511

7. Philippe M, Esling C, Hocheid B (1988) Role of twinning in texture development and in plastic deformation of hexagonal materials. Textures and Microstructures 7:265–301. 10.1155/TSM.7.265

8. Philippe M, Serghat M, Vanhoutte P, Esling C (1995) Modelling of texture evolution for materials of hexagonal symmetry—II. application to zirconium and titanium α or near α alloys. Acta Metallurgica et Materialia 43:1619–1630. 10.1016/0956-7151(94)00329-G

9. Philippe M, Wagner F, Mellab F et al (1994) Modelling of texture evolution for materials of hexagonal symmetry—I. Application to zinc alloys. Acta Metallurgica et Materialia 42:239–250. 10.1016/0956-7151(94)90066-3

10. Kelley EW, Hosford WF (1968) Transactions of the metallurgical society of AIME., pp 242:654–660

11. Philippe M (1994) Texture formation in hexagonal materials. Mater Sci Forum 157–162:1337–1350

12. Hutchinson B, Barnett MR, Ghaderi A et al (2009) Deformation modes and anisotropy in magnesium alloy AZ31. Int J Mat Res (formerly Z Metallkd) 100:556–563. 10.3139/146.110070

13. Krajewski PE, Ben-Artzy A, Mishra RK (2010) Room temperature tensile anisotropy of extruded magnesium plates. In: Agnew SR, Neelameggham NR, Nyberg EA, Sillekens WH (eds) Magnesium Technology. TMS, Warrendale, PA, pp 467–472

14. Xiong F, Davies CHJ (2003) Anisotropy of tensile properties of extruded magnesium alloy AZ31. Mater Sci Forum 426–432:3605–3610

15. Barnett MR (2003) A taylor model based description of the proof stress of magnesium AZ31 during hot working. Metall and Mat Trans A 34:1799–1806. 10.1007/s11661-003-0146-5

16. Reed-Hill RE, Robertson WD (1957) Deformation of magnesium single crystals by nonbasal slip. Transactions of the Metallurgical Society of AIME 220:496–502

17. Reed-Hill RE, Robertson WD (1958) Pyramidal slip in magnesium. Transactions of the Metallurgical Society of AIME 221:256–259

18. Ward Flynn PW, Mote J, Dorn JE (1961) Transactions of the Metallurgical Society of AIME 221:1148–1154

19. Agnew SR, Duygulu Ö (2005) Plastic anisotropy and the role of non-basal slip in magnesium alloy AZ31B. Int J Plast 21:1161–1193. 10.1016/j.ijplas.2004.05.018

20. Barnett MR, Ghaderi A, Sabirov I, Hutchinson B (2009) Role of grain boundary sliding in the anisotropy of magnesium alloys. Scr Mater 61:277–280. 10.1016/j.scriptamat.2009.04.001

21. Stanford N, Sotoudeh K, Bate PS (2011) Deformation mechanisms and plastic anisotropy in magnesium alloy AZ31. Acta Mater 59:4866–4874. 10.1016/j.actamat.2011.04.028

22. Foley DC, Al-Maharbi M, Hartwig KT et al (2011) Grain refinement vs. crystallographic texture: Mechanical anisotropy in a magnesium alloy. Scr Mater 64:193–196. 10.1016/j.scriptamat.2010.09.042

23. Krajewski PE, Hector LG Jr, Du N, Bower AF (2010) Microstructure-based multiscale modeling of elevated temperature deformation in aluminum alloys. Acta Mater 58:1074–1086. 10.1016/j.actamat.2009.10.023

24. Schroth JG (2004) General Motors' Quick Plastic Forming Process. In: Taleff EM, Friedman PA, Krajewski PE (eds) Advances in Superplasticity and Superplastic Forming. TMS, Warrendale, PA, pp 9–20

25. Barnes AJ (1994) Superplastic forming of aluminum alloys. Mater Sci Forum 170–172:701–714

26. Cipoletti DE, Bower AF, Krajewski PE (2011) A microstructure-based model of the deformation mechanisms and flow stress during elevated-temperature straining of a magnesium alloy. Scr Mater 64:931–934. 10.1016/j.scriptamat.2010.12.033

27. Bower AF, Wininger E (2004) A two-dimensional finite element method for simulating the constitutive response and microstructure of polycrystals during high temperature plastic deformation. J Mech Phys Solids 52:1289–1317. 10.1016/j.jmps.2003.11.004

28. Freund L, Suresh S (2003) Thin film materials: stress, defect formation, and surface evolution. Cambridge University Press, Cambridge [England]; New York

29. Frost H, Ashby MF (1982) Deformation-mechanism maps: the plasticity and creep of metals and ceramics, 1st edn. Pergamon Press, Oxford [Oxfordshire]; New York

30. Obara T, Yoshinga H, Morozumi S (1973) {1122}(-1-123) Slip system in magnesium. Acta Metall 21:845–853. 10.1016/0001-6160(73)90141-7

31. Yoshinaga H, Horiuchi R (1963) Deformation mechanisms in magnesium single crystals compressed in the direction parallel to hexagonal axis. Japan Institute of Metals 4:1–8

32. Yoo M, Agnew S, Morris J, Ho K (2001) Non-basal slip systems in HCP metals and alloys: source mechanisms. Mater Sci Eng A 319–321:87–92. 10.1016/S0921-5093(01)01027-9

33. Cahn J, Taylor J (2004) A unified approach to motion of grain boundaries, relative tangential translation along grain boundaries, and grain rotation. Acta Mater 52:4887–4898. 10.1016/j.actamat.2004.02.048

Permissions

List of Contributors

Jenny Rudnizki
Department of Ferrous Metallurgy, IEHK, RWTH Aachen University, Intzestr 1, 52056, Aachen, Germany

Ulrich Prahl
Department of Ferrous Metallurgy, IEHK, RWTH Aachen University, Intzestr 1, 52056, Aachen, Germany

Wolfgang Bleck
Department of Ferrous Metallurgy, IEHK, RWTH Aachen University, Intzestr 1, 52056, Aachen, Germany

Michael A Groeber
Air Force Research Laboratory, 2230 Tenth St, 45433, WPAFB, Ohio, USA

Michael A Jackson
BlueQuartz Software, 400 S. Pioneer Blvd, 45066, Springboro, OH, USA

John M Sosa
Center for the Accelerated Maturation of Materials, The Ohio State University, 1305 Kinnear Rd., Columbus, OH 43212, USA

Daniel E Huber
Center for the Accelerated Maturation of Materials, The Ohio State University, 1305 Kinnear Rd., Columbus, OH 43212, USA

Brian Welk
Center for the Accelerated Maturation of Materials, The Ohio State University, 1305 Kinnear Rd., Columbus, OH 43212, USA

Hamish L Fraser
Center for the Accelerated Maturation of Materials, The Ohio State University, 1305 Kinnear Rd., Columbus, OH 43212, USA

Adam Creuziger
National Institute of Standards and Technology, 100 Bureau Dr., 20899, Gaithersburg, MD, USA

Lin Hu
Carnegie Mellon University, 5000 Forbes Avenue, 15213, Pittsburgh, PA, USA
IBM Semiconductor Research and Development Center, 2070 Route 52, 12533, Hopewell Junction, NY, USA

Thomas Gnäupel-Herold
National Institute of Standards and Technology, 100 Bureau Dr., 20899, Gaithersburg, MD, USA

Anthony D Rollett
Carnegie Mellon University, 5000 Forbes Avenue, 15213, Pittsburgh, PA, USA

Kannan Subramanian
Stress Engineering Services, Inc., 3314 Richland Ave, 70002 Metairie, LA, USA

Harish P Cherukuri
Department of Mechanical Engineering and Engineering Science, University of North Carolina at Charlotte, 9201 University City Blvd, 28223 Charlotte, NC, USA

Veera Sundararaghavan
Aerospace Engineering, University of Michigan, Ann Arbor, MI 48109, USA

Sai Kiranmayee Samudrala
School of Mechanical Engineering, Georgia Tech, Atlanta, GA 30332-0405, USA

Prasanna Venkataraman Balachandran
Department of Materials Science, Drexel University, Philadelphia, PA 19104, USA

Jaroslaw Zola
Rutgers Discovery Informatics Institute, Rutgers University, Piscataway, NJ 08854, USA

Krishna Rajan
Department of Materials Science and Engineering, Iowa State University, Ames, IA 50011, USA

Baskar Ganapathysubramanian
Department of Mechanical Engineering, Iowa State University, Ames, IA 50011, USA

Carelyn E Campbell
Materials Science and Engineering Division, National Institute of Standards and Technology, Gaithersburg, MD 20899, USA

Ursula R Kattner
Materials Science and Engineering Division, National Institute of Standards and Technology, Gaithersburg, MD 20899, USA

Zi-Kui Liu
Department of Materials Science and Engineering, The Pennsylvania State University, University Park, PA 16802, USA

Sudhanshu S Singh
Materials Science and Engineering, Arizona State University, Tempe, AZ 85287-6106, USA

Jason J Williams
Materials Science and Engineering, Arizona State University, Tempe, AZ 85287-6106, USA

Peter Hruby
Materials Science and Engineering, Arizona State University, Tempe, AZ 85287-6106, USA

Xianghui Xiao
Advanced Photon Source, Argonne National Laboratory, Argonne, IL, USA

Francesco De Carlo
Advanced Photon Source, Argonne National Laboratory, Argonne, IL, USA

Nikhilesh Chawla
Materials Science and Engineering, Arizona State University, Tempe, AZ 85287-6106, USA

McLean P Echlin
University of California Santa Barbara, Materials Dept. Building 503 Santa Barbara, CA, 93106-5050, USA

William C Lenthe
University of California Santa Barbara, Materials Dept. Building 503 Santa Barbara, CA, 93106-5050, USA

Tresa M Pollock
University of California Santa Barbara, Materials Dept. Building 503 Santa Barbara, CA, 93106-5050, USA

David E Cipoletti
School of Engineering, Brown University, Providence, RI 02912, USA
Present Address: College of Engineering, Bucknell University, Lewisburg, PA 17837, USA

Allan F Bower
School of Engineering, Brown University, Providence, RI 02912, USA

Paul E Krajewski
General Motors Company, 30001 Mound Rd, Warren, MI 48090, USA

www.ingramcontent.com/pod-product-compliance
Lightning Source LLC
Chambersburg PA
CBHW080258230326
41458CB00097B/5137